THE INSECT FACTOR IN WOOD DECAY

THE RENTOKIL LIBRARY

The Woodworm Problem
The Dry Rot Problem
The Insect Factor in Wood Decay
The Conservation of Building Timbers
Pests of Stored Products
Household Insect Pests
The Cockroach (Vol. I)
Wood Preservation—A guide to the meaning of terms
Termites—A World Problem
Pest Control in Buildings—A guide to the meanings of terms

The adult beetle of *Anobium punctatum* emerging from a flight-hole.
Water-colour by Norman Weaver

The Insect Factor in Wood Decay

AN ACCOUNT OF WOOD-BORING INSECTS WITH PARTICULAR REFERENCE TO TIMBER INDOORS

Norman E. Hickin

Scientific Consultant, Rentokil Limited

Third Edition, edited and revised by

Robin Edwards

Entomologist, Rentokil Limited

Associated Business Programmes London

HN

ASSOCIATED BUSINESS PROGRAMMES LTD
17 Buckingham Gate, London SW1

First published by Hutchinson & Co. (Publishers) Ltd 1963
Second edition 1968
Third edition (revised) 1975

Set in Monotype Times
Printed in Great Britain by The Anchor Press Ltd
and bound by Wm Brendon & Son Ltd
both of Tiptree, Essex

ISBN 0 85227 101 8

Contents

Preface to Third (revised) Edition

The Insect Factor in Wood Decay was first published twelve years ago, and the second edition appeared five years later. Only minor additions were made to the second edition, although the index was extended considerably.

A large amount of new information has become available in the intervening years, some of which has now been incorporated into this third edition. The selection of what to include proved difficult: the book, already running to 344 pages, could not be made much larger without becoming unwieldy or altering the character of the work. Dr. Hickin and I have chosen those items which we believe will increase the usefulness of the book without changing the presentation.

In previous editions, Dr. Hickin's excellent drawings have suffered from over-reduction in size; a feature which we have taken the opportunity to correct. Many of the insects are now reproduced much larger, with a consequent improvement in detail.

The text has been extensively re-arranged to bring sections into a more logical order. Keys for the identification of the Cerambycidae (and some other groups) have been brought together at the start of each sub-family, rather than being scattered through the text. Notes on the species now follow the sequence in which they appear in the keys.

The index has again been revised, with particular attention to the listing of host timbers. Both scientific and common names are given, even though in some cases, only one is quoted in the text. These names are becoming increasingly important as more and more unfamiliar timbers are being employed in the building industry.

I would like to thank Dr. Hickin for his part in selecting new items and for checking the revisions. Mr. E. A. J. Duffy of the C.I.E. has kindly checked the names of Cerambycids and Scolytids to bring them into line with current usage, and Miss Evans of the Royal Entomological Society of London has agreed to the use of a drawing from one of the Society's 'Handbooks for the Identification of British Insects'. Finally, I should like to reiterate Dr. Hickin's thanks to all those who helped in the production of the first two editions of this book.

R.E.
1975

Introduction

In planning this book, the chapter sequence gave a little trouble. It was thought desirable, before dealing with the insect factor in detail, to give a brief review of the nature of wood itself and then to give a catalogue of the various factors causing wood to decay. The chapter on the relative importance of the various wood-boring beetles was placed, after some discussion, *after* the accounts of the various biologies of these insects, since it was thought that it would then be more readily understandable. The sequence of the chapters on the wood-boring beetles has no relationship to classification. The most important wood-boring beetle in Britain—*Anobium punctatum* and its family, the ANOBIIDAE—is dealt with first. This is followed by the Death Watch Beetle in the same family, which in turn is followed by an account of the Powder Post Beetles, so that the order of chapters is more or less in order of importance of the beetles concerned. Towards the end of the book will be found chapters on termites and on the role of wood-boring insects in laboratory techniques for evaluating wood preservatives.

Termites do not occur in the United Kingdom, except as accidental importations or as laboratory animals, but manufacturers are having to take more and more notice of these most destructive wood-pests if they export their goods to countries where termites are a hazard. Even if the goods themselves have no wooden parts, the packaging materials are often destroyed by termites, with consequent damage to the contents. A brief general account of these insects is therefore given, together with the history of their accidental introduction into Britain.

Perhaps it would not be out of place to say here something about the sense in which the word 'decay' is used, not only in the title, but in several places in the text. In all cases the word is used in the wide sense as given in the *Shorter Oxford English Dictionary*—the process of falling off from a sound condition, a falling off or wasting or wearing away, dilapidation, or the wasting of organic tissue. It is seen from this that the word decay is not confined to a dilapidation association with any specific physical or biological factor and the word is never used in this narrower sense in the present work.

CHAPTER ONE

The Nature of Wood

The importance of wood as a constructional and decorative material: the national investment in wood—A general account of wood; its origin, chemical, physical and some mechanical properties.

The importance of wood as a constructional and decorative material

Wood is a very desirable and attractive substance. It is produced by approximately 44,000 different species of woody plants in a great variety of physical forms and properties. In colour it varies from black to white, and in range of density some timbers will sink in water and others will float with as much as nine-tenths of their volume out of the water. Its gross annual increment of growth is of staggering proportions and it has been used by man for almost every conceivable purpose. The ease with which it can be fashioned set man upon the ocean and every land mass was first visited by men in wooden boats. Man appears to be at his happiest with a piece of wood in his hands, whether it be the bowl of a pipe, butt of a gun, neck of a fiddle, tiller of a boat, or handle of racket or bat. Wood gives a sense of physical well-being. In colder climates it is warm to the touch, and where it is hot, wood is cool. It is used in such engineering projects as harbour works, railway works, sleepers and sea-defence groynes. The sixteen and half million dwellings in Great Britain are predominantly timber-framed and the floors and joinery are of the same material. The greater part of the furniture on which the population sits, sleeps, and eats is of wood, selected for its aesthetic appeal, its durability and fitness for purpose. Some of the sweetest music man has made is produced by instruments of wood, and his artistic sense is satisfied by carving it into shapes.

The national investment in wood

From the record given above of the occurrence of wood in large-scale engineering projects, buildings and homes, the tremendous national investment in wood can be realized. In its immensity it can only be measured in thousands of millions of pounds. If we take away wood from these aspects of living, not much is left. But perhaps in discussing wood, we ought also to mention two very important points concerning the growth of trees. In the first place, the presence or absence of trees influences the nature of the climate of a place; increasing or reducing the numbers of trees changes it. Secondly, but not less important, is the great effect which the presence of trees has on the productivity of the surrounding earth, by reason of the leaf fall and tree decay which bring great masses of cellulose and other substances back into the earth again. The presence of trees also ensures the conservation of water.

Unfortunately the forests of the world are receding, not advancing, so that timber becomes more expensive to cut and transport as the more accessible forests are worked out. The replanting of forests is looked upon as a long-term project and in many countries is therefore placed low in order of priority for national investment income. Moreover, the not unnatural desire of the peoples of the world to attain a higher and higher standard of living means that every operation in cutting, conversion, transportation, and the ultimate manufacture of the wooden article costs more and more.

It seems an obvious inference to draw that we must do more to preserve wood that is already in man's service. The day of periodic replacements of wood in buildings and elsewhere is passing. Timber that has served man hitherto for one generation, must serve now for two or more. Many thousands of tons of timber in buildings have been carrying out their original function for 300 years or more. The rapidly developing technology of *in situ* timber conservation is meeting an obvious need, and it is hoped that the point has been made that this work of caring for wood which man has employed to such a vast extent in his economy and his culture, is a very worth-while task.

A general account of wood; its origin, chemical, physical and some mechanical properties

In dealing with such an abundant product there is danger that familiarity may breed contempt. We may not be altogether certain about the origin and properties of the material wood, so that space is devoted to a consideration of *wood* itself and some prominence is given to its biological origin and to its chemical and physical properties. It must be emphasized, however, at the outset, that the descriptions which are employed are purely introductory, and may lead to over-simplification unless the references are consulted. It is earnestly hoped that this will be done, and that far from the study of wood in all its aspects being considered a 'fringe subject' by those employed on the technicalities of wood decay, it may be approached as an important part of their training.

Certainly wood appeals to our aesthetic senses. Whether this is due to some inherent feeling for the material which, in the form of branches, was used to construct the first dwellings of the human race (although those who work in stone aver that the first human dwellings were caves, and the constriction of the mouth of the caves with blocks of stone was man's first essay at building a home), or whether its intrinsic beauty of grain or colour, its warmth to the touch, or the endless shapes in which it can be fashioned for man's use endears it to the human heart—none can tell.

If what is said in the following pages encourages any of my readers to increase their interest in, and enjoyment of, wood, as a material whose life *must be prolonged*, then my task will not have been in vain.

What is wood?

Wood is the product of trees and a part of their substance. It consists of (a) tissues which carry out vital functions for the tree as a living plant, and (*b*) to a very much greater extent, those same tissues after their death.

Trees are plants which have adopted a special form of growth. By the production of 'woody' tissues, the plant is enabled to reach towards the light to a much greater height than can the non-woody plants or 'herbs', as they are generally called. It follows from this that trees are long-lived plants—able to withstand seasonal change, storm and tempest, and to use the same structure for carrying out vital functions of the plant often for as much as several hundred years. It is this tissue, the dead tissue, that gives structural rigidity to the main stem or trunk and the principal branches.

Classification of woody plants

The 'woody' or tree habit is shown by a wide range of plant types. The most primitive plants adopting the tree form are the PTERIDOPHYTA (Horsetails, Ferns, etc.), and although present-day examples are rare and of no economic importance, in former geological epochs they were the dominant plant forms, and their compressed and modified substances constitute the present-day coal-measures. It is interesting to consider the first deposits of carbon as being won from carbon dioxide of the air by the action of the sun on the leaves of these woody plants (*photosynthesis*). The SPERMATOPHYTA are more advanced and are divided into two groups, the GYMNOSPERMS—characterized by having naked seeds, and the ANGIOSPERMS, with seeds enclosed by an ovary. (This last group includes the flowering plants of our gardens.) From the evolutionary point of view, the GYMNOSPERMS are the older group, and although there are botanically important orders such as the GINKGOINAE (containing *Ginkgo biloba*, the maidenhair tree), the CYCADINAE and the GNETINAE, the most important order in this group from the woody habit point of view, is the CONIFERAE. The CONIFERAE are usually characterized as bearing cones and having needle-shaped leaves. This is a very general description, but the timber converted from the woody tissues of this group is known as SOFTWOOD. This type of wood is, without a doubt, the most important on a gross tonnage basis. The family TAXACEAE (Yew, *Podocarpus*, etc.) is contained in the CONIFERAE, although the fruit is not a cone, nor are the leaves needle-shaped.

The ANGIOSPERMS are divided into two well-defined groups, MONOCOTYLEDONS and DICOTYLEDONS. The MONOCOTYLEDONS have one seed-leaf and linear venation in the leaves, and in this group the production of woody tissue is rare, and if produced is of no value as timber (e.g., bamboo). The DICOTYLEDONS are characterized by usually having two seed-leaves and the venation is mostly reticulate. Timber converted from woody examples of this group is known as HARDWOOD. The woody habit has sometimes been expressed as being a primitive character of plants, but a large number of families among the ANGIOSPERMS have the woody habit. Whilst some families, e.g., CRUCIFERAE and UMBELLIFERAE have no woody members, the families BETULACEAE and CUPULIFERAE contain most of the forest trees of the northern temperate zone.

Production of wood by the plant

The manner in which the plant produces woody tissue is of exceptional interest, but only the briefest outline of the process can be given here. In Fig. 1 is shown an outline plan of the tissues in transverse section as shown in a lime stem at the end of the first year. The vascular tissue is first

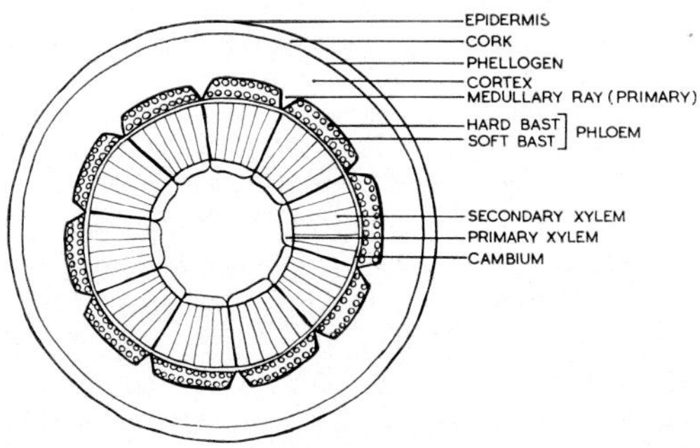

Fig. 1. An outline plan of tissues in a transverse section of a one-year stem of lime (*Tilia*). The cambium is shown having joined into a complete circle and secondary xylem is being produced.

produced as discreet areas or 'bundles' with the xylem tissue towards the pith cavity (medulla) and the phloem or bast towards the outside. These are separated by the cambium, which is shown joined into a complete circle. The separate bundles, however, are still shown by the primary medullary rays, which will keep their identity in the wood even after many years. The cambium is the *meristemmatic* tissue, constantly dividing and pushing out new cells to the inside—the secondary xylem and associated tissue—and new cells to the outside—the various constituent elements of the phloem and associated tissues. Many of these latter cells become impregnated with suberin and become cork cells. The process continues year after year, the diameter of the cambial layer increasing with each succeeding year, the active inner elements of the bark serving to conduct food materials downwards to all the actively growing parts of the tree, whilst the active wood elements—the SAPWOOD—conduct water and nutrient materials upwards. After a period of activity, the older woody elements die, as they are forced by the peripheral growth into the central area of the trunk or branch. This is the HEARTWOOD. As the change takes place, waste products may be forced into the dying cells and other changes may take place which give specific timbers their characteristic properties of colour, durability, etc.

In temperate climates, the period of active growth is confined to the spring and summer months, which has the effect of giving a characteristic pattern to the timber if it is viewed in transverse section. This is because the size of the individual elements, and sometimes their nature, varies with the seasonal needs of the tree. The concentric rings in these trees are known as 'growth rings'. In certain tropical areas many trees of the rain forests do not show this growth pattern, although it is present in trees growing in areas where wet and dry seasons alternate.

The Microstructure of wood

Every plant is composed of a number of box-like elements known as cells (the most primitive plants consisting of only a single cell), and in the woody stem the cell shows its simplest construction in the cambial layer when freshly split off from the actively dividing tissue. The cell wall at this stage is thin and composed entirely of cellulose, and contains the vital living material—protoplasm. The latter is divided into two important components, the general mass of protoplasm lining the cell wall, known as the cytoplasm, and a darker body apparently much concentrated, known as the nucleus. The subsequent behaviour of the cell is initiated by changes which take place in the nucleus. In this book only the cells formed inwards from the cambium giving woody tissue will be considered.

Softwoods

The elaboration of the cells in softwoods and hardwoods to form the wood elements differs considerably. In softwoods, those cells developing into tissue for conduction and for supplying mechanical strength become elongated and needle-like. They are known as tracheids (Fig. 2), and vary

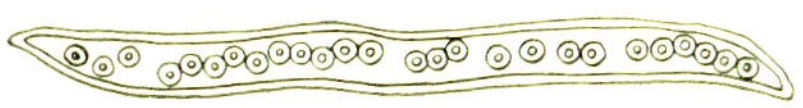

Fig. 2. A softwood tracheid with bordered pits. Much magnified.

in length from 2–10 mm, although the majority occur at the shorter end of the length range.

The cell is capable of considerable modification once the final size has been attained by the extension of the cell wall (the primary cell wall). The latter is thickened by the deposition of further cellulose and other cell wall substances secreted from inside the cell. Certain areas, however, are not thickened, and these are known as pits. It is through the pits that conduction of nutrient materials and other substances from cell to cell, and ultimately from one part of the tree to another, takes place.

The tracheids are packed tightly together so that in transverse section they appear like honeycomb, but more detailed examination reveals alternating zones of smaller thicker-walled tracheids and longer ones with thinner walls. The former are concerned less with sap conduction (and are furnished with fewer pits), but more with mechanical rigidity of the tissue, whilst in the latter the converse is true. There is often an abrupt change in the tracheid zone which can be seen with the naked eye. The thin-walled tracheids represent the early wood, whilst the late wood is represented by the thick-walled tracheids. The complete band represents a growth ring.

In addition to tracheids, another type of cell is present in the wood, which undergoes less modification than the tracheid when split off from the cambium. This is known as the wood parenchyma cell, and has the important function of food storage. Carbohydrate food material not required immediately by the growing tissues is usually elaborated to a more complex form and deposited in the wood parenchyma. This tissue takes the form of long strands or rays, and is visible to the naked eye. The amount of wood parenchyma in softwoods is usually small.

Resin, which is a sticky, gummy, aromatic secretion characteristic of many softwoods, is produced from special parenchymatous cells lining resin canals, which are just visible to the naked eye when the wood is carefully sectioned.

Hardwoods

It is a relatively simple task to differentiate hardwoods from softwoods, as in the former there is a high degree of cell specialization which can be easily recognized. The cells having the function of conduction are quite separate from the fibre members, which give rigidity to the wood structure. An important difference, of great economic significance in the microstructure of softwoods and hardwoods, is the very much smaller fibre length in hardwoods compared with the tracheid length in softwoods. The wood pulp made from softwoods is almost always superior to that made from hardwoods.

The conducting tissue in hardwoods consists of vessels or pores (Fig. 3). These consist of longitudinal rows of special cells, elaborated from the cambium, which lose their connecting walls and thus construct a pipe or tube. (In some timbers the vessel members still retain the end walls as perforated oblique plates, e.g., birch and alder.) The vessels can be seen in transverse section as holes in characteristic patterns, and in longitudinal section as long thin furrows. These open-ended vessels are manifestly more efficient in conduction than the tracheids of softwoods, but some vessels are furnished with pits and indeed tracheid-like cells are found in some hardwoods (oak and sweet chestnut).

When the vessels of hardwoods are no longer able to carry out the function of conduction, and thus turn from sapwood to heartwood, adjoining parenchymatous cells often grow into the vessels through the pits. These growths are known as tyloses, and assume a foam-like and often glistening appearance.

It is probably due to the presence of tyloses that certain timbers resist the penetration of wood preservatives, and also resist the spread of fungal decay by making the ingress of fungal hyphae very much more difficult.

As has already been mentioned, the strengthening tissue of hardwoods consists of fibres. These cells have pointed ends, and are usually furnished with a few pits. The important strength characters of hardwoods are usually dependent on the wood fibre construction. The storage tissue of hardwoods is often very abundant, and has a wide range of forms and distribution. It consists of wood-parenchyma and also rays. The varied patterns of the vessels, fibres, parenchyma, and rays are of diagnostic

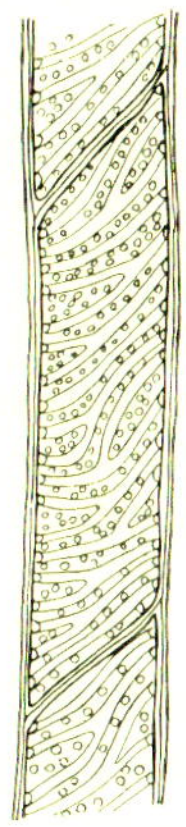

Fig. 3. A vessel of lime partially in transverse section; showing reticulate thickening, pits and oblique vessel end walls.

value in timber identification, and indeed make possible positive identification of many thousands of species of woody plants by the examination of a small transverse section of a piece of wood with a × 10 lens.

Wood as a chemical substance

It will have been seen from the description of the micro-structure of wood that it consists of a mass or aggregate of hollow pipe-like structures cemented together. The chemical composition of the walls of these elements differs widely amongst the thousands of tree species, but in the main, three substances are always present. Cellulose and lignin (or lignone) are the two main chemicals present, but a smaller quantity of a gum-like substance, hemicellulose, is present also. Much is known of the structure of cellulose, but that of lignin is little understood. It is known that lignin helps to prevent the degradation of cellulose by enzymes associated with fungal attack. Chemically, however, lignin is less stable than cellulose, and can be removed from wood pulp by processing with steam, or by the use of warm, weak acids. This is how lignin is removed during the process of paper manufacture.

In reality, cellulose is a collective name for a number of different substances, but they have the common property of being polymerized from simple sugars. Wood cellulose is based upon glucose, and the way in which this is built up is shown in Fig. 4. If cellulose is treated with nitric or acetic acid, it produces a soluble compound, and many industries utilize this substance as a basic material. Cellulose in various forms can

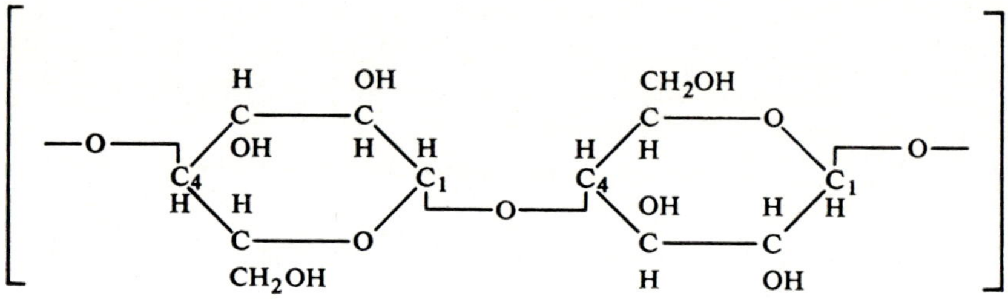

Fig. 4. The empirical formula for cellulose is based on polymerized glucose units $(C_6H_{10}O_5)n$, where *n* may be up to about 3,000. The structural formula is given here. The molecular chains are unbranched, but appear to be bound laterally into fibrils. Lignin has a somewhat similar polymer system which is interlinked.

be precipitated from solution. Explosives can be made from a solution of cellulose in nitric acid.

If wood is distilled destructively, a mass of tar derivatives and methyl alcohol is produced.

What has been said above is general for practically all timbers, but in addition, individual species have a host of other substances present, including resins, ethereal oils, colouring substances of various kinds, and tannins.

Wood as a physical substance

It is often convenient to consider wood as a gel, having properties very similar to glue. For instance, wood is hygroscopic: as it absorbs water, which it does readily, it swells; as it loses water so it shrinks. The amount of water which wood substance can absorb is definite and constant for any particular species, and lies near 30 per cent of the dry weight. This is the amount of moisture which would be taken up when the wood is in a saturated atmosphere.

The water so taken up would be found in the wood substance and not in the cavities. Such a condition is known as the *fibre saturation point.* Here, the timber has reached its maximum volume, and further absorption of water does not increase it.

The amount of water in a piece of wood (provided its surfaces are not protected in any way) is constantly changing with the humidity of the atmosphere. Thus, when a tree is felled, the water content in the sapwood is about 100 per cent of that of the dry weight of the wood. The water content of the heartwood is usually much less than this. Timber is then usually cut into planks and allowed to dry out in a current of air. The amount of water in the wood drops fairly rapidly at first, and then more slowly until the water content at the surface of the wood has a vapour pressure equal to that of the atmosphere. When this occurs the water content is somewhere between 15 and 20 per cent of the dry weight. When such timber is brought indoors into heated rooms, a further drying ensues, so that furniture in a heated room has a water content usually about 8 per cent. If the room becomes unheated and damp, the wood will absorb more moisture, or if it becomes very dry and hot, then moisture is lost, and this process is going on continually in all wood, whether in furniture or in the structural timbers of a house. It was not specifically mentioned above, but it is generally known, that as the water content of the wood decreases, there is a reduction in volume of the wood often accounting for splits during the seasoning process.

Another property which shows the similarity of wood to a gel is the decrease in hardness and stiffness when water is added to it; but a rise is shown in flexibility and extensibility. These are well-known properties of a gel such as glue or gelatine, and wood behaves in exactly the same way. The property of glue of melting when heat and water are applied to it (that is, loss of stiffness) is exactly the same property as the bending of wood for the manufacture of furniture by the application of moist steam.

The actual specific gravity of wood substance does not vary much for various timber species, and it is approximately 1·55. Wood substance is therefore heavier than water and would sink, but (as has been shown in the microstructure) wood contains a very large number of air spaces in the various elements. The specific gravity of wood is a measure of the cellular structure, and timbers differ very widely in their densities. Balsa wood floats with a large proportion of its volume out of the water, but ebony and box sink.

The mechanical properties of wood are remarkable, especially the combination of strength with lightness. If the crushing strength is divided by the density, wood gives a higher value than iron, steel and other metals. The grain structure of wood, however, must be taken into consideration when its mechanical properties are tested; for instance, in compression and tension, wood is strongest along the length of the grain, and is always weakest at right angles to it, but in shear the highest value is obtained at right angles to the grain.

REFERENCES

ANON 1953. The Timber Industry in the U.S.A. *IV. The Treatment and Preservation of Timber*. T.A.M. 59B. O.E.E.C. Paris.

COWLING, E. B. 1958. *A Review of Literature on the Enzymatic Degradation of Cellulose and Wood*. Report 2116, Forest Products Laboratory, Forest Service, U.S. Dept. of Agriculture.

DESCH, H. E. 1948. *Timber—Its Structure and Properties*. Macmillan. London.

HENDERSON, F. Y. 1939. *Timber—Its Properties, Pests and Preservation*. Crosby Lockwood. London.

HUNT, G. M. and GARRETT, G. A. 1953. *Wood Preservation*. 2nd Edition. McGraw-Hill. New York.

PHILLIPS, E. W. J. 1948. The Biology and Properties of Wood. *New Biology*. **4.**

SNYDER, T. E. 1948. *Our Enemy the Termite*. Comstock. Ithaca.

CHAPTER TWO

The Factors of Wood Decay

Fungi; animals; mechanical wear; chemicals; heat

Definition of wood decay

Dictionary definitions of the word decay all refer to deterioration, and although not specifically saying so, obviously refer to comparison with the original properties of the material. The dictionaries also seem preoccupied with the end-product of decay; although the *process* of decay is referred to, it is obvious from examples given that the *degree* of decay envisaged is considerably advanced. A few flight holes of *Lyctus* appearing in a piece of sapwood incorporated into an oak table gives the owner an uncomfortable feeling that his furniture is decaying, although the process of decay due to the depredation of a wood-boring insect may only be in the primary stages.

Decay of wood, then, occurs when any deterioration in any property of the wood takes place. This may be in appearance only, when the offence is to the aesthetic sense. On the other hand, the deterioration may be due to alteration in physical properties such as hardness, tensile strength, or load-carrying capacity, with dangerous consequences.

The factors of wood decay

The main groups of causative agents producing deterioration in wood have been listed a number of times by various authors. The individual groups have been given greater or lesser emphasis according to the eventual purpose of the author. Modern research, however, suggests that several of the factors of decay, previously thought quite distinct as decaying organisms, are indeed dependent on each other.

A convenient main grouping is as follows:

I. *Attack by wood-rotting fungi.* This is perhaps the most important factor of wood-decay. If the world's total cellulose crop is envisaged, it is seen that for conditions of equilibrium to exist, there must be an equal amount destroyed, and it is mainly through the medium of the wood-rotting fungi that this takes place.

II. *Attack by wood-boring animals.* Many different orders of animals utilize wood as a foodstuff, either directly or indirectly, although it appears that very few animal groups can digest cellulose unless in the presence of wood-destroying fungi or other plant forms. Fungal attack may already be in process before the animal arrives at the scene, or it may be brought with the animal—the Ambrosia Beetles, PLATYPODIDAE, etc., being examples of the latter. Or the cellulose fragments may be torn off and passed into the gut of the animal, where digestion takes place through yeast-like cells which secrete the enzymes which break down the cellulose. This is an example of symbiosis—an association of differing organisms to their mutual advantage.

Several wood-boring animals previously thought to have no connection with wood-destroying fungi are now known to derive important benefit from their presence in the superficial layer of wood, e.g. *Teredo* and soft rots.

III. *Mechanical wear or breakdown.* Wood which has broken down or failed, through loading past the failure point, has obviously suffered deterioration, and is a factor in wood decay. This may occur during end-compression (Fig. 5), tension or shearing, or on account of mechanical

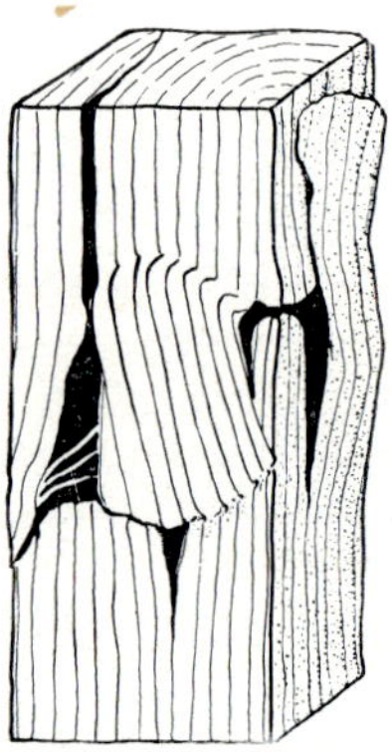

Fig. 5. Characteristic failure of a brittle wood with straight grain under end-compression. The force applied from the top. (After Henderson.)

wear caused by some element of some or each of these factors, with or without cutting action (abrasion). Two everyday examples of this are (*a*) wooden stairs unprotected by coverings and subject to heavy use, which show a gradual wearing away of the used surface; (*b*) drawers in furniture used continually and not subject to lubrication which often abrade and throw off a fine dust, sometimes confused with the bore dust of *Lyctus*.

IV. *Chemical degradation or decomposition of wood.* Wood is decayed by many chemical substances, although wooden vessels are often used to contain corrosive fluids which would attack metals. Individual timbers, however, vary widely in their resistance to chemical reagents. Softwoods, which have a high content of cellulose and lignin, and a low pentosan content, are more resistant to acids, and the hardwoods, teak, jarrah and abura are similarly resistant. The rate of attack on these timbers by mineral acids of below 25 per cent concentration is slow at ordinary temperatures. Acids which act as strong oxidizing agents, such as nitric acid, have a violent reaction when in contact with wood.

Wooden vats and processing equipment, however, are employed in the food and beverage industries where the fluids may be mildly corrosive, even though the wood employed (usually pitch pine and Douglas fir) is in constant decay. They are preferred because the products of such decay are more desirable than the corrosion products from steel or other metal vats.

An example of chemical degradation of wood which can often be observed is the 'woolliness' of timber caused by sulphur dioxide released into the atmosphere from a faulty flue. This is the process of defibration.

So far, only acid degradation of wood has been considered. Alkalis decay wood at a much faster rate than acids. Sodium carbonate solution, for example, which is only a weak alkali, corrodes wood, as can be seen in draining boards and other wooden objects coming into contact with washing solutions.

When many timber species are wet they are distinctly acidic; about half of a comprehensive range of hardwoods and softwoods showed a pH value of below 4·5 in a recent series of tests. This degree of acidity is often sufficient to corrode iron nails, but in the case of oak, corrosion is due to the vapour of acetic acid which is given off, and indeed copper and lead fittings are often corroded, even when not in actual contact with the oak.

Electro-chemical decay is best described in this group. It often occurs in boats when an electrolytic cell effect may take place, causing extensive damage to wooden hulls, which may look like fungal decay. Dissimilar metals must be present, such as copper and steel bolts, with the brine-soaked wood acting as the electrolyte between them. The metals corrode, and electrolysis of the brine causes sodium to accumulate at the cathode (simultaneously forming caustic soda), and chlorine at the anode. Both accumulations attack the wood locally, and can cause it to proceed to an advanced stage of decay.

V. *Decay by heat.* Wood which has been seasoned to normal moisture content limits is not harmed by prolonged exposure to low temperature. On the other hand, continued exposure to temperatures near the boiling point of water may damage the wood. The appearance of wood so exposed is remarkably like that of fungal decay.

The inflammability of wood is a subject best briefly discussed under this heading. If the temperature of wood is raised to about 250° C, inflammable gases are given off, which, if ignited by a flame, will burn. This further increases the temperature, and the wood 'burns' until all is consumed. The inflammable gases burn first, leaving a mass of carbon; if the temperature is raised sufficiently, the carbon will also be consumed. Timbers have differing properties in this regard, but the size of the timber is of great importance. A piece of burning timber will become extinguished when the external source of combustion is removed, if the thickness is greater than about 12 mm. The charcoal accumulation is one of the factors tending to reduce combustibility.

The special chemical formulations for preservation of wood against combustion act by preventing flaming when the inflamable gases are produced. They are known as fire retardants.

How wood-rotting fungi decay wood

The ability of fungi to decay wood is due entirely to their ability to break down cellulose into simple degradation products which can be utilized for respiration. This is brought about by the secretion of suitable enzymes and is known as enzymatic degradation. Enzymes are catalytic substances produced by living cells which can cause chemical changes to take place without suffering any change themselves. Enzymes are specific in their action, one particular enzyme producing one particular reaction. In this and other respects they differ from chemical catalysts, which they resemble in function. Enzymes are named by adding 'ase' to the root of the substrate on which they act. The enzyme cellulase is the one responsible for the hydrolysis of cellulose, and acts both outside and inside the fungal cells. Outside the cell wall, cellulose is broken down to low molecular weight sugars by enzymes which are secreted by the actively growing hyphae onto the water film coating the wood fibre. The intermediate stage from cellulose to glucose is known as 'cellobiose'.

The enzymatic effect on lignin has not been studied and all that is known is that lignin forms a barrier to enzymes normally acting on cellulose. In other words, without the presence of lignin, the cellulose would be degraded much more readily in the presence of the appropriate enzymes. The products pass through the cell wall to the inside, where final breakdown takes place (the related enzyme β glucosidase is responsible for this).

The wood-rotting fungi

The origin of the extraordinarily large group in the vegetable kingdom known as the fungi is shrouded in mystery. Most authorities consider that they have been derived from algae-like plants, but that their parasitic or saprophytic mode of life has caused profound changes in their morphology and physiology. The common character of all fungi is the absence of chloroplasts—the small mobile bodies in the cells containing chlorophyll and concerned in the process of photosynthesis. Fungi are therefore unable to build up carbohydrate food material from carbon dioxide from the air and water, and must in consequence obtain this essential nutrient from other organic matter, animal or vegetable, living or dead. There is hardly a group of living organisms that is not associated with fungi, either living on or in it as a parasite when it is alive, or helping to split up or decay it, when it is dead.

The group of the fungi most concerned in the decay of wood is the BASIDIOMYCETES, with a few species in the ASCOMYCETES. Recently, a great deal of attention has been given to a group of fungi known as the

'Fungi Imperfecti', which is concerned with the decay of wood known as soft rot.

One of the most important features of fungi is the nature of the cell wall. In almost all cases, this is not cellulose as in all other plants, but a substance very similar to chitin—from which a large part of the insect skeleton is composed. On reflection it is hardly surprising that an organism secreting a cellulose-digesting enzyme should not have cell walls of cellulose. In the evolution of the fungi, the change in the basic chemical substance of the cell wall must have preceded the cellulose-feeding habit. The thallus of the wood-rotting fungi is often not a gross feature of the wood decay it causes. It consists frequently only of threads one cell in thickness (the hyphae), which proliferate inside the wood, attacking the wood substance by boring holes in it just large enough for the hypha to pass through (Fig. 6). Sometimes, however, the hyphae are present as a

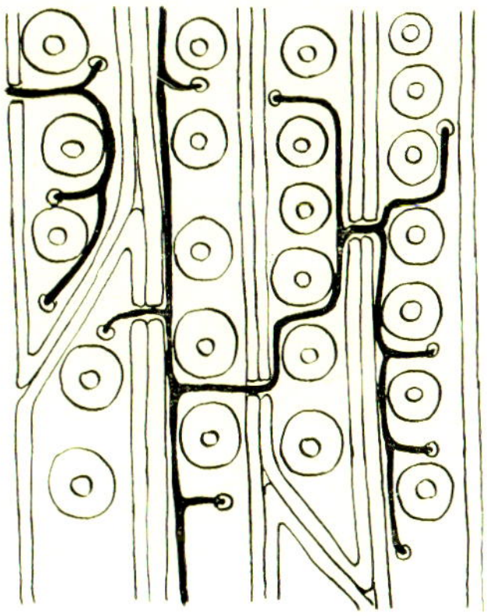

Fig. 6. Softwood shown in longitudinal section highly magnified. The hyphae of a wood-rotting fungus are shown entering the tracheid walls in section and in surface view. Diagrammatic.

dense mat (mycelium) covering the surface of the wood, and invading as a sheet over intervening brick, stone, or plaster. Other hyphae sometimes form aggregates or strands for conveying nutrient materials or water.

Distribution of the wood-rotting fungi is usually effected by spores. These are minute—a single spore not being visible to the naked eye—but they are produced in very large quantities. Being so light, they are blown about by the wind, and their presence can be demonstrated in almost every conceivable situation.

Little is known of the fungi responsible for soft rot. A species which has had some study in the United Kingdom is *Chaetomium globosum*. It appears to be present in almost all very wet situations, an example being the cooling towers of electricity generating stations.

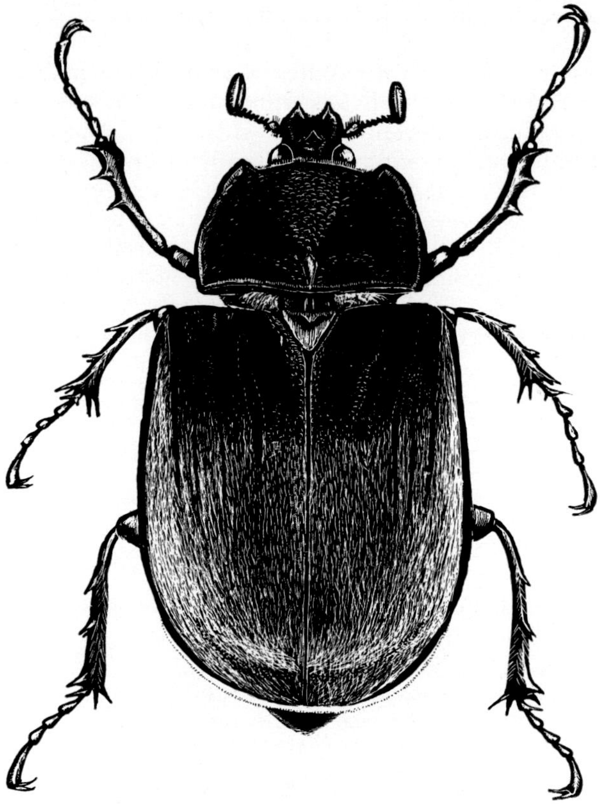

Fig. 7. *Megasoma gyas* Herbst. Length: 60 mm. The specimen illustrated was shown, set and carded, at a meeting of the Royal Entomological Society in London on 2nd November 1949, and made history by unshackling its tarsi of gum and walking off the card! It was referred to in *Proc. R. ent. Soc. Lond.* (C) **14**:42 as a *living* female adult. The specimen together with a larva was collected by Mr. M. G. Fraser on 19th October, 1949, at Seaforth, Lancs, from Brazilian rosewood. This timber species—*Dalbergia nigra*—is a hard, heavy, durable wood, but the heartwood is often decayed, thus accounting for the presence of this member of the SCARABAEIDAE. Surely this species is the bulkiest beetle ever to have been accidentally introduced into Britain.

Wood destroying animals

Many diverse groups of animals destroy wood. This is always done by burrowing within it either by the immature and adult stages, or by the immature stage alone. At each generation, means exist for distribution by the resulting adults leaving the infested wood (*Anobium punctatum*) or by young larvae being ejected from it (*Teredo*).

Crustacea

This animal class comprises crabs, lobsters, crayfish, prawns, shrimps, sand-hoppers, woodlice, barnacles and water-fleas. The CRUSTACEA differ from insects with which they (with some other groups) form the phylum ARTHROPODA, by possessing two pairs of antennae and at least three pairs of jaw-like appendages, behind the mouth. A number of crustaceans are known as wood-borers in floating or sunken wood in the sea, the most important being the Gribble (*Limnoria lignorum*). This is a member of the group ISOPODA and resembles a small woodlouse about 3 mm in length and dark grey in colour. It has seven pairs of walking legs. Where found it is usually present in very large numbers, and bores into the surface of the wood, the gallery in the first place running obliquely until it is 10–15 mm from the surface, then parallel to it. From time to time a 'manhole' is constructed, giving direct access to the fresh sea-water.

When the outer layer of timber has been completely honeycombed, it is washed away by wave action, and the tunnelling then starts in the next layer. Each gallery contains a pair of animals, but only the female does the burrowing, and she produces about twelve larvae at a time. The latter make burrows sideways from the parental burrow, and when they obtain access to the sea, they swim off and find other timber to infest. Another crustacean inhabits the *Limnoria* burrows named *Chelura terebrans*, belonging to the group AMPHIPODA. It is thought that this is an example of symbiosis—two animals living in close association with each other from which they derive mutual benefits.

No cellulose-digesting enzymes are secreted by the *Limnoria*, so that cellulose itself is not being utilized by the animal as a foodstuff. The suggestion has been made, however, that *Limnoria* is actually feeding on soft rot fungi, which would be present in the superficial layer of the wood.

Mollusca

This group contains among many other forms, snails, slugs, mussels, oysters, and whelks, and anatomically contains many diverse elements. The group EULAMELLIBRANCHIATA contains a number of animals with the wood-boring habit. The best known is the ship-worm *Teredo*, a bivalve which, with related genera such as *Martesia*, is of great importance in many seas of the world as a destroyer of wooden-hulled ships, and of wooden harbour works. The free-swimming larvae are produced in large quantities, and actively swim about in search of wood. Those that are fortunate enough to find it, first attach themselves to it by a thread, then start to burrow into it after a metamorphosis. The gallery in the wood is always lined with a calcareous material which serves to make identification a simple operation.

The animal can attain a length of 45 cm or so, and its burrow can be up to 25 mm in diameter. The only signs of external are damage the small openings in the wood made by the larvae when they first bore in. Two siphons about 35 mm in length emerge from this hole, and they can be quickly retracted, whilst at the other end of the elongated body, two small shells rasp away the wood by their continual rocking motion. Cellulase is secreted by *Teredo* so that some digestion of cellulose takes place, but some planktonic organisms sucked in through the inhalant siphon are also used as nutrient material. Again, it is believed that soft rot fungi play some part in the boring relationship of this animal, but precise details are unknown. *Xylophaga dorsalis* is another wood-boring bivalve mollusc. The shells are much larger and the body smaller than the previous species, the galleries seldom exceeding 50 mm in length.

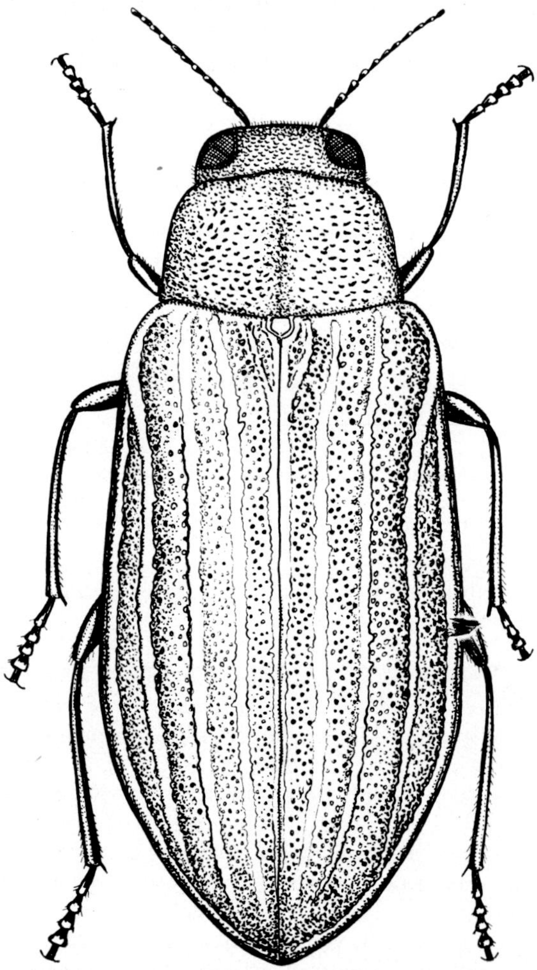

Fig. 8. *Buprestis aurulenta*. Length: 17 mm. The larva of this bright green shining beetle bores into Douglas fir in Canada. The timber, containing larvae, is often converted and imported into the British Isles. The beetles emerge many years afterwards—the decrease in moisture content of the timber prolonging the larval life.

Insects

By far the most important animal group causing destruction of wood is the class INSECTA, which is divided into twenty-nine orders, of which five have members known to bore into wood.

A species with wood-boring larvae is found in the order EPHEMEROPTERA (Mayflies). This is *Povilla adusta*, whose larvae burrow into wood floating or submerged in the fresh water of East and Central Africa. It is almost certain that aquatic wood-rotting fungi (probably species of soft rot fungi) cause the initial infestation of the wood, enabling the *P. adusta* larvae to commence burrowing operations.

The next order, the ISOPTERA (Termites), is an extremely important wood-destroying group. These insects have a social organization, and live in large communities consisting of several castes. In the termitarium, often deep in the ground in a special cell, live the royal pair. These are usually the original founders of the colony, having now lost their wings.

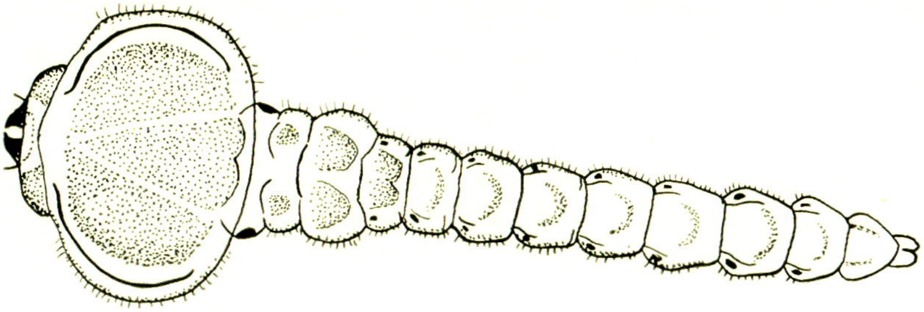

Fig. 9. Larva of *Chalcophora mariana* (L.). A member of the beetle family BUPRESTIDAE. Larvae of this family usually feed in the phloem when young before entering the wood.

The abdomen of the female swells to an immense size, and often up to 1,000 eggs per day are laid over relatively long periods (several or many years). In addition to the reproductive castes, there are the sterile castes—workers and soldiers. The former, as their name suggests, carry out the duties of foraging for food, chewing up wood on which to grow fungus, cleaning the nest, nursing, etc., whilst the soldiers defend the nest from invaders and marauding insects. There are over 1,700 different species known, and they are extremely common in the tropics and warm temperate countries. Only two species are commonly found in Europe, *Kalotermes flavicollis* and *Reticulitermes lucifugus*, but neither of these occurs in the British Isles. There have been many studies on the nutrition of termites and their method of feeding. It has been discovered that the insects do not secrete cellulase, but the digestion of cellulose is carried out by intestinal protozoa—primitive one-celled animals inhabiting the gut. This is yet another case of symbiosis—animals of different species associating together to their mutual benefit. Each species of termite carries its characteristic species of flagellate protozoa, although some termites harbour as many as ten different protozoan species.

About 100,000 different species of LEPIDOPTERA (Butterflies and Moths) have been described, but relatively few members of this order have adopted the wood-boring habit. However, one rather primitive family, the

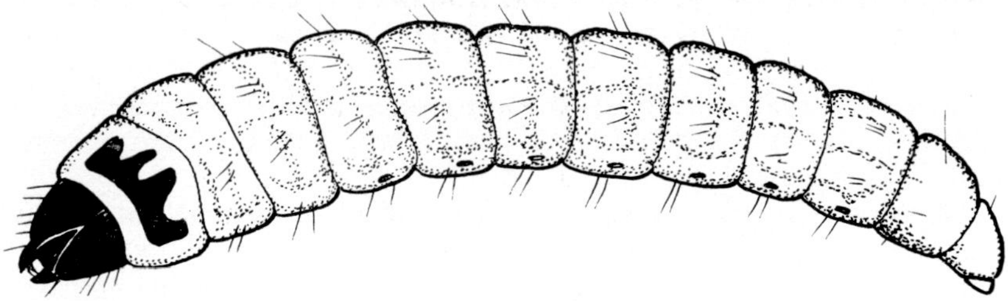

Fig. 10. Larva of the Goat Moth *Cossus cossus*. It is found boring in the trunks of oak, ash and willow. It takes three to four years to complete growth.

COSSIDAE (Goat and Carpenter Moths) consisting of large or very large species, has wood-boring larvae. These infest fruit and other trees, making very large galleries.

The HYMENOPTERA (Ants, Bees, and Wasps) contains a primitive family, the SIRICIDAE, known as Wood-wasps or Horntails. These are large insects and the female has a long stout ovipositor, by means of which she drills a hole through the bark of a tree and passes a single egg into the new wood. When the larva hatches from the egg it bores into the heartwood for about two years before completing the life cycle.

Perhaps the most highly developed of all insects are the COLEOPTERA (Beetles). They number about 220,000 different species, and are therefore the most numerous order in the animal kingdom. They outnumber in species all other insects. The biology of beetles is certainly most diverse, and no other insects have invaded land, water, and air to anything like the same extent. Almost every form of organic matter except living animals is utilized as food by beetles. It is not surprising, therefore, to find many families of beetles using wood as a foodstuff. Many thousands of species of beetle larvae bore into the wood of living trees, and wood in the last stages of fungal decay is likewise attacked.

The most important families of beetles with wood-boring habits are: BUPRESTIDAE, ANOBIIDAE, BOSTRYCHIDAE, LYCTIDAE, LYMEXYLIDAE, OEDEMERIDAE, CERAMBYCIDAE, CURCULIONIDAE, SCOLYTIDAE and PLATYPODIDAE.

REFERENCES

CARTWRIGHT, K. ST. G. and FINDLAY, W. P. K. 1945. *Decay of Timber and Its Prevention*. Forest Products Research Laboratory. H.M.S.O. London.

CHRISTENSEN, C. M. 1951. *The Molds and Man*. University of Minnesota Press. Minneapolis.

FINDLAY, W. P. K. 1953. *Dry Rot and Other Timber Troubles*. Hutchinson. London.

FITTING, H., KARDER, R., SIERP, H., and KARSTEN, G.—*Strasburger's Textbook of Botany*. Sixth English Edition, Macmillan. London.

GRAHAM, S. A. 1952. *Forest Entomology*. 3rd Edition. McGraw-Hill. New York.

HICKIN, N. E. 1954. *Woodworm, Its Biology and Extermination*. London.

HICKIN, N. E. 1967. *The Conservation of Building Timbers*. Hutchinson. London.

HICKIN, N. E. 1971. *Wood Preservation—A Guide to the Meaning of Terms*. Hutchinson. London.

HICKIN, N. E. 1971. *Termites, A World Problem*. Hutchinson. London.

HICKIN, N. E. 1972. *The Woodworm Problem*. 2nd Edition. Hutchinson. London.

HICKIN, N. E. 1972. *The Dry Rot Problem*. 2nd Edition. Hutchinson. London.

IMMS, A. D. 1957. *A General Textbook of Entomology*. 9th Edition, revised by O. W. Richards and R. G. Davies. Methuen. London.

SKAIFE, S. H. 1955. *Dwellers in Darkness*. Longmans, Green & Co. London.

VITÉ, J. P. 1952. *Die Holzzerstorenden Insekten Mitteleuropas*. Musterschmidt Wissenschaftlicher Verlag. Göttingen.

CHAPTER THREE

The Anobiidae. I

An account of Anobium punctatum*—A brief description of all the species of the British* ANOBIIDAE, *including* Ernobius mollis *and* Ptilinus pectinicornis

---

An account of the biology of *Anobium punctatum*

Introduction

This chapter represents an attempt to put together all the known information, both morphological and biological, concerning this wood-boring beetle. Critical comment, however, has not been excluded where appropriate, and suggestions are made concerning those aspects of which our knowledge is meagre with regard to the research which should be carried out in order to remedy such a state of affairs. A long bibliography is included for which no claim is made of completeness and all the cited papers have not been seen in original where abstracts have given the desired information. It has not been possible to include in this chapter a detailed account of much of the experimental work which has been carried out. In particular the section on nutrition, an understanding of which is so important if we are to learn very much about this insect and its impact on our economy, has been dealt with, of necessity, only in brief. From several points of view, *Anobium punctatum* is an enigma. Several aspects of its biology require to be confirmed, and there still remain secrets to be unravelled. It has, in addition, been the subject of intense controversy and litigation.

The technology of wood conservation

One fact concerning this insect, however, is incontrovertible. It has been the cause of the rapid development of the technology of WOOD CONSERVATION.

The attack of *Anobium punctatum*, at least to the extent that commercial extermination treatment is recommended, occurs only when the property has been erected some 20 to 30 years. Therefore, the treatment can only be given to the timber in the permanently fixed condition and often when it is partially or wholly hidden. This calls for special and often elaborate techniques.

This technology is based on a broad spectrum of knowledge. The botany of timber must be understood. The multiplicity of timber species, although a joy to the user, presents a difficult exercise for the technologist.

The structure of the woody plant, aspects of the physiology of the growing tree, felling, conversion, and dehydration are subjects with which the wood conservationist must have some familiarity.

In addition, timber as a structural material must be studied because it is at this point that the greatest economic losses are likely to occur. It is axiomatic that a wide knowledge of the biological factors of wood decay and of the principles of ecology are drawn on daily by the technologist. When one further considers the chemical and physical properties of the liquids, gases, and solids used in conserving wood, the engineering problems of application of these chemical products, the handling of timber already *permanently* fixed, then it will be realized that the wood conservation technologist must be a man of broad view, with a fund of detailed knowledge, but with a resilience such that developing techniques can be grasped, distributed and applied with skill.

During the past 28 years it has been the writer's task to play a part in this work, and to induce a reluctant public to recognize the scale on which infestation by *Anobium punctatum* was developing.

Thirty years ago, this insect was associated only with old houses and old furniture, yet today its incidence is so serious as to be of national concern, doing much damage in property only constructed some 20 to 30 years ago. In the opinion of the writer, this change was fairly abrupt, coinciding approximately with the 20-year period after the great upsurge of softwood timber imports which came soon after the First World War.

Economic status of Anobium punctatum

In the United Kingdom there are 600 or so companies carrying out treatments against wood-boring insects as a major part of their work, 120 of whom are members of the British Wood Preserving Association. These vary enormously in size, from the national company with 200 trained timber infestation surveyors, at the time of writing, and a large fleet of elaborately equipped vehicles, to the man with a couple of employees, with relatively simple equipment, carrying out work on a local scale. About 80 per cent of their work is known to be due to *Anobium punctatum* attack. In addition there are several companies who manufacture chemical products for application to timber by builder, general contractor, and the householder himself.

A high proportion of property owners, having obtained a report and estimate for *Anobium* extermination, will not face the expense involved. A very much higher number of property owners either know well that their property is *Anobium* infested, but because it is leased to a tenant (or for some other reason) do nothing about it, or are in ignorance of the situation. In both cases the adult beetles in the summer are free to distribute themselves in the district.

Most owners, when they come to sell their property, have the attack of *Anobium* pointed out to them by the prospective purchaser, and have to accept a figure much less than that asked because of this.

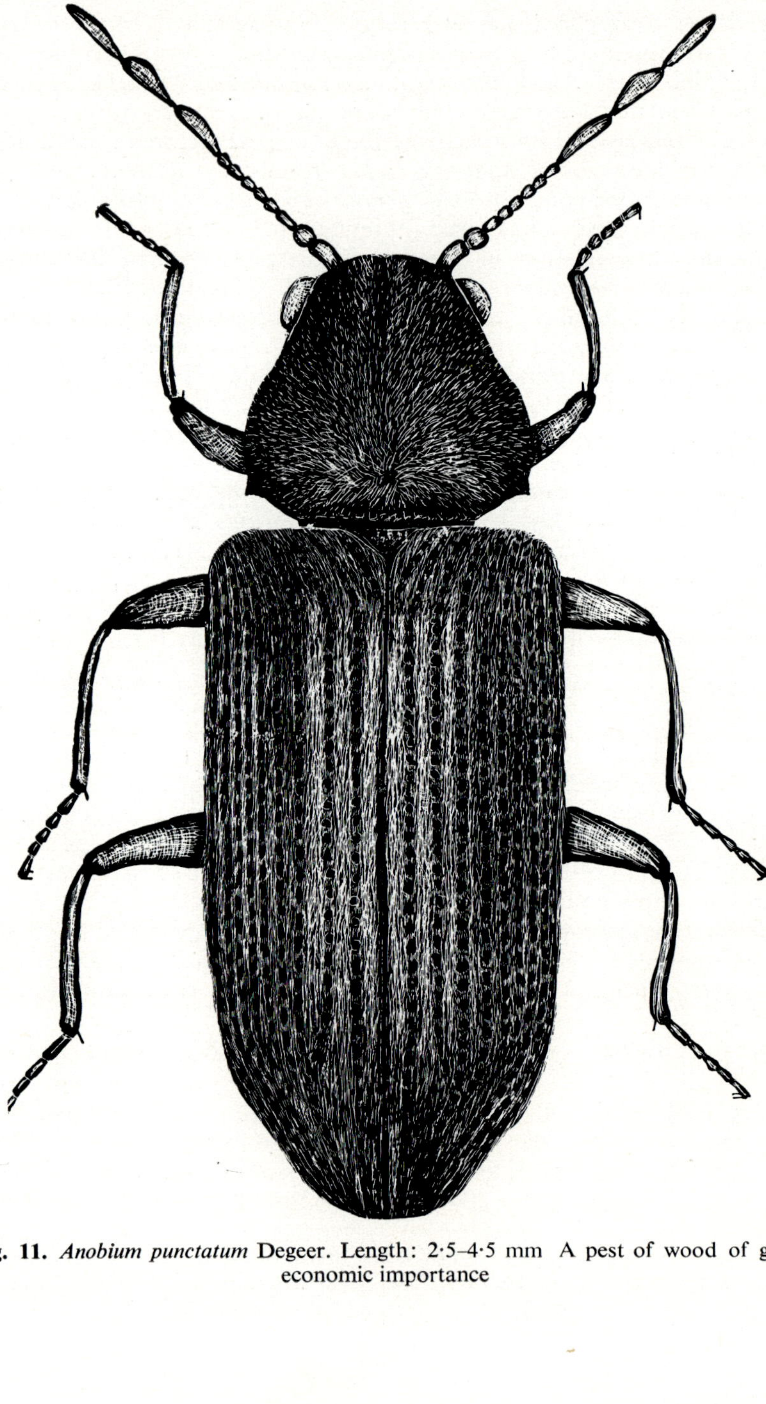

Fig. 11. *Anobium punctatum* Degeer. Length: 2·5–4·5 mm A pest of wood of great economic importance

Distribution of Anobium punctatum *in the United Kingdom*

Stephens (1839) stated that *Anobium punctatum* was abundant in old houses throughout the country, and today there can be no doubt that it is found in every parish in the United Kingdom. It is a common insect out of doors attacking the dead parts of trees, such as where branches have been cut off, or any situation where the bark has been accidentally removed thus killing the underlying sapwood. It appears never to be found where the adult would have to make its flight hole through bark.

Since about 1945 this species has become exceedingly common as a pest of wood indoors in furniture, musical instruments, tools and implements, and on a very much greater scale in the joinery timbers, flooring and structural timbers of buildings (Hickin, 1957). *Anobium punctatum* is today a pest of major economic importance and a nationally marketed product for woodworm extermination sells to half a million customers every year in the United Kingdom from the Shetlands, to the Scillies, and the Channel Islands.

There have, however, been a number of opinions expressed as to the differing geographic incidence of infestation. 'It is more prevalent and damage is more extensive in Devon, Cornwall and the West Country generally, as well as in the West of Scotland. The mild and wet climate in these parts of the country appears to be particularly suitable for the insect, and for spread of infestation' (Fisher, 1949). 'Without a doubt much more common in those counties with seaboard (but including Surrey) and appears to be much less common in Stafford, Warwick, Nottingham, and Leicester' (Hickin, 1954). 'Commonly thought to be especially prevalent in areas of greater humidity' (F.P.R.L. Leaflet No. 9, 1959).

The commercial servicing operations for *Anobium* extermination are now on such a scale that a study of the distribution of the incidence of these operations has given some valuable information which is summarized in Chapter XI.

World distribution of Anobium punctatum

The distribution of *Anobium punctatum* through the world can be considered in three parts. Firstly, the area in which this species is thought to be indigenous. Secondly, those areas to which it has spread by commerce; in this case it would be by means of 'woodworm' infested furniture taken from Europe by early settlers. Lastly, the third area—where the insect is absent.

There seems little doubt that *Anobium punctatum* is indigenous to the great mass of temperate Europe. In the absence of evidence to the contrary, it can be assumed that it is indigenous also to temperate Asia. It was first described by Degeer from Swedish specimens. Becker states that it is the commonest and most destructive of the ANOBIIDAE of Central, North, and East Europe. He goes on to say that the distribution is so extensive in Germany that it is possible to find the animal in nearly every building.

Every kind of processed or other dead wood, of every kind of tree, used in any way at all, is attacked by it. It is known also from Russia, although its precise status as an economic pest is not known.

Through commerce, *A. punctatum* has been introduced to and now thrives throughout temperate Australasia. In Tasmania, Evans states that *A. punctatum*, although not attacking the timber of the eucalyptus, infests blackwood (*Acacia melanoxylon*) and particularly New Zealand white pine (*Podocarpus dacrydioides*). In New South Wales, old, well-seasoned timber, mainly softwoods (*Pinus* spp.), are attacked and *Podocarpus dacrydioides* is again specially mentioned as being particularly susceptible. Incidence of attack in eucalyptus and hardwoods is comparatively slight.

In New Zealand *Anobium* was mentioned by Kirk (1905), Speight (1918) and Cockayne (1919). Spiller has given an account of the great damage to buildings in a country where wood is the main building material. He says that every house that has been erected for more than 15 years is infested with the Common House Borer, *Anobium punctatum*. New Zealand has a population of only about 3 millions, but there are several dominion-wide and numerous local firms specializing in borer control. The amount spent in the 20-year period 1930–50 was more than £2 million. Although the native longhorn beetle *Ambeodontus tristis* F. and the native termite *Kalotermes brouni* Frogg cause some damage, the main income is from dealing with *Anobium punctatum*.

In North America it appears to have a wide distribution on the eastern side, although it does not occur in such spectacular proportions. According to Simeone, *Anobium punctatum* only became established in the United States of America in the last 100 years.

In South Africa, Tooke (1949) says *Anobium punctatum* is of European origin, but is now widely distributed throughout the world, and, being the commonest of the anobiids, is responsible for the greatest damage by this group of beetles, particularly to wood used inside buildings. He then goes on to say that *A. punctatum* is a modern introduction into house timbers in South Africa, and was not noticed in infested furniture until about 1939. It is now widely distributed throughout all four provinces of the Union, but the most serious and extensive damage is to be found in the coastal areas. He concludes his remarks on the distribution by saying, from the nature and extent of the damage, it is apparent that the attacks are of several years' standing, and that the insect was introduced into the Union quite a considerable time ago.

With regard to the areas where *Anobium punctatum* has not been recorded, these include tropical Africa and tropical Asia (Beeson says that it seems that *A. punctatum* has not been found in India). There do not appear to be records of its occurrence in tropical or South America.

In conclusion, it can be stated that *Anobium punctatum* is a serious pest to timber indoors in the greater part of the temperate world. It remains a potential hazard to building timber and furniture in areas where as yet it has not been found. According to Becker, it is not likely to become a pest to timber in tropical areas, except where altitude brings about conditions more suited to it.

Historical classification and synonymy of Anobium punctatum *Degeer*

The first description of this insect was given by Degeer in 1774, and the title page of the fourth volume of his *Memoires pour servir a l'Histoire des Insectes* is shown in Fig. 12.

The original description is as follows:

Vol. 4. Page 230. MEMOIRES POUR SERVIR A L'HISTOIRE DES INSECTES.

'Vrillette pointilleé. — 3. VRILLETTE d'un brun griseâtre, a étuis canéles a points concaves.
Ptinus (punctatus) griseo-fuscus elytris striatis punctisque excavatis.

MEMOIRES
POUR SERVIR A
L'HISTOIRE
DES
INSECTES.

Par M. le Baron CHARLES DE GEER,
Marechal de la Cour, Commendeur & Grand-croix de l'Ordre Royal de Waja, & Chevalier de celui de l'Etoile du Nord; |Membre de l'Académie & de la Société Royales des Sciences de Suede, & Correspondant de l'Académie Royale des Sciences de Paris.

TOME QUATRIEME.

A STOCKHOLM.
De L'IMPRIMERIE de PIERRE HESSELBERG.
M. DCC. LXXIV.

Fig. 12. Title page of Degeer's *Memoires.*

Elle est des plus petites, sa longueur n'est que d'une ligne & demie, de sorte qu'elle est beaucoup plus petite que la Vrillette *opiniârte*, à laquelle elle ressemble au reste dans toute sa figure, ayant, le corps allongé & cylindrique. Sa couleur est entierement d'un brun un peu griseâtre, & les étuis, qui couvrent les ailes, ont des canelures longitudinales, marquées par points concaves. Le corcelet a une élévation bossue au milieu.'

It will be seen that the relationship with the genus *Ptinus* in the PTINIDAE was even then realized, but it has often been a source of wonder to the writer that this species was apparently not known to Linnaeus.

The genus *Anobium* was erected by Fabricius in 1775 in his work the *Systema Entomologiae* published in Flensberg and Leipzig. He includes six species in his new genus (pp. 62–3), none of which appears referable to *punctatum* Degeer. In the *Entomologia Parisiensis*, edited by de Fourcroy, in which the classification of Geoffroy is followed, the species *Birrhus domesticus* is described, which is referable to *punctatum*. This work was published in Paris in 1785. The insect is given the common name of *La Vrillette des tables*, and is described as 'B. testaceo-niger, thorace subhirsuto. *Loc.* Ligneas suppelectiles foraminibus rotundis terebrat.'

Kloet and Hincks (1945) give Olivier (1790) as the author of the species *striatum*, a synonym of *punctatum* which has been widely used. This would be in the *Encyclopedie Methodique* published in Paris, in which the classification of Fabricius given in 1775 is used. The date 1790, however, would appear to be incorrect, as only Volume V was published in that year, and any species referable to *punctatum* could not be found in it.

Classification

The family ANOBIIDAE, together with the families, PTINIDAE, BOSTRYCHIDAE and LYCTIDAE comprise the superfamily, BOSTRYCHOIDEA. This is referable to the suborder TEREDILIA. Crowson (1950) characterizes the superfamily by the prothorax nearly always being produced forward over the head in a hood-like fashion; tarsi with segment 1 very small or trochanters elongate and with an almost transverse junction with the femora; dorsal ocelli absent; meta-endosternite nearly always with anterior tendons close together, and borne on a median projection. Larva soft and usually C-shaped, without distinct tergal sclerites and with the gula usually indistinct.

The LYCTIDAE, which are without the hood-like projection of the prothorax, have previously occupied a somewhat anomalous position, but the work of Cymorek has shown that the anobiid *Ptilinus pectinicornis* has some important similarities to the LYCTIDAE in its biology.

Morphology and identification

A detailed morphological study of *Anobium punctatum* appears to be lacking and only a very brief incomplete description is given here. Adult beetles vary in length from 2·5 to 5·0 mm, and the colour, according to

Stephens (1839), is pitchy-brown, with legs and antennae dull red. Hickin (1957) says variable in colour, from light reddish yellow through dark chocolate brown to pitchy-red; some of the variation in colour is due to the pubescence rubbing off. The legs are usually reddish-brown.

Anobium fulvicorne Sturm may be identified from *punctatum* by being dull blackish in colour with the antennae, tibiae, and tarsi reddish. The antennae are shorter than in *punctatum*, and the elytra are less rounded at the apex. The tarsi are more distinctly thickened towards the apex. The most common insects with which *punctatum* is confused by the public are *Ptinus tectus* Boieldieu in the PTINIDAE and *Stegobium paniceum* (L.) and *Ernobius mollis* (L.) in the ANOBIIDAE.

Pronotum

This is almost as wide as the elytra. Viewed from above (Fig. 11), the pronotum is pear-shaped with prominent sharp angles at the base. It is covered with pubescence. Viewed from the side, the front margin is sinuate rising to a prominent crest (Fig. 39). At each side of the crest there is a fovea which appears to vary considerably in extent. A shallow depression occurs in the mid line from the front of the pronotum to the crest. This also varies somewhat in depth and width. The acute angles at the base of the pronotum, as seen from above, are actually thickened ridges which run almost vertically downwards, just anterior to the hind margin of the pronotum.

Wings

Wings are well developed (Fig. 13) and the beetle is capable of sustained flight. The wing-beat is very rapid and the flight is somewhat reminiscent of that of the common house fly. After flight the wings are not covered completely by the elytra and the wing-tips protrude.

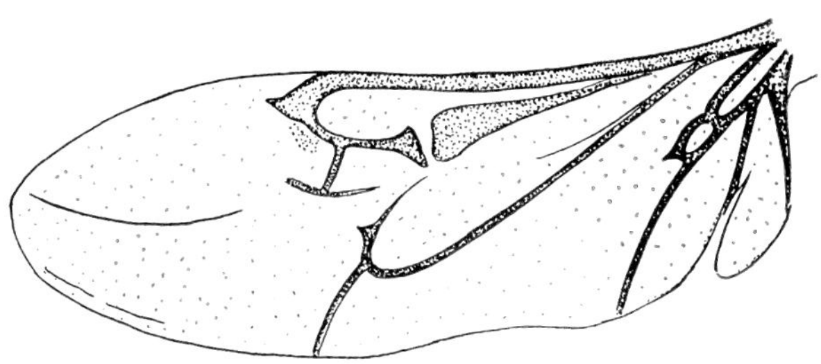

Fig. 13. Left wing of *Anobium punctatum*.

Legs

These are approximately equal in length and covered in fine short hair (Fig. 14). Each leg has five tarsal segments, of which the proximal segment is the longest, and in each case the segments become shorter as they

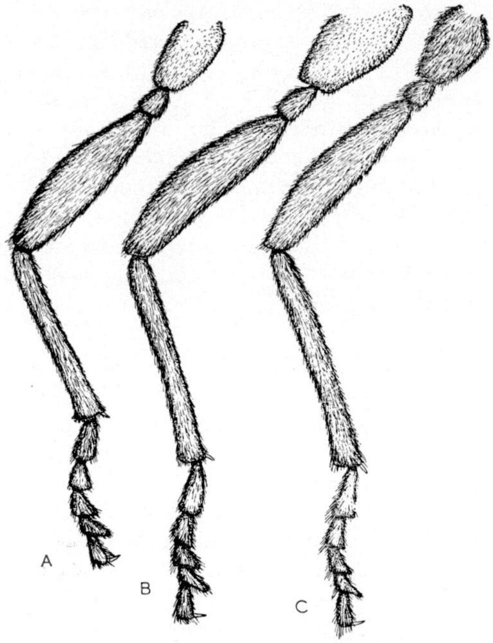

Fig. 14. Legs of *Anobium punctatum.*

progress distally (with the exception of the fifth tarsal segment). In addition, the distal lobes of the segments become narrower and more attenuated as they progress distally. *Fore leg:* this is only slightly smaller than the middle and hind legs, with the tibia only slightly less in length than the femur and trochanter together. *Middle leg:* the tibia of this is as long as the femur and half the length of the trochanter together. The coxa is larger than that of the fore and hind leg and the femur is the same length as that of the hind leg. The trochanter is longer than that of hind leg. *Hind leg:* the tibia is just as long as the femur and trochanter together, and only half as wide as the femur.

Elytra

The characteristic feature of the elytra is the series of nine rows of sculptured pits on each elytron. They vary from subcircular to bean-shaped. The thin web of material left at the base of the pit appears almost black in the living or dead insect, but if the elytron is detached and light passed through, it appears pale lemon, and is so thin that certain colours, such as green, can be passed through. The elytra are dilated at the apex.

Identification of Sexes

Kelsey *et al.* (1945) have described differences between the male and female beetles that allow for easy identification (Figs. 15 and 16). The

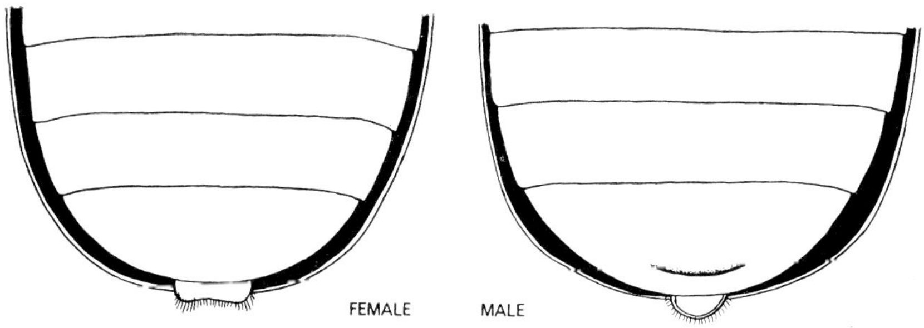

Fig. 15. Terminalia of adult *Anobium punctatum* from beneath.

'telson', the apical sclerotized lobe that covers the genitalia from above, has the apical margin sinuate in the female, whereas in the male it is strongly convex. In addition, the last visible sternite bears a depression in the case of the male which is absent in the female.

Number of Matings

The same authors consider that there is evidence that the females may mate three or more times during the course of egg-laying.

Sex Ratio and Mating

Kelsey *et al.* thought that the sex ratio was approximately equal, although in most circumstances the number of males found exceeds the number of females. This is on account of the habit of the female of hiding or laying eggs, in old flight holes. Mating takes place soon after emergence from the wood, and often occurs with the female in the flight hole and the male on the surface (Hickin), or they retreat into the flight holes whilst still *in copula* (Kelsey *et al.*). This usually takes an hour or so, and on disengagement, the females spend some time laying eggs, whilst the male returns to the surface of the wood. This may be repeated several times. Kelsey *et al.* record 'swarming' in this species which they observed five times, and on sexing one swarm they found 167 males and three females.

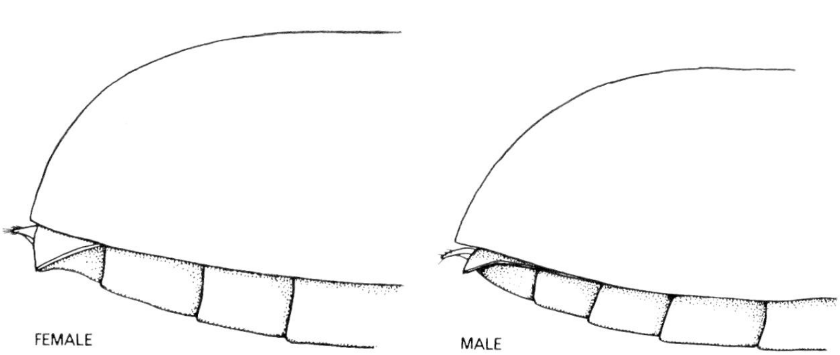

Fig. 16. Terminalia of adult *Anobium punctatum* from the side.

Becker (1940) found that adults lived for about 20 days, but some lived as long as 30 days. They possess the habit of shamming death; the legs and antennae are held close to the body much as in the pupal stage.

The immature stages of *Anobium punctatum*

Eggs

The eggs of *Anobium punctatum* are white or whitish and ellipsoidal; they have been described as lemon-shaped. Gahan pointed out that there was a fine honey-comb-like sculpturing at one end of the egg, extending to about one-third of the length, the rest of the egg being smooth. Kelsey, in his description of the transmission of symbiotic yeast cells from the reproductive organs of the female to the gut of the young larvae, says that the eggs resemble in shape a perfect acorn and its cup, the latter being alveolate with minute pits. The alveolate end is slightly wider than the distal two-thirds, which is perfectly smooth and terminates in a short, blunt point. The eggs are 0·35 mm wide and 0·55 mm in length. The site for egg-laying must incorporate a crack, groove or rough surface in which the egg can be partially wedged or anchored. The eggs seem never to be laid indiscriminately on smooth or otherwise unsuitable surfaces, except where the natural oviposition site is simulated in the laboratory. The normal oviposition sites indoors are cracks and crevices in wood and

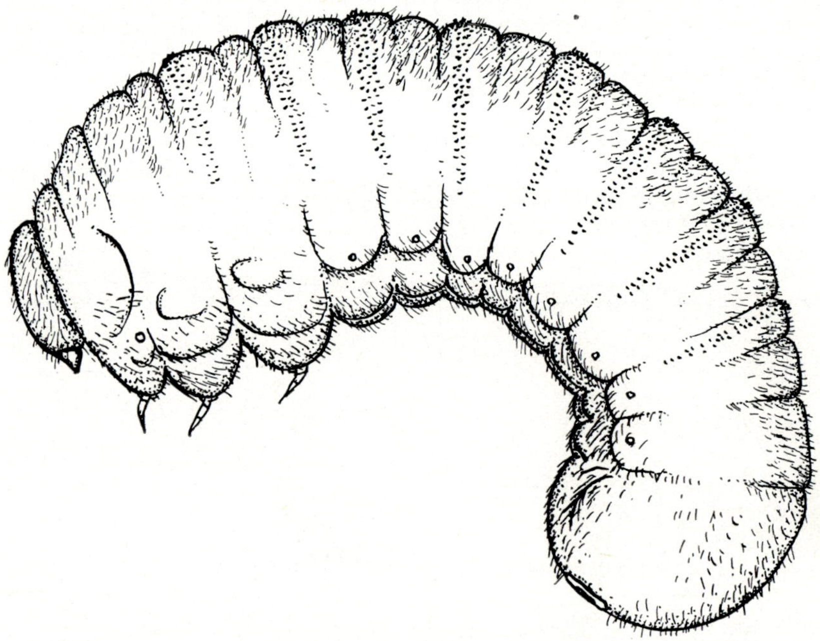

Fig. 17. Larva of *Anobium punctatum*.

rough end grain, rough-sawn timber, and unplaned wood where the grain is open. Unfinished wooden surfaces of various kinds, open joints between two pieces of wood, and the rough edges of plywood are commonly utilized.

The laboratory simulation of an egg site already mentioned consists of stretching organdie muslin or other types of pile-less muslin with adhesive (Kelsey) or cotton (Bletchly) over a block of wood. The *Anobium* female is stimulated to lay her eggs on this material on account of its mechanical suitability.

In an experiment it was shown that 83 per cent of eggs were laid out of doors during the period 10 a.m. to 6 p.m., whilst 72 per cent were laid during this same period under laboratory conditions. On the other hand, 50 per cent were laid during this period in an incubator, and it is suggested that temperature is an important factor in determining the time of

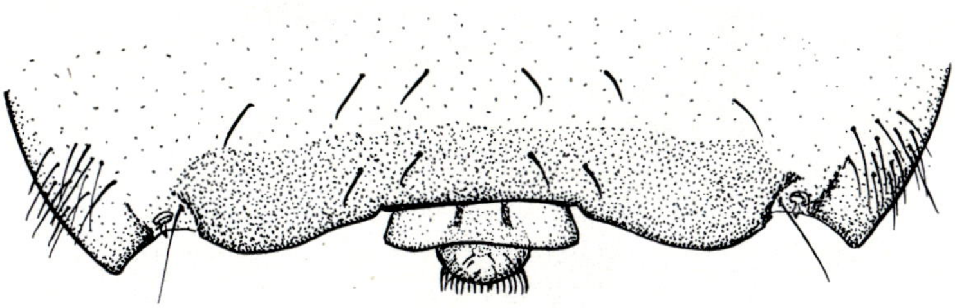

Fig. 18. Epistomal area (after Parkin).

egg-laying, although the effect of light intensity and humidity fluctuations may also be significant.

The eggs are laid singly or in small rows of two, three or four, and altogether a female is capable of laying up to eighty. Fisher had previously put this figure at considerably less (twenty to forty). Bletchly puts this as upwards of sixty, but it is unusual to obtain an average of more than twenty-eight even from individuals taken after emergence early in the season. It is stated in a number of accounts that the egg 'under normal conditions' takes from four to five weeks to hatch, but Bletchly found that at 20°C the incubation period varied from 15·5 days (87 per cent relative humidity) to 23·3 days (43 per cent relative humidity).

Emergence of larva from the egg

Kelsey has described in detail the emergence of the young larva from the egg. The alveolate end of the egg is the last to leave the ovipositor, and the yeast cells already referred to become attached to the rougher, pitted alveolus. The developing embryo within the egg is normally orientated with the head end into the alveolar portion of the chorion, so that when the larva bites its way out, it takes into its gut a number of yeast cells, usually eating a portion of the chorion in the process. The young larva, however, always leaves the egg at the point which is adajcent to the wood. Usually the eggs are wedged firmly into the end grain of the wood, or lie on their sides in similar positions, but the emerging larvae bite their way directly into the wood, normally starting from the narrow end of the cracks in which the eggs are fixed. Altson, however, says that when a number of eggs are laid close together, the ensuing congestion of young larvae all trying to bore into the same place, causes some to wander off to seek another crack. Sometimes these wandering larvae will return after the first larvae have disappeared into the wood.

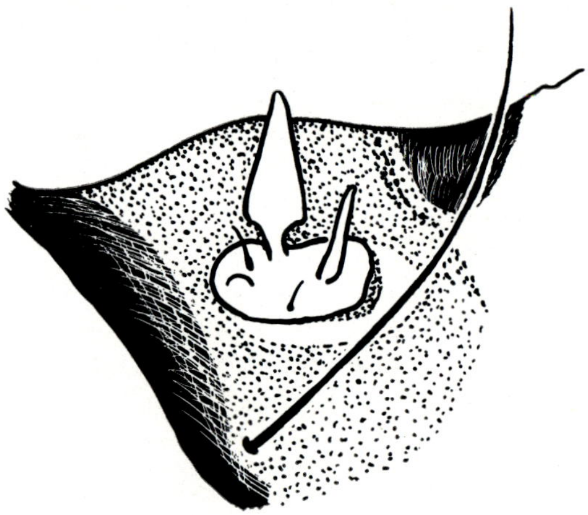

Fig. 19. Right antenna of larva.

The young larvae are straight at this stage—not strongly curved as during later instars—and the dorsal spinules are absent. As the larva commences to tunnel, the frass collects in the empty egg-shell, which often becomes tightly packed, and it may even be forced off its place of attachment. The young larva can withstand some degree of starvation and Kelsey *et al.* record larvae alive nineteen days after removal from wood blocks.

Kelsey *et al.* carried out experiments on the hatching of eggs under controlled relative humidity and temperature, and they found that at 22·5° C, no eggs hatched below a relative humidity of 45 per cent. Although the embryonic larvae developed, they were not able to break through the shell. Below 65 per cent relative humidity the percentage hatching was always below 80 per cent. Within the range of 65 per cent to 95 per cent relative humidity more than 80 per cent of the eggs invariably hatched, and the maximum hatching of eggs (from a large group of block experiments) was 93 per cent.

Larvae

Parkin (1933) and Böving (1954) have made detailed studies of the larva of *Anobium punctatum*. In the work of Parkin, seven British species of

ANOBIIDAE with wood-boring larvae, and one with a non-wood-boring larva, are described, and Böving's work includes sixty-six Nearctic and Palaearctic species. In Böving's paper only one reference to previous work is made—to the work of Parkin—which is quite remarkable, but in the list of references at the end of this chapter, given specifically in connection with work on the larva, seven are cited from Parkin's paper.

Before giving a description of the *Anobium punctatum* larva, it is of interest to note that Parkin found that the larva of *Ptilinus pectinicornis* did not fit into the pattern of the generalized anobiid larva. Cymorek has found that this insect is more closely lyctid in its biology than anobiid. Böving from his study and his wide experience of coleopterous larvae was able to comment concerning the classification of the BOSTRYCHOIDEA. He thought that this was a very well-defined family group, although its

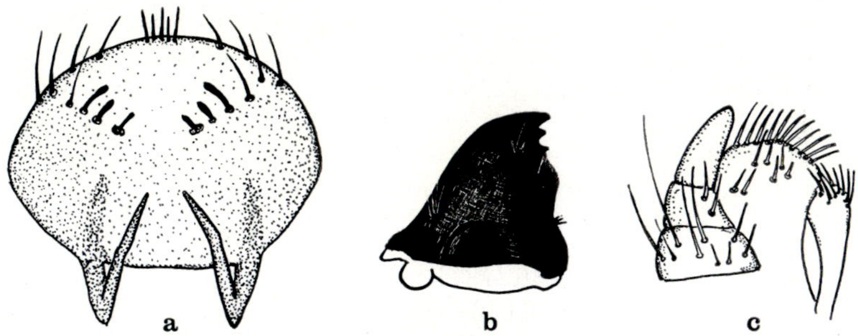

Fig. 20. Larva of *Anobium punctatum*: a, labrum; b, left mandible; c, right maxilla from below.

overall position in the order was not certain. (This was on account of the possession of the so-called nates, a specialized structure just anterior to the anus.) In addition, the larvae of PTINIDAE are so like those of ANOBIIDAE that the validity of the former family is in question.

Larvae of ANOBIIDAE can be distinguished from all other coleopterous larvae by a combination of the following characters:

1. The mandibles are normally dentate with at least two apical teeth, and never possess a true mola grinding against an anvil-like hypopharyngeal sclerotization.

2. A single ocellus with convex, round cornea is located on each side in many species.

3. The body is more or less curved, soft and subcircular in cross-section. The thoracic and abdominal segments are generally of nearly equal thickness. Each segment has not more than two distinct folds.

4. Legs are always present.

5. The spiracles are annular, but often peculiarly aberrant. There are no airsacs from main tracheal branches.

6. The mesothoracic spiracle has moved into the prothorax—but not as far as the anterior margin.

General description of larva of Anobium punctatum

The larva of *Anobium punctatum* (Fig. 17) spends the whole of this stage in a gallery or tunnel which it has constructed by boring with its mandibles within wood, so that to examine a larva it must be obtained by cutting open or slicing the infested wood. The larva is strongly arched (crescentic) when taken from the wood, but this, obviously, is not its natural position. Except for the head capsule, the skin is soft and varies in colour from creamy-white through greyish-white to yellow, or may even have a pinkish tinge. The head is usually a light brownish-yellow, with the more highly sclerotized epistomal area and mandibles chestnut coloured to black. In general type the larva is scarabaeoid and eruciform, and its six legs are very small. Although a transverse section is subcircular, the dorsa surface is strongly curved and the ventral surface flattened. The body, in general cylindrical, is enlarged in the region of the eighth and ninth abdominal segments. Due to the crescentic arching of the body, the last three abdominal segments point anteriorly. The length may be up to 7 mm and the head and body are covered with fine erect setae of a light golden colour.

Head. The head is orthocentrous, light yellow and almost circular. The sclerotized epistomal area (Fig. 18) is narrow, transverse and dark chestnut brown in colour. Munro figures frontal sutures, but Parkin could determine only an epicranial and a fronto-clypeal suture. The latter also states that he could find no sutures on the underside of the head capsule, except the continuation of the epicranial. The epicranial suture runs forwards over the vertex to approximately half-way down, to what Parkin calls the 'face', where the termination is seen as a small depression. Except for the epistome and the area which would normally be covered by the forward projecting folds of the prothorax, the surface of the head is covered with fine erect setae of a light golden colour. The epistome is about five times as wide as long, deeply segmented, and clearly demarked from the fronto-clypeal area. Two setae arise from the anterior margin, but their exact origin appears to vary. There is a long seta by the antennal pit, but no setae elsewhere on the anterior half, although there are scattered setae on the posterior half. The clypeus has the anterior margin slightly sinuous, is four times as wide as long, and two setae are located in each posterior angle.

The antennae (Fig. 19) are very small and membraneous, situated in pits and surrounded by a sclerotized area. There is some confusion concerning the nature of the microscopic parts. Kemner considered that there were two segments and Parkin one, whilst Böving refers to the antenna being without distinct articles. There are two appendages, one large and stalked and the other short; perhaps the former represents the apical segment of Kemner. There is, in addition, a small hemispherical tubercle. There are four setae situated on the antenna, whilst at a short distance from the base is a long stout seta twice as long as the antenna. Böving shows, in addition, another seta about as long as the antenna, and just anterior to the long seta. Parkin and Böving mention two clear oval and margined areas at the antennal base.

Eyes. Parkin made out a pair of degenerate ocelli, one on each side of the head behind the lower angle of the mandible. They consist of minute spots of greyish pigment covered by a lens-like projection. These are

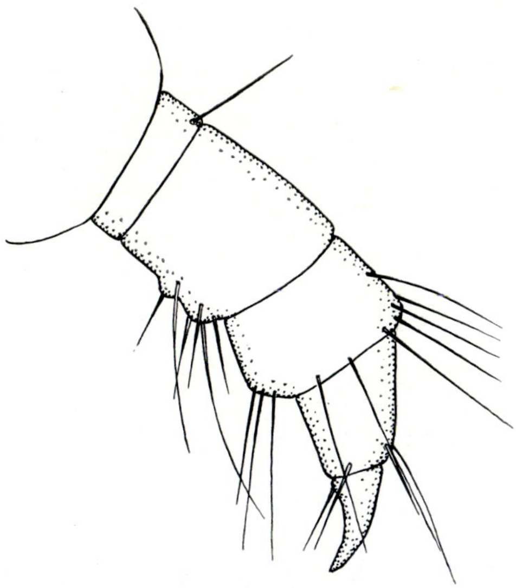

Fig. 21. Larva of *Anobium punctatum:* leg.

recorded also by Altson, Kemner, and Gahan, although Böving, in his specific description of the *Anobium punctatum* larva makes no mention of ocelli.

Mouthparts. The labrum (Fig. 20a) is oval, but more nearly approaching sub-circular than the more transverse labrum of *Xestobium rufovillosum.* The posterior angles extend backwards to form a pair of highly sclerotized projections (labral hooks of Parkin). On the dorsal surface in the anterior half are situated a number of setae and adjacent to the position where the clypeus overlaps it are three small pits. On the ventral surface (the epi-pharynx) there are about twenty stout spines, of different shapes.

The mandibles (Fig. 20b) each have three teeth (Parkin), but Böving draws a distinction between the two short apical teeth and a subapical edge developed into a tooth-like projection which, in turn, is connected to a minute pointed tip (four teeth). There are three brushes on the mandible;

two groups of setae on the dorsel surface, and a small patch of backwardly directed hairs on the ventral surface. All are inconspicuous.

The maxillae (Fig. 20c) are noticeably joined by the submental membrane of the labium. The cardo is emarginate and strongly sclerotized (Parkin). The lacinia is only about a third as large as the galea. There is no terminal spine, but six strong setae are present near the apex and finer setae are present on the dorsal surface. The galea has about twenty strong setae similar in size to those on the apex of the lacinia, and mostly on the anterior margin. The maxillary palp has three segments,the proximal being broad and transverse, and bearing six, large, strong setae and a few shorter ones. Böving describes the prementum, mesomentum, and submentum as having fine long setae on each side, with a few fine setae scattered on the surface. Parkin describes the labium as membraneous (a feature shared with *Xestobium*), and the submentum is large with two posterior groups of about ten setae each on the ventral surface. There are no distinct anterior groups as seen in *Xestobium*, but one large group of twenty to twenty-five rather scattered setae. The ligula, according to Parkin, is rather conical, round apically and furnished with a double row of strong setae extending

Fig. 22. Larva of *Anobium punctatum*: a, spinule from prenotal fold; b, spiracle.

posteriorly into the mentum. The bases of these setae are set on a rather more deeply pigmented band. The mentum also has a few setae laterally below the palps. The latter are two-segmented and lacking in setae. There is no bulbous sensory papilla at the tip of the apical segment of the labial palp as exists in *Xestobium*.

Thorax. The description by Parkin is mostly followed here. The thoracic segments are larger than those of the abdomen, giving a distended appearance. Each segment is divided into a number of folds as shown in Fig. 17.

Prothorax. The notal fold is large, not divided into pre- and post-notal folds but the hypopleural folds are small and incompletely separated from notal. Pedal folds are small, lying somewhat posteriorly to the hypopleural. A large spiracle lies in a depression between the prothorax and the mesothorax.

Mesothorax. The notal fold is divided into pre- and post-notal fold. The prenotal fold is divided into two halves by a transverse furrow. Hypopleurals are large, and pedals small.

Metathorax. This resembles the mesothorax except that the hypopleurals are smaller and the pedal folds larger. On the prenotal fold there is a band of spinules which are in rows two or three deep, near the mid dorsal line.

Legs. These are well developed (Fig. 21) and five segmented (Houlbert says three segments). The coxa is short and broad, the trochanter somewhat triangular, stout, roughly twice as long as broad, and the tibia is about the same length as the femur, but is not so stout. The tarsus bears a slightly curved claw, about as long as the proximal part of the tarsus, and also a pair of short spine-like setae opposite the base of the claw. The setae are disposed as in the figure. The legs of *Xestobium* can be differentiated from those of *Anobium* by their possession of shorter claws, and relatively larger femur and tibia, and also by being more setose.

Abdomen. This is divided into ten segments, the first eight of which are divided into folds in a similar manner to that of the thorax, except that there are no pedal folds. Segments nine and ten are not folded. The hypopleural folds are especially setose. The prenotal folds are relatively larger in relation to the postnotal folds than in *Xestobium*. Spinules (Fig. 22a) occur on the prenotal folds of the first seven abdominal segments only, and are absent on the eighth, ninth and tenth segments. The bands on the first segments have two rows of spinules, and on the seventh only a single row.

The spiracles (Fig. 22b) have a tubular projection (air tube) about equal in length to the spiracle itself.

Biology of larvae

The larval tunnels contain rather loose aggregations (frass or bore-dust), consisting partly of rejectimenta—small wood particles torn off, but not taken into the mouth—and faecal pellets. The latter (Fig. 23) are ellipsoidal and consist of a number of wood particles cemented together that have passed through the gut.

The number of larval instars seems not to have been settled. Kelsey *et al.* say that the first moult occurs when the young larva has tunnelled into the wood to a distance equal to its own length, and the larval head at the moult is usually directed outwards.

Becker (1942) has made a number of ecological and physiological studies on the larvae of *Anobium punctatum* which perhaps should be mentioned at this stage. He thought that, by analogy with *Hylotrupes* and *Lyctus*, a high mortality is probable among very young larvae when maintained under controlled conditions. Becker was able to remove the complicating influence of the variable foodstuff factor by using special wood pulp made from spruce and pine sapwood. He found that the speed of growth of young larvae was greater than that of older larvae. Larvae weighing from 0·5 to 1·5 mg grew fastest. Growth acceleration under favourable environmental conditions was greatest for this weight group. Speed of growth was influenced by temperature in all weight groups, and it was also markedly influenced by relative humidity, although above 70 per cent this had no appreciable effect on the older larvae.

Becker thought that larvae of *Anobium punctatum* developed best at a relatively low temperature compared with other wood-boring larvae. He found this to lie between 22·0 and 23·5° C. On the other hand, the temperature range allowing constant growth, lay within relatively narrow limits. At a relative humidity of 90 per cent these limits for younger larvae were from 12° C to about 29° C, and for older larvae, from approximately

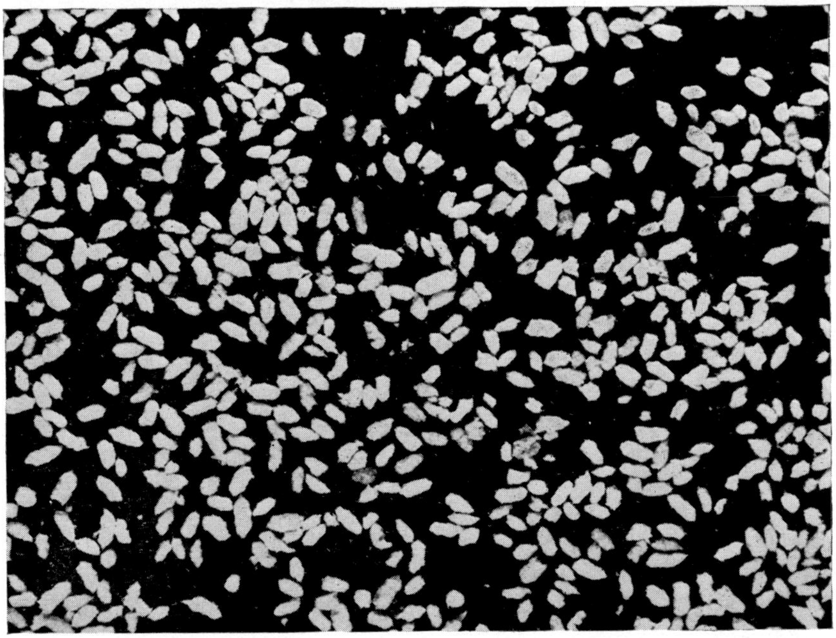

Fig. 23. Faecal pellets in frass of *Anobium punctatum*. Magnified 10 times.

17° C to about 27° C. Temperature influence was considerable, especially so in the case of the younger larvae. With regard to the actual moisture content of the wood, larvae grew most rapidly at the fibre saturation point of the wood, the younger larvae growing most rapidly at just above this amount. Growth was found to be retarded at relative humidities giving less than this amount of moisture in the wood.

At a temperature of 20° C larvae under 3·5 mg failed to grow at a relative humidity below 55–60 per cent and larger larvae, below 60–65 per cent. In the case of newly hatched larvae the lower limit for development lay between 65–70 per cent relative humidity.

Becker goes on to say that these laboratory experiment results completely confirm practical experience. The wood in cellars and ground floors and in unheated buildings with higher moisture is more seriously affected than elsewhere. In the United Kingdom this remark would be challenged. Whilst it is true that attack by *Anobium punctatum* is common and widespread in such situations, in the United Kingdom the great economic problem is *Anobium* damage in the roof void.

Pupae

There does not apparently exist a detailed description of the pupa of *Anobium punctatum*, but Bletchly has pointed out that it bears a superficial resemblance to that of *Xestobium rufovillosum*, which has been described by Fisher, and also to that of *Stegobium paniceum*, which has been

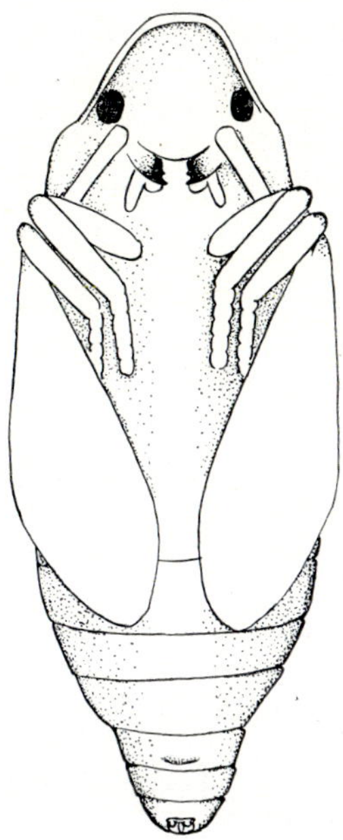

Fig. 24. Pupa from below. Male.

described by Metcalfe. The external sex organs which Bletchly has described differ considerably, however, from those of the above two insects. The pupa of *Anobium punctatum* (Fig. 24) is exarate with a thin soft cuticle enveloping all the appendages. This cuticle covering is almost complete although the elytra and wings of the pupa can easily be moved to a position at right angles to the body. There are no spines, nor is there pubescence. The pupa can move only the last few segments of the abdomen. The pupa is at first milky-white, but the eyes are the first to become pigmented, then the mandibles, followed by the tarsal claws. Lastly, just before metamorphosis, the adult colouring on the whole body can be easily seen immediately underneath the now completely transparent pupal cuticle.

The female pupa (Fig. 25) is more pointed posteriorly, and, visible to the naked eye, there is a pair of genital processes corresponding to those of *Xestobium* and *Stegobium*. There is a gonopore at the base of these appendages. An oval depression with a median furrow occurs in the centre of the eighth sternite at a distance of about one-third the width of the sternite from the posterior margin. In addition, there is a crescent-shaped depression on the seventh sternite. These features correspond closely to similar features in *Xestobium* pupae as described by Fisher.

In the male pupa (Fig. 25) the tip of the abdomen is rounded, and situated on the last sternite there is a pair of processes joined anteriorly by a 'protuberant strip'. Bletchly discusses the relationships of *Anobium* to *Xestobium* and *Stegobium* on the basis of the pupae.

Preparation for pupation first becomes evident by the larva commencing

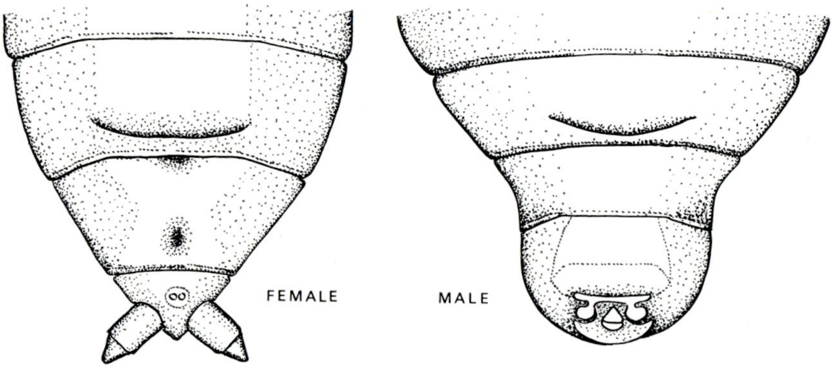

Fig. 25. External sex characters of female and male pupae.

to bore its tunnel towards an outside surface of the wood it has been infesting. The tunnel is then driven in a more or less straight line to this outer surface, but the larva finishes its tunnelling by making a large rounded and smooth oval chamber just underneath the surface of the wood.

The pupal chamber is free from rejectimenta and faecal pellets. These are utilized as a plug to the chamber, the wood fragments stuck together with the aid of a secretion.

At about this time, the thoracic and abdominal musculature, which hitherto has maintained the larval trunk in the strongly arched position, becomes relaxed and the larva straightens. The alteration from a narrow tunnel to the larger, oval pupal chamber is probably a function of this change in the muscle action. Bletchly endeavoured to find whether the sex of the resulting adult could be determined by examination of the prepupae, especially in regard to weight, but he came to the conclusion that the sexes cannot be separated with accuracy at this stage on a weight basis. The prepupae in Bletchly's work varied from 2·3 mg to 5·9 mg in the case of those that subsequently became male adults, and from 2·7 mg to 8·6 mg in the case of those which subsequently became female. The weights obtained gave an average of 3·9 mg for males, and 5·0 mg for females. Some confusion existed, however, in this work, due to the death of two very heavy prepupae which weighed 8·8 mg and 9·2 mg respectively, and which in each case appeared to be male, although lacking confirmation. Becker has stated that pupation may occur after a weight of 3 mg has been reached, but that the largest larva weighed between 13 and 14 mg. This is an extraordinarily wide variation in weight, especially as male and female weights overlap obviously to such a large degree.

Bletchly, in his study of the external characters of pupae of *Anobium punctatum*, found that the female pupa on the average is heavier than the male. The average weight of nineteen males was 2·4 mg, and twenty-five females was 3·8 mg.

The length of the pupal stage is given as six to eight weeks (F.P.R.L. leaflet No. 8, 1959), and two or three weeks (B.M.(N.H.) Economic leaflet No. 11). In South Africa three to four weeks is given by Tooke.

After rupture of the pupal cuticle, the adult beetle remains for a time in the pupal chamber, presumably for hardening of the exoskeleton. It then bores out of the wood, making a circular flight hole, 1·1–2·0 mm (average, 1·5 mm) in diameter. This hole is made by the adult using its mandibles whilst rotating on its axis. If the adult beetle is prevented from leaving the wood by the intervention of a covering, such as glass, it will bore a considerable distance in a straight line in an endeavour to get round the obstacle. In a piece of plywood backing to a mirror, straight tunnels made by adult beetles of 2·1, 2·2, 3·3, two of 4·2, and two of 5·0 cm were found. Not only were the tunnels in a straight line, but they were entirely superficial, the response to the direction stimulus obviously being very strong.

Many materials are recorded as being bored by the emerging adult beetle, a common material being leather, but there are instances of lead, 3·2 mm in thickness, being perforated. In the roof of an old lead-covered building, such perforation from underlying timber structures allowed water to percolate into the timber and thus permitted the spread of fungal decay. By this means *Xestobium* attack had been made more widespread.

A point of interest which might be of considerable economic value, if the answer were known, concerns the particular wood surface towards which the pupating larva chooses to direct itself. In a wooden board it is commonly seen that a preponderance of flight holes is situated on one side of the board. This commonly occurs not only on one or two boards, but sometimes this phenomenon is shown by a whole floor. Any one of a number of factors could account for this. The larva could perceive a moisture gradient, directing itself towards the damper wood, and this would account for the large number of flight holes occurring in such places as the 'cupboard under the stairs', where the exit holes predominate on the inside. On the other hand, it is possible for the larva to follow the nitrogen gradient in the wood, or perhaps some other nutritional factor is involved. The perception of the actual growth ring curvature by the larva should not be ruled out. In a state of nature, the larvae would require to direct their tunnels towards the naturally occurring outside of the dead branch or trunk.

Nutrition and its significance in timber infestation

A study of the literature on this subject shows a wide divergence of opinion. A view most widely held is perhaps best given in F.P.R.L. *Bulletin* 19 (1940): 'The Anobiids occur in both coniferous and hardwood timbers, in heartwood as well as sapwood. They seem to prefer well-seasoned, older wood, and belief in this preference is so common that the presence of worm-holes in furniture is frequently accepted evidence of age. How far such evidence is trustworthy is uncertain, and what changes occur in the ageing of timber to make it suitable for infestation by these wood-borers are still incompletely understood,' and then later on when *Anobium punctatum* is being discussed in particular: 'The damage caused by the common furniture beetle, occurring as it does in nearly all private houses containing old furniture is familiar to everyone. In the course of a few years a chair leg, or some other part of an article of furniture may be completely reduced to powder. Apart from changes which may take place during the ageing of wood, making it more attractive to the insect, two factors render old furniture liable to attack. As the result of ordinary wear and tear, joints open and varnish and glue chip off, thus affording convenient sites for egg-laying. The second factor is that during the many years old furniture remains in use, or is stored away in a lumber room, it is frequently left in close proximity to worm-eaten wood for lengthy periods, and it is not surprising that it becomes infested. The removal of attacked furniture from one place to another obviously tends to spread the area of infestation. Plywood and wickerwork are liable to severe damage, and softwoods (pine, larch, spruce, etc.), as well as hardwoods, are attacked; indeed, there are few timbers, if any, which are immune from infestation by this insect.'

Tooke and Scott found in South Africa that '*Anobium* beetles attack both sapwood and heartwood, hardwoods as well as softwoods. Peculiarly, the beetles do not appear to attack timber freshly seasoned, but prefer it to be some years old. As a result, most damage occurs in private homes, museums, etc., and in furniture stores and other warehouses where wooden articles may be retained over periods of some years, or where antique furniture is stored.' Fisher (1949) also mentioned the preference for old furniture, and added 'old floor joists, floor boards, joinery work and plywood', but later on he adds, 'Although this beetle is well known as a pest of old and apparently dry furniture the indications are that in such wood the rate of larval development is very slow. Where the insect occurs in structural woodwork, its development is speeded up when the moisture content of the wood rises. In this connection, the insect is more prevalent and causes greater damage in timber in cellars where wood is damp.'

Fisher continues: 'It should be emphasized, however, that both solid wood and plywood appear to have to be in use for a period of years before they become suitable for infestation; new timber and recently manufactured plywood are not liable to attack.'

It was in 1949 that attention was first drawn to the great increase in *Anobium punctatum* by Hickin, who stated that there appeared to be no doubt that *A. punctatum* was increasing extremely rapidly in Great Britain, and had to be reckoned with as a major household pest. It was liable to cause serious damage not only to furniture, but to all types of building construction where wood was employed. The same author said that *A. punctatum* was found in 1949 causing extensive damage to a very large number of houses built within the last 25 to 30 years. He then gave his experience that it was rare for softwood to be infested until the timber had been cut about 20 years. The sapwood of oak was seldom attacked until it had been cut for 60 years, but birch plywood was remarkably susceptible to attack, instances of flight holes having occurred five or six years after utilization.

Again Fisher (1952) stated that they (*Anobium punctatum*) are seldom, if ever, found in new timber, but require some condition for their natural development which is not present in wood until it has 'aged' or in some other way differs from freshly felled, converted, sound timber.

There appear to be several points of importance, when examining the food relationships of a wood-boring insect such as *Anobium punctatum*. There is firstly the point made by Parkin (1940) when he stated: 'The demands of the larva for a supply of suitable food material may be reflected in the tropisms of the beetle when selecting a site for oviposition.' He goes on to cite the example from his own work of the female *Lyctus* beetle being able to recognize the presence of starch in wood so that the eggs are not laid in starch-free wood. One is led to ask whether the female *Anobium*, apart from seeking the purely mechanical requirement of a crevice in which to cement the eggs, carries out any test by taste or smell in order to satisfy herself that nutritional requirements for the resulting larvae are adequate. In addition, if, as seems most probable, timber such as softwoods and oak require to be 'aged' in some way before eggs are laid on it, how does the female beetle detect this ageing? Kelsey *et al.* found that from an egg-laying point of view, results indicated that the physical nature of the wood surface is as important as the timber species. On one occasion more eggs were laid on roughened *Podocarpus totara* (normally non-susceptible to *Anobium* attack) than on finely sandpapered *Podocarpus dacrydioides* (normally susceptible).

The second point is that from a consideration of the very wide range of timber species attacked including both softwoods, hardwoods, their heartwood, sapwood, as well as pulp boards, and cardboard, at least in some respects it is certain that the nutritional requirements can be met by an extremely wide and varied range of woody substance.

The third point, an extremely important one from the standpoint of the economic importance of *Anobium*, is that it must be borne in mind when considering and analysing experimental work on *Anobium* nutrition, that the insect does not necessarily find its *optimum* nutritional requirements (i.e., the nutrition giving the most rapid life cycle or the maximum population increase) in the conditions where it is a pest.

The experimental work on *Anobium* nutrition can be divided into two main parts. Firstly, an examination of the feeding and digestive system of the larva and secondly, an examination of the development behaviour of larvae when forced to feed on various wood samples.

Parkin (1940) gave a short summary of the literature of the nutrition of wood-eating insects, and mentioned the work of Mansour and Mansour-Bek (1934b), and by Uvarov (1928). These authors divided wood-eating insects into cellulase producers, or cell wall eaters, and non-cellulase producers, or cell content feeders. Many years previously, Mer (1893–1903) had shown the dependence of some beetle larvae, such as *Lyctus*, on starch (therefore cell content feeders). Hopf (1938) had divided wood-feeding beetle larvae into three divisions according to the complexity of the wood components which they hydrolyse during digestion. Earlier (1937), he had demonstrated that, during its passage through the gut of *Anobium punctatum*, wood lost some of its protein nitrogen. Parkin, by testing for digestive enzymes, found that *A. punctatum* obtained its nourishment not only from the cell contents, but from the carbohydrates of the cell wall, including cellulose. This author also demonstrated the presence of a proteinase in the gut of this insect, and thus he thought that the wood and not symbiotic organisms, was the source of nitrogen. He also refers to the work of Koch (1934), and thought it possible that the symbiotic organisms produced vitamins of importance to the larval nutrition. Bletchly and Farmer (1959) gave a review of the literature bringing work on nutrition of *Anobium* up to date from Parkin's work in 1940.

Turning now to work on the development of larvae in wood blocks, pulp blocks and chemical products, Becker (1942) found that *Anobium* larvae could gain weight on a diet of cellulose over a considerable period. On the other hand, he found that growth was accelerated by adding simpler, more easily digested glucose and other hexose sugars. Spiller (1951b) suggested that alpha-cellulose comprised the bulk of the larval food, but only one-third of the available amount was actually digested. Müller (1934) appeared to confirm this. Falck (1930) demonstrated that larvae of *Anobium* removed about 21 per cent of the cellulose and hemicellulose and he concluded that only the hexosan and not the pentasan part of the hemicellulose was digested in addition to a part of the cellulose and simple carbohydrates.

Turning again to the nitrogen content of the nutritional requirements, Uvarov had drawn attention to the fact that the nitrogen content of wood was low, and Becker (1942) showed that if peptone and diastase were added to wood, they had a favourable effect on larval growth. However,

Becker had already shown that larvae could develop in pure cellulose, and therefore in the complete absence of nitrogen except as the gaseous element, so Toth (1952) considered that the symbiotic organisms must be able to fix atmospheric nitrogen.

Becker (1949) thought that protein was the critical material for growth in *Hylotrupes bajulus* (L.), and he put forward the explanation that the species had no symbiotic organisms, and was thus more dependent on the wood protein than *Anobium*, which possesses symbionts. Bletchly (1959) discussed the work of Bollard (1953, 1957, *a*, *b* and *c*) on the nitrogenous compounds present in tracheal sap, and made an effort to relate this to literature concerning the age of wood which various investigators thought rendered the wood more susceptible to attack than at other times. He quotes Becker (1940), who considered that old timber is less attacked than wood only a few years old, and Tragärdh (1938), who showed that attack could commence in houses less than 10 years old, the incidence of attack reaching a maximum in houses 30 to 40 years old, then subsequently declining. Becker stated that larvae could develop in hardwoods more than 20 years old. In New Zealand, Mason (1952) said that few houses more than 5 years old are free from infestation, and after 10 years the infestation is severe.

Bletchly and Farmer draw attention to a small local survey (D.S.I.R. 1958), which showed that a slight attack was already present in houses seven to nine years old, and they discussed the view that *Anobium punctatum* only attacks wood which has been in service for some years. They thought the view that wood does not become susceptible to attack until it has matured in some way was too facile. Bletchly (1957) gave his view that the true explanation seemed to be the slow build-up of an infestation on account of the small number of eggs laid and the long life cycle. Bletchly and Farmer (1959), working with Scots pine, confirmed Becker's contention that the outer zone of sapwood was more suitable for larval development than the inner, and this had also been shown by Spiller (1951a), with *Pinus radiata*. Weiss (1905) had shown a nitrogen gradient, the highest value being at the outside of the sapwood.

Bletchly and Farmer also found that extraction of Scots pine and oak with acetone increased the suitability for larval development. Extraction of Scots pine sapwood with water rendered it unsuitable for larval development, and few larvae survived in it. Extraction of oak sapwood with ether rendered it more suitable for larval development. These authors believe that their work confirmed that *Anobium* is, to a large extent, a cell content feeder, and they state that it follows that freshly seasoned wood is likely to be of a higher nutritive value than wood which has been in service for a number of years.

This statement, however, is not in accord with Bletchly (1953), when he stated that decayed wood is more suitable than sound wood for egg-laying, and for initial larval penetration. The larvae grow more quickly in the former though fewer may survive. He goes on to say that exploratory experiments on walnut inoculated with the moulds *Paecilomyces varioti* and *Chaetomium globosum* have indicated that although larvae do not appear to develop more rapidly than in the controls, yet the percentage surviving is greater.

From the foregoing it can be seen that there has been a great amount of speculation concerning the nature of the wood which *Anobium punctatum* will infest and that there is a certain amount of experimental evidence to show the rates of larval development under various experimental conditions, and the digestive capabilities of the larva.

Host timbers

A large number of timber species is liable to attack by *Anobium punctatum*, and the following list has been drawn up by Linscott (1967) from the literature and from practical experiments. Those hardwood timbers shown by her to be susceptible to attack are indicated by an asterisk (Linscott 1971).

Abies alba	Silver fir
grandis	Giant fir
nobilis	Noble fir
Acacia melanoxylon	Australian blackwood
Acer platanoides	Norway maple
pseudoplatanus	Sycamore
saccharum	Rock maple
other spp.	Maples
Aesculus hippocastanum	Horse chestnut
Agathis australis	New Zealand kauri
robusta	Queensland kauri
Alnus glutinosa	Alder
Araucaria cunninghamii	Hoop pine
Aristotelia racemosa	Makomako
Beilschmiedia tawa	Tawa
Betula pubescens	Hairy birch
verrucosa	European birch
other spp.	Birch
Buxus sempervirens	Box
Carpinus betulus	Hornbeam
Castanea sativa	Sweet chestnut
Cedrus atlantica	Atlas cedar
deodara	Deodar cedar
libani	Cedar of Lebanon
Chamaecyparis lawsoniana	Lawson's cypress
Coprosma divaricata	Mingimingi
Cotoneaster spp.	—
Crataegus spp.	Hawthorn, Whitethorn

Dacrydium colensoi	Silver pine
cupressinum	Rimu
Duboisia myoporoides	Corkwood
Elaeocarpus dentatus	Hinau
obvatus	Whitewood
Eucalyptus maculata	Spotted gum (rare)
Euroschinus foliatus	Port Macquario beech
Evodia sp.	Soapwood
*Fagus sylvatica**	European beech
other spp.	Beech
Fraxinus excelsior	European ash
sp.*	American ash
Fuschia excorticata	Kotukutuku
Griselinia littoralis	Papauma
Hedera helix	Ivy
Hedycaria arborea	Porokaiwhiri
Hoheria populnea	Ribbonwood, Lacebark
Ilex aquifolium	Holly
Juglans regia	Walnut
Khaya spp.*	African mahogany
Larix decidua	Common larch
kaempferi	Japanese larch
Laurelia novae-zelandiae	Pukatea
Libocedrus bidwillii	Kukawaka
Liriodendron tulipifera	Canary whitewood
Lonicera sp.	Honeysuckle
*Magnolia acuminata**	Magnolia
Myoporum laetum	Ngaio
Myrsine chathamica	Mapu
salicina	Toro
Nothofagus fusca	New Zealand red beech
menziesii	Silver beech
solandri	New Zealand black beech
Ochroma spp.*	Balsa
Olearia argophylla	Muskwood
aricenniaelobia	Akeake
traversii	Akeake
Panax colensoi	—
elegans	Silver basswood
murrayii	Mutari
Paratrophis heterophylla	Turepo
Phyllocladus alpinus	Mountain-toatoa
trichomanoides	Toatoa, Tanekaha
Picea abies	European spruce
sitchensis	Sitka spruce
other spp.	Spruces

Pinus halepensis	Aleppo pine
nigra var. *calabrica* (*P. laricio*)	Corsican pine
radiata	Radiata pine
strobus	Yellow or White pine
sylvestris	Scots pine
other spp.	Pines
Piper excelsum	Kawakawa
Platanus acerifolia	London plane
Podocarpus dacrydioides	New Zealand white pine
spicatus	Matai
totara	Totara
other spp.	Yellow woods
Populus alba	White poplar
canescens	Grey poplar
nigra	Black poplar
serotina	Black Italian poplar
tremula	Aspen
other spp.	Poplars
Prunus avium	Gean
Pseudopanax crassifolium	New Zealand lancewood
Pseudotsuga taxifolia	Douglas fir (rare)
Pyrus communis	Pear
malus	Apple
Quercus cerris	Turkey oak
ilex	Evergreen oak
petraea	Sessile oak
robur	Pedunculate oak
other spp.	Oaks
Robinia pseudoacacia	Robinia
Salix alba var. *caerulea*	Cricket bat willow
fragilis	Crack willow
other spp.	Willows
Taxus baccata	Common Yew
Thuja plicata	Western red cedar
Tilia vulgaris	Common lime
*Triplochiton scleroxylon**	Obeche
Tsuga canadensis	Eastern hemlock
Ulmus glabra	Wych elm
hollandica	Dutch elm
procera	Common elm
Vitis sp.	Vine

The species found by Linscott (1971) to be moderately resistant to attack, i.e. those which do not appear to be able to support the complete development of the beetle, are:

Dipterocarpus sp.	Keruing
Gonystylus warburgianus	Ramin
Gossweilerodendron balsamiferum	Agba
Khaya sp.	African mahogany
Quercus sp.	Japanese oak
Shorea pauciflora	Red meranti
Swietenia humilis	Central American mahogany

Six timbers were classified by Linscott as very resistant or immune to attack by *A. punctatum*:

Afrormosia elata	Afrormosia
Chlorophora excelsa	Iroko
Entandrophragma cylindricum	Sapele
Lovoa klaineana	African walnut
Mitragyna ciliata	Abura
Terminalia ivorensis	Idigbo

Plywood

Birch or alder plywood made up with glue of natural origin (e.g. blood casein) is particularly susceptible to attack by *Anobium punctatum*. The larvae obtain extra nourishment from the nitrogenous compounds present in the glue.

The use of artificial resin glues (e.g. urea-formaldehyde or phenol-formaldehyde) is said to confer immunity to attack. However, this has been shown to be incorrect in some circumstances (Hickin 1972, Taylor 1968), and the use of an insecticide (particularly dieldrin or lindane) in the glue is recommended (Taylor 1968).

Parasites and Predators

There is great difficulty in giving definitions of 'predator' and 'parasite' as applied to insects and other arthropods which are not mutally inclusive. However, the idea generally accepted by the term 'Predator' implies an arthropod actively seeking out, then devouring, its insect prey, and then moving on to seek out another. The word 'Parasite' is usually reserved in this connection for an arthropod which spends the whole, or at least the greater part, of its life (in the case of an Arachnid, such as a mite, the nymphal and adult stages) intimately associated with the body of its single host.

Parasites

I. ACARI

Certainly the most frequently occurring parasite of the larvae of *Anobium punctatum* is the mite *Pyemotes ventricosus* (Newport). This mite was first placed in the genus *Heteropus* by Newport in 1850, but as this name was preoccupied, it was renamed *Pyemotes* by Amerling in 1862. However, the name *Pediculoides* given by Targioni-Tozzeti in 1878 has been widely used. *Pyemotes* is one of eight genera comprising the PYEMOTIDAE of Oudemans, 1937, which was formerly the PEDICULOIDIDAE of Berlese of 1907.

Linscott, in an unpublished report, counted the number of larvae of *Anobium punctatum* in six plywood boards. These had been kept in an unheated and well-ventilated wood storage building. This was in connection with a count of larvae suitable for the larval transfer technique, but the number of larvae parasitized by mites of the species *Pyemotes ventricosus* was also noted. The table is given in full below:

TABLE I

Board No.	1	2	3	4	5	6	*Average* %
TEST SIZE (1·5 mg–3·0 mg)	160	67	24	170	20	11	
%	8·4	15·1	7·9	12·8	8·9	4·4	9·6
TOO BIG (above 3·0 mg)	452	114	49	477	42	55	
%	23·7	25·7	16·1	36·0	18·7	22·2	23·7
TOO SMALL (below 1·5 mg)	411	153	50	334	51	36	
%	21·6	34·5	16·4	25·2	22·8	14·5	22·5
DAMAGED	362	57	58	89	25	19	
%	19	12·9	19	6·7	11·2	7·7	12·8
MORIBUND	184	10	44	151	46	46	
%	9·7	2·3	14·4	11·4	20·5	18·5	12·8
MITES	334	19	80	107	40	81	
%	17·6	4·3	26·2	7·9	17·9	32·7	17·8
HYMEN. PARASITE [prob. *Spathius*]	0	23	0	0	0	0	
%	0	5·2	0	0	0	0	0·8
TOTAL	1,903	443	305	1,328	224	248	

It is seen that the percentage of parasitization by *Pyemotes ventricosus* varied from 4·3 to 32·7%, but of six boards examined only one contained hymenopterous parasites.

Pyemotes ventricosus undergoes an extraordinary life cycle. The eggs hatch and the nymphs carry out the whole of their development within the abdomen of the female, both sexes being produced as sexually mature individuals. (Perhaps this is not quite so extraordinary as the species *Siteroptes graminum* (Reuter), where fertilization takes place between the males and females of the brood whilst still within the body of the parent.) Only a small number of males (Fig. 26) are produced, amounting to about 3–4 per cent, which emerge first and never leave the surface of the distended abdomen of their mother. They move about over the surface (Fig. 27), sometimes clustering around the genital orifice awaiting the emergence of females, which they fertilize immediately. From time to time they pierce the abdomen of their mother, and suck fluid for nourishment and presumably for water replenishment, thus becoming ectoparasitic on the parent. They live for 20–30 days, and in this time each male mates with about 30 females. Up to 300 offspring are said to be produced from one female. The young females after being fertilized (Fig. 26), drop off the parent and seek another insect larva to parasitize. If they do not find one within about 48 hours, they die. Unfertilized females sometimes produce a brood parthenogenetically, when only males are produced.

A wide variety of insect larvae are parasitized by this mite, perhaps the best known host being the Angoumois grain moth—*Sitotroga cerealella* (Olivier). When a suitable host larva is encountered by the young female, the stylet-like chelicerae are thrust into the body and the fluid contents

Fig. 26. *Pyemotes ventricosus*. Left, male; right, female before finding host.

gradually extracted. The tip of the abdomen distends, becoming relatively enormous, and it may reach a diameter of about 1 mm.

Often several such mites (and sometimes many) feed on the same larva, when they may easily be seen with the naked eye as glistening pearl-like spheres. It is apparent that parasites of the larvae of *Anobium punctatum* occur more frequently when the habitat is damp, and seem especially common in damp plywood and in timber taken from out of doors.

On a few occasions, members of the public who have been handling woodworm-infested wood have enquired whether woodworm bite, having experienced a number of itching weals on the arms and neck. The condition is a well-known phenomenon when grain containing *Pyemotes* parasitized larvae of *Sitrotoga cerealella* is being handled. It is known as 'grain itch', and is produced by the bites of *Pyemotes* mites. The lesion produced consists of a whitish raised area, in the centre of which is a small vesicle, which is the site of the actual bite. The area immediately surrounding the white area is rosy-red. The itching, which may last for two days, is intense, especially when the victim is warm in bed. *Pyemotes* cannot burrow under the skin, nor can the mite live for long on the human body, but a thousand bites have been recorded on one person at one time. In the past, some severe cases of very high temperature (99–102°, and in one case 103·8°), and exceptionally fast pulse rates (110, in one case 130) have been attributed to bites by *Pyemotes* mites. Albumen in small quantities is found in the urine. General joint pains, backache, intense headache,

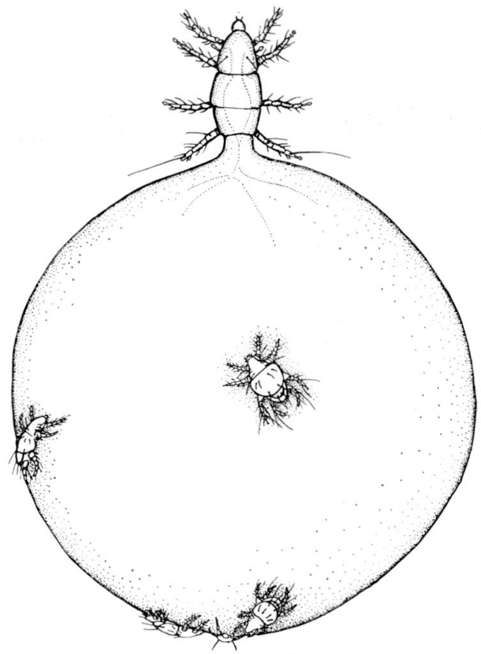

Fig. 27. *Pyemotes ventricosus*. Female in distended condition with five attendant parasitic male offspring. Length: 0·75 mm. Partly after Vitzhum.

nausea, and loss of appetite, are other symptoms associated with the bites of *Pyemotes*, according to Webster. It has also been recorded that asthma has been caused by this mite.

II. BRACONIDAE

With the exception of the mite *Pyemotes ventricosus* described above, all known parasites of *Anobium punctatum* belong to the HYMENOPTERA PARASITICA. No dipterous parasites are recorded. Amongst the PARASITICA, the braconid *Spathius exarator* (L.) (Fig. 28) is one of the most important parasites of *Anobium*. Becker considered that, from his many years' experience in North Germany, Berlin, and Brandenburg, this species was the commonest and most regular parasite of *Anobium*, occurring in most infested timbers. In other parts of Germany it has been recorded by J. Ratzeburg (1837), C. Brischke (1881), N. A. Kemner (1915), F. Zacher (1933), and J. Fahringer (1934). Becker goes on to give an example where 65 per cent of *Anobium* larvae were destryoed by *Spathius exarator*, and he did not consider this unusual. It is worth while recording the details quoted. In a plank of pine 40 × 15 × 1·5 cm from a lighted cellar there were 153 *Anobium* and 286 braconid flight holes. In addition, many living *Anobium* larvae still in the board were parasitized. It is assumed that *Spathius* lays one egg after stinging the *Anobium*, so that each braconid flight hole is equivalent to one *Anobium* larva destroyed. *Spathius exarator* shows a very marked preference for parasitizing larvae on the top and edges of boards, and egg-laying takes place in the light.

Another point of interest, for which the reason can be clearly seen, is that *Spathius* parasitization is more nearly complete in thin boards than in thicker boards. The length of the ovipositor is obviously the determining

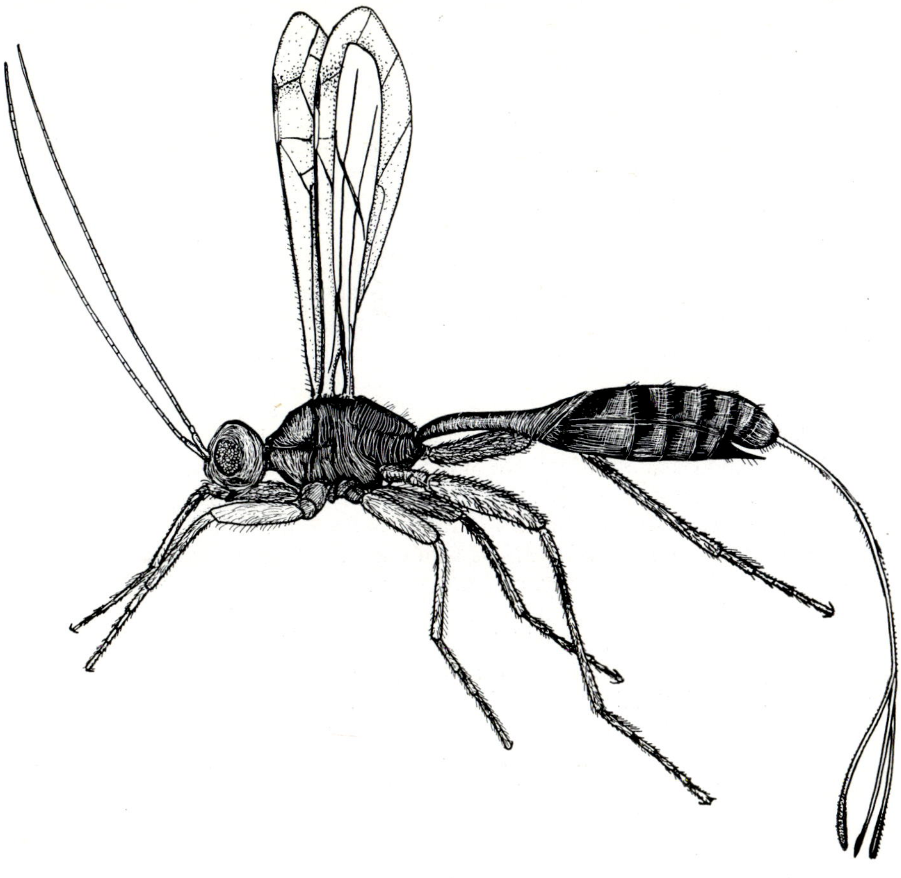

Fig. 28. *Spathius exarator*. Female partly in perspective. Length: 5 mm without ovipositor.

factor. More *Anobium* larvae are safer from attack if they are deeper in the wood than the *Spathius* ovipositor length.

Becker thinks that, in cases where certain conditions are favourable, *Spathius* can very nearly obliterate an *Anobium* attack. Unfortunately, little is known of the biology of *Spathius*; what we do know is mainly due to Becker's observations. About equal numbers of males and females of *exarator* are produced, but Krishna Ayyar (1940), studying the related *Spathius critolans* Nixon, found that parthenogenesis could occur, but unmated females laid fewer eggs. The latter author also found that there was a period of 7–10 days before the female sought out, then stung, a host larva.

A single egg was then laid on any part of its body, and hatched within one or two days. After piercing the skin of the host, the larva immediately commenced to suck its juices. The larva of this species of *Spathius* is fully grown in from three to eight days. Somewhere adjacent to the remains of its host, a white cocoon is spun of silk, and 7–10 days after pupation, the adult emerges and may live for as long as 162 days.

Returning to Becker's account of *Spathius exarator*, he states that egg-laying takes place through the surface of the wood, that damper situations are favoured, such as cellars, and that these parasites are much rarer in furniture. Presumably, the female would find much more difficulty in probing hard wood than soft. At room temperature the males survive on the average for 13 days, and the larger females for an average of slightly more than twice as long. The size of *Spathius exarator* varies within wide limits, apparently according to the size of its host. The male varies from 3 to 6 mm in length, whilst the length of the female, without ovipositor, is 4–7 mm. The ovipositor and its sheath is about the same length as the whole body.

Nixon has monographed the SPATHIINAE of the Old World, and he considers that they are distributed throughout the world with the exception of South America, but they are poorly represented in Australia (three species) and New Zealand (one wingless species). They are almost entirely parasitic on wood-boring beetles of various families. It does not appear, however, that the value of *Spathius* as a control of *Anobium* is as great in Great Britain as in Germany, as shown by Becker. Only a few specimens at any one time have been observed by the present author, and nothing like the mass development which Becker describes, although mention is made in official British literature (1951) of mass emergences in May, causing complaints of plagues of 'small flies'.

Synonomy of *Spathius exarator* is as follows:

Ichneumon exarator L., 1758; *Ichneumon mutillarius* F., 1775; *Ichneumon mystacatus* Schrank, 1781; *Ichneumon innaturus* Grav., 1807; *Cryptus clavatus* Panz., 1809; *Ichneumon attenuator* Thunb., 1822. Besides *Anobium punctatum*, it has been recorded as parasitizing *Anobium pertinax* L., *Ptilinus pectinicornis* (L.), *Leperesinus fraxini* (Pz.), *Cryphalus tibiae* Pz., and *Scolytus scolytus* F.

A number of other braconids have been listed as parasites of *Anobium*. Kemner (1915) mentions *Wachsmannia spathiiformis* (Ratz.), a species of *Bracon* and a species of *Doryctes*, and this is repeated by Ruhl (1918) and again by Fahringer (1934). *Rogas collaris* is mentioned by Kemner and Ruhl. Picard also found the species *Spathius pedestris* Wesmael, *Spathius rubidus* (Rossi), another unspecified *Spathius* and a species of *Dendroster* parasitizing *Anobium*. Nixon also mentions *Spathius pedestris* as parasitizing *Anobium*.

Thompson, in his host parasite catalogue, gives *Doryctes leucogaster* (Nees), as a parasite of *Anobium punctatum*, and *Hecabolus sulcatus* Curt., as a parasite of *Anobium rufipes* F.

It appears that with world-wide commerce in timber, especially the increase in the overall importation of Malayan timbers into the United Kingdom, more species of Braconids, especially of the SPATHIINAE, will inevitably appear here. Nixon considers that it is in Malaya that this group has its origin, being represented there by a great proliferation of species.

III. BETHYLIDAE

Sclerodermus domesticus Klug has been reported a few times as an *Anobium punctatum* parasite. It appears normally to parasitize cerambycid larvae. Becker was not certain whether the species *Sclerodermus unicolor* Westw. was an *Anobium* parasite in Berlin, but the species *Sclerodermus brevicornis* Kieff. was once found. Two species of *Cephalonomia* are known as parasites of *Anobium punctatum*. The biology of *Cephalonomia rufa* Kieff. has been studied by van Emden. Small *Anobium* larvae were picked up by the adult *rufa* and carried about, whilst the larger ones were paralysed, and eggs laid inside the body. The complete life cycle took 10 weeks at 20–24° C. *Cephalonomia formiciformis* Westw. was stated by Kunicke (1938) to be a parasite of *Anobium punctatum*.

IV. ICHNEUMONIDAE

Two ichneumonids have been recorded as *Anobium punctatum* parasites by Kemner. In the case of one of the cited species, *Pimpla flavicoxis* Thoms. however, Becker considers this improbable. The other species mentioned is *Hemiteles modestus* Grav.

Thompson gives *Hemiteles cinctus* L. and *Dimophora robusta* Brischke, but Kerrich, in a letter to the author, states that *Hemiteles* is assuredly a secondary parasite.

V. CLEONYMIDAE

Kerrich has also called the author's attention to two species, *Plutothrix coelius* (Walker), which is found in southern England, is widespread in continental Europe and is known as a parasite of *Anobium punctatum*, and *Plutothrix scenicus* (Walker). The latter is found as far north as Cumberland. The former species has been studied by Lyngnes under the name *Trigonoderus tristis* Walker.

VI. EUPELMIDAE

Calosota vernalis Curt. is also known as a parasite of *Anobium punctatum.*

VII. PTEROMALIDAE

The species *Ceraphala cornigera* Westw. has been observed by Becker parasitizing *Anobium* in Berlin. This author also states that *Theocolax formiciformis* Westw. has been found once in Berlin as an *Anobium* parasite, and calls attention to the records of Zacher and Kemner (1942).

The most frequently encountered hymenopterous parasite of *Anobium punctatum* in the United Kingdom is *Theocolax formiciformis* Westwood. It is quite commonly found running on or near *Anobium* infested wood, especially plywood, in an ant-like manner. Although this species is rare in Berlin, Becker and Weber have given a detailed account of the insect, which they reared in laboratory cultures.

Description of adult of Theocolax formiciformis

In general, the body is cylindrical, the head cuboidal, and the abdomen barrel-shaped (Fig. 29). In colour it usually appears blackish. The antennae are conspicuously angled, and consist of nine segments in the female, and ten in the male. The last three segments of the antennae are black, whilst the remainder are light in colour. A feature of interest in the legs is the presence of a cleaning spur at the distal extremity of the tibia of each front leg—see figure. This is in the form of a hinged comb with very small teeth on the spur-comb, and larger teeth on the edge of the tibia, where the spur-comb fits. Its function is to clean the antennae. When the insect has been burrowing in the soft bore-dust of an *Anobium* larva, it pauses from time to time and backs out to have a thorough toilet, when all parts of the body are dusted by the legs or mandibles—the tibial spur-combs being used for the antennae, as already mentioned. The wings are usually vestigial, but if developed, are without veins. The ovipositor is protected by two sheaths, which are covered by short hairs, and it projects a distance equal to about the last two abdominal segments (less when the female is gravid). Becker found that the total length of the female varied from 1·3 to 3·8 mm, but the usual body-length was between 2·8 and 3·1 mm. On the other hand, males varied from 1·6 to 3·0 mm, without any particular size range being much more frequent than others. During breeding experiments, fully winged individuals occurred to the extent of 4 per cent of females and 6·7 per cent of males. Only one winged specimen had previously been recorded. The winged insects, however, could not, nor did they ever make the slightest attempt to, fly. Winglessness is, of course, a great advantage when searching out *Anobium* larvae, and making their way through the packed frass.

In the same experimental work, the length of adult life at 20° C and at

24° C (at 70 per cent relative humidity in each case) was ascertained. At 20° C, females lived from 3 to 54 days, and the males from 17 to 50 days, whereas at 24° C, the females lived from 5 to 42 days, and the males from 14 to 34 days. The frequency peak for length of life of the female adult at 20° C lay at 20–30 days, and at 24° C about 5–6 days less. In the case of

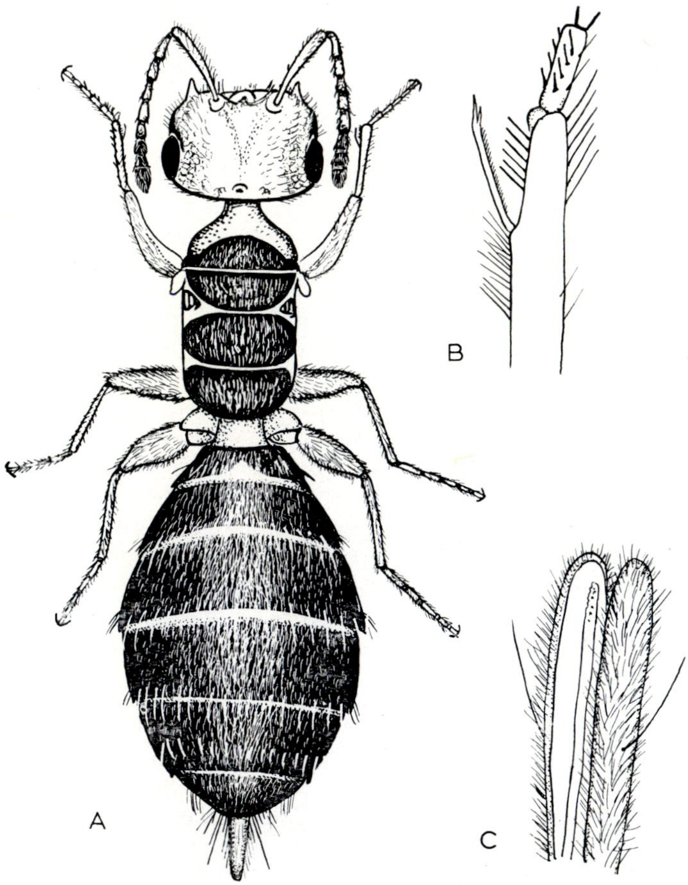

Fig. 29. *Theocolax formiciformis*. A. Female. Length: 3·75 mm. B. Distal end of tibia of front leg. C. Ovipositor and its sheaths.

the males, there was a pronounced frequency peak only in the experiments carried out at 24° C, and here the length of adult life lay between 20 to 25 days. Expressed as arithmetic means, the length of adult life for females at 20° C was 25 days, for males 34 days. At 24° C, for females it was 20 days, and for males 24 days. Two interesting points emerge. Firstly, this length of adult life is rather long compared with other hymenopterous parasites, and, most unusual, the males live longer than the females.

The immature stages of Theocolax formiciformis

Egg-laying takes place about eight to nine days after copulation. It is not clear, however, whether or not the parasite stings the host to paralyse it. The 'stung' *Anobium* larvae that Becker and Weber observed did not show a disability until shortly after the eggs had been laid. Usually they could still move two or three days after the *formiciformis* larvae commenced to suck. The tables can be turned however, and an *Anobium* larva was observed, by Becker and Weber, to squash and eat four parasite eggs.

Usually the parasite deposits its eggs on the gallery wall of the *Anobium* larva, but sometimes, although more rarely, on the *Anobium* larva itself. As a rule, the eggs are laid near the rear end of the larva. Becker found that from one to nine eggs were laid by the parasite, which can also deposit eggs a second time on another larva, but he never observed this happen more than twice. The eggs of *Theocolax formiciformis* are of a relatively large size compared with the body length of the adult. They are about 0·5 mm in length, and opalescent, the chorion being quite smooth.

After 3 days at 20° C (1½ days at 24° C), the eggs hatch, which is rather a short development period, and is probably related to the fact that the *Anobium* larva is by no means paralysed by the 'sting'. After emergence, the young larvae wander about until they come into contact with an *Anobium* larva blocking the tunnel, when they fasten on to it and suck, moving their position from time to time. At first the larvae are conical, the largest dimension being at the head end, and there are twelve segments in addition to the head. On developing, the thickest part moves to the region of the fifth to sixth segment.

When fully fed the larvae crawl a short distance from the now brown and shrivelled host, and without any cocoon or even the slightest preparation, change into the pupal stage. The length of the sucking stage of the larva at 20° C varies from 12 to 19 days, averaging 14–15 days, and at 24° C it varies from 7 to 10 days, averaging 8–9 days. The resting stage, before actual pupation, lasts at 20° C from 11 to 16 days, and at 24° C from 3 to 7 days.

The pupa is very much like the adult, except that the head is depressed towards the ventral surface of the thorax. The pupal stage lasts at 20° C from 17 to 29 days, averaging 22–23 days, but the most common lengths were 21 and 22 days. At 24° C, the period varied from 15 to 22 days, the mean value being 18 days, and the most common value 17 to 19 days.

When the parasites emerge, brothers often mate with sisters, but sometimes a considerable period may elapse before the sexes meet. Becker found that only relatively rarely did these parasites bore their way out of the wood, but the present author has observed three specimens all boring their way out of a tube through the cork!

Theocolax formiciformis is known from England, Germany, France, Sweden, and New Zealand. In England it is by far the most common hymenopterous parasite of *Anobium punctatum*. Its occurrence in New

Zealand is doubtless due to the importation of *Anobium*-infested furniture with the belongings of the early settlers, but it is difficult to reconcile this argument with its absence from North America. Furniture imported into Virginia in the early days of the colony was certainly infested with *Anobium*. Becker and Weber considered that the distribution of *formiciformis* is restricted to countries of moderate temperature.

Fig. 30. *Korynetes caeruleus* (Degeer). Length: 4 mm.

Predators

CLERIDAE

The most important predator of *Anobium punctatum* in England is the Clerid beetle *Korynetes caeruleus* (Degeer) (Fig. 30). The genus *Korynetes* of Herbst 1792 is often referred to in the literature in the spelling of Paykull (1798) as *Corynetes*. Becker considers that the most important enemy of *Anobium punctatum* (in Germany) is, without a doubt, *Opilo domesticus* Sturm. This member of the CLERIDAE, however, does not occur in Britain. The species *Opilo mollis* L. (Fig. 31) is found, but it cannot be considered anything but a rare insect in this country, and very rare indoors—the specimen figured came from Rottingdean, Sussex, being the only example the writer has collected in 15 years' search. The *Opilo* larvae crawl around the tunnels excavated by *Anobium*, and when they encounter a larva, they consume it. Sometimes these predators consume their own kind, so that *Opilo* larvae are usually found living solitarily. A characteristic of larvae of *Opilo* is that they push out the frass of the *Anobium* in front of them, and this is often ejected in piles out of old flight holes, and Becker gives an illustration of this. *Opilo* larvae can also, if necessary, bore through softwood without using the tunnels of *Anobium*, and they can successfully undergo periods of starvation if *Anobium* larvae are not found.

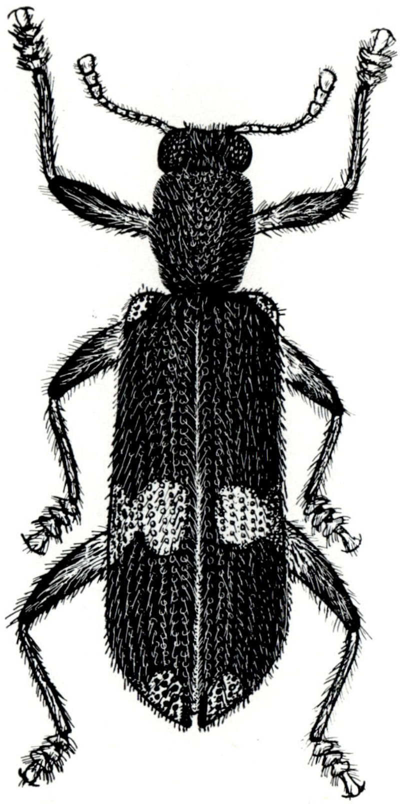

Fig. 31. *Opilo mollis* L. Length: 10 mm.

Opilo larvae consume a large number of *Anobium* larvae. The adult *Opilo* also attacks and devours adult *Anobium* beetles, and eggs are not laid by the female until this has been done.

The British CLERIDAE consists of twelve species, of which two, *Paratillus carus* (Newman) and *Thaneroclerus buquetii* (Lefebvre), have been introduced. *Korynetes caeruleus* is said by Joy to be local, and found only as far north as Lancashire, but there is no doubt that it has become very much more common in the last 20 years or so. As well as *Anobium*, it preys also on *Xestobium rufovillosum*. If egg-laying by *Korynetes caeruleus* must be preceded by a meal of its adult host, then this would be an explanation of the phenomenon of its emergence time recorded by the author. This concerns the observation that when it is preying on *Xestobium rufovillosum*, the adult emerges at the same time as its host, and when *Anobium* is its host, then its period of emergence is timed to coincide with that of *Anobium*. This is obviously of value if it must consume its host in the adult stage. The mechanism of the timing of emergence would be of interest to study.

The family CLERIDAE is contained within the suborder MALACODERMATA. The tarsi are characterized by having five segments, the first of which is minute, but segments 2–4 (in some cases 4 only) are bilobed, and they are usually hairy. The elytra are somewhat softer than is normal. The antennae are hairy and clubbed (in *Korynetes caeruleus* they are strongly serrate, but in *Opilo mollis* much less so). The beetles are pubescent, but the covering of fine hairs is not readily visible to the naked eye. *O. mollis* varies from 8 to 11 mm in length, and is pitchy-brown to pitchy-black, with yellow marks at both ends of the elytra, and at a little nearer the posterior than the middle. In addition, the bases of the femora are yellow. *Korynetes caeruleus* is bright steely blue, with the legs and antennae bluish-black. The elytra and thorax are bright and shining in spite of the covering of fine black pubescence. The length varies from 4 to 6 mm. *K. caeruleus* appears to be more resistant to insecticides than its hosts.

Biological control

It is difficult to assess the overall part played by parasites and predators in controlling infestations of *Anobium punctatum*. The individual counts of *Pyemotes, Spathius*, etc., have been made in special circumstances. It is not valid to equate such cases with the level of parasitism that would be expected in normal circumstances. It has often been recorded that great difficulty accompanies the carrying out of the larval transfer technique in wood preservative bioassay owing to the fact that a very high proportion of the *Anobium* larvae used succumb to the attacks of *Pyemotes* mites. In this case, under the laboratory conditions of constant relative humidity of a high degree, and constant temperature, also at a level probably near to the optimum for the parasite, the mites quickly reach the maximum level of parasitism. The small blocks of wood, short galleries, and high concentration of larvae, are factors all tending to promote this same result. Although Becker refers to *Spathius exarator* as a parasite which can nearly obliterate an *Anobium* infestation, and to *Opilo domesticus*, which he states can effectively reduce the attack of wood pests, and sometimes even bring it to a halt, he has listed the various factors which lead him to conclude that the natural enemies of wood pests (he had been referring to *Hylotrupes bajulus* in addition to *Anobium*) are not able to prevent, effectively, a complete destruction of the wood. The reason for this (with which the present author concurs) was that the population development of the parasite or predator species does not get fully under way until the population of the host species has reached such a numerical preponderance that wood damage has already reached considerable proportions. It must be the experience of all those whose task it is to inspect damage by *Anobium punctatum* that the greater number of *Korynetes caeruleus* beetles is found where the greater number of living adults of *Anobium punctatum* occurs also, and where the wood damage is considerable.

Becker, in listing the factors which, in his opinion, made biological control of *Anobium punctatum* slow, unrewarding and uneconomic, stressed particularly the difficulty in rearing *Spathius*, each specimen of which would require an *Anobium* larva upon which to subsist. Attempts to rear it on other larvae have failed. *Opilo* and *Korynetes caeruleus* can be reared on other small larvae, but in this case the larvae and adults are so voracious that they consume each other when reared together in numbers —they would require to be reared separately. The speed of development of a culture of *Theocolax formiciformis* is so slow that this would be a very long-drawn-out procedure.

However, even with all these disadvantages in mind, the idea of biological control in some special circumstances should not be lightly thrown aside. Timber of very large dimensions, perhaps with an internal cavity, or inaccessible timber, could have some measure of *Anobium punctatum* (or *Xestobium rufovillosum*) control, by the use of a relatively large number of *Korynetes* larvae.

The British Anobiidae

A general description of all the species follows, giving a picture of the ANOBIIDAE as a whole, containing insects of no economic importance but of great interest, together with insects of profound economic significance.

An interesting point about the ANOBIIDAE is that the total number of insects known to Linnaeus in 1758 was 1,929 and of the twenty-eight species in the British ANOBIIDAE Linnaeus knew four, but neither the Death-Watch Beetle (*Xestobium rufovillosum*) nor the Furniture Beetle (*Anobium punctatum*) was included in these four.

The family ANOBIIDAE is contained within the superfamily BOSTRYCHOIDEA together with the families PTINIDAE, BOSTRYCHIDAE and LYCTIDAE. This approximates to the suborder TEREDILIA. The characteristics of the superfamily are as follows: The integument of the adult is generally hard and brittle (but may be very soft—*Ernobius mollis* (L.)) and the pronotum is well developed, often shield-like or hood-like, sometimes with lateral projections.

The larvae of the BOSTRYCHOIDEA are soft and hook-shaped, lacking both sclerotized dorsal plates and setae, and many are wood-borers, spending the whole of the larval life within wood. None of the PTINIDAE, however, bores into wood except that a few species will hollow out a pupal chamber in wood adjacent to material in which the larvae have been feeding.

The ANOBIIDAE are distinguished from the other families of the superfamily by having the prosternal intercoxal process not received into the mesosternum and the antennal sockets are more or less under the sides of the frons and separated by more than the length of segment one. The hind coxae are contiguous. Altogether, seven of the recognized subfamilies containing fifteen genera are present in the British fauna, with a total of twenty-seven species.

In the following description of species the author has followed the order adopted by Kloet and Hincks although from time to time criticisms are made.

HEDOBIINAE

Hedobia

The first subfamily, the HEDOBIINAE, contains only Laterille's genus *Hedobia* (1829) in which the sole British representative is the Linnaean species *imperialis* (1767). This handsome and characterstic insect (Fig. 32) was classified by Joy in the PTINIDAE and, having worked for a number of years with this family the author can understand why it could be placed close to the genus *Ptinus*. Its correct place, he believes, is in the ANOBIIDAE but in such a position that it becomes adjacent to the PTINIDAE and not, as in Kloet and Hincks' arrangement, next to *Bostrychus capucinus* (L.) in the BOSTRYCHIDAE.

Hedobia imperialis (L.) has a background colour of dark pitchy-red and where rubbed may be mahogany, but a distinctive pattern of scale-like pubescence gives the insect its characteristic appearance. The scutellum is almost white. The pronotum is much less broad than the rest of the body and is divided into two distinct areas each with lateral scale-like pubescence. The antennae are almost as long as the elytra and the last three segments are no larger than the others. The length of the insect is 3·5–5·5 mm. It is local in England and Ireland and Joy gives it as 'in old hedges'. The author has found one only, about 20 years ago in a piece of walnut furniture of Italian origin.

DRYOPHILINAE

This subfamily contains three species divided between *Grynobius* Thomson (1859), one species, and *Dryophilus* Chevrolat (1832), two species, which share the character that the pronotum is not as wide as the elytra, but is much wider than that of *Hedobia*. Again the pronotum is divided into two areas, the oral area being much narrower than the aboral. In *Grynobius* these areas are very distinct but in *Dryophilus* they are much less so.

Grynobius

G. excavatus (Kugelann 1792) (Fig. 33) has short antennae only a little longer than the pronotum and the last three segments are longer and more robust than segments 2 to 8. The elytra are covered with a series

Fig. 32. *Hedobia imperialis* (L.). Length: 5 mm.

of longitudinal ridges—seven or eight on each elytron are usually visible from the top surface. These ridges alternate with lines of short fine pubescence. The beetle varies from a dark pitchy colour to reddish and is from 4 to 5·5 mm in length. It is considered to be common in old wood, but the author has never found it indoors. There is a rare aberration, described by Seidlitz (1889) and named *eichoffi*, in which the ridges on the elytra are more distinct and shining, and the third segment of the antenna is distinctly longer than the second and twice as long as the fourth. It is interesting to note that Strand considers it probable that the genus *Grynobius* consists only of a single species, *planus* F. and that *excavatus* is a synonym. The same author states that this species is frequently a pest in houses in Norway in timber of birch, alder, and willow. With regard to *G. kiesenwetteri* Edwards 1921, Buck has shown that this is, indeed, a

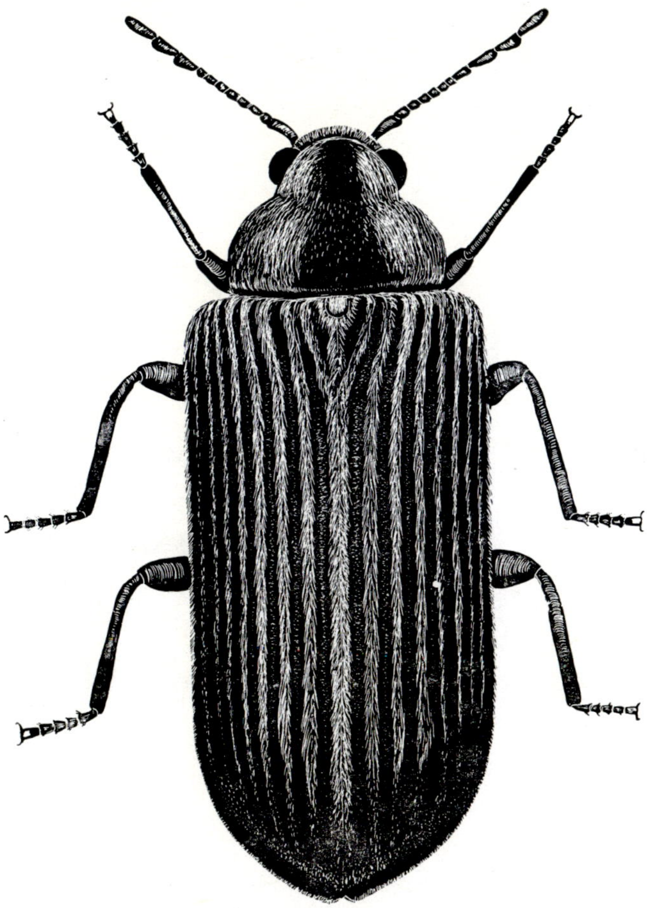

Fig. 33. *Grynobius excavatus* (Kugelann). Length: 5 mm.

large female of *excavatus* with some varietal characters that often accompany excessive size. It is 6–8 mm in length and is distinguished by the pronotum being more sharply contracted at the front and the alternate ridges on the elytra being more pronounced than the others along their whole length.

Dryophilus

Both species of *Dryophilus* are characterized by the antennae having their bases fairly close together—they are separated by less than the length of the first antennal segment. The eyes are prominent when viewed from above and the antennae are long—about as long as the elytra—with the last three segments very long. The species *pusillus* (Gyllenhal 1808) (Fig. 34) is a small beetle 1·3–2·3 mm in length and the scutellum is small and without pubescence. Elytra are reddish in the male, black in the female;

Fig. 34. *Dryophilus pusillus* (Gyll.). Length: 2·3 mm.

the head and pronotum being black in each case. It is found locally in England and Ireland, and is collected by beating fir trees.

Dryophilus anobioides Chevrolat 1832 is a larger insect, measuring from 2·5 to 4 mm in length and is very rare, being found in old stumps of broom and bramble in southern England. The scutellum has white scale-like pubescence and segments 4–8 of the antennae are transverse whilst those of *pusillus* are elongate to quadrate.

ERNOBIINAE

The subfamily ERNOBIINAE consists of three genera: *Ochina* Stephens (1830) with one species; *Xestobium* Motschulsky (1845) with one species; and *Ernobius* Thomson (1859). Kloet and Hincks list four species of *Ernobius* as British, Johnson (1965) in his recent revision of part of the genus shows it to contain five species, but variation within the species makes this genus rather more difficult to identify than others in the ANOBIIDAE. In Joy's work *Ochina* is found in the XYLETINIDAE but seems to be in its rightful place in Kloet and Hincks' list, following *Dryophilus*.

Ochina

Ochina ptinoides (Marsham 1802). This is the *hederae* of Mueller 1821 (Fig. 35). It is a fairly small insect, 2·5–3·5 mm in length. The pronotum is slightly less in width than the elytra, and, as seen from the top, is rounded at the front with two lateral projections behind. The elytra are elongate and somewhat shining with two transverse bands of white pubescence. One band is situated near the middle legs and the second band behind the third pair of legs. It is local in England, Scotland, and Ireland and is found around old ivy, the larvae boring in its stems. The author found it flying at dusk in August around an ivy-covered window in Northumberland and indeed it flew into the room.

Xestobium

X. rufovillosum (Degeer 1774) (Fig. 36) is the *tessellatum* of de Villers 1789, found in many works. It is the largest species in the ANOBIIDAE, being from 5 to 7 mm in length. The antennae are short, only as long as the pronotum with the last three segments enlarged. The pronotum has three regions when seen from above, shown by two constrictions and is as wide as the elytra. Two prominent longitudinal ridges are situated on each elytron and extend backwards usually nearly as far as the hind legs. The colour is somewhat variable, pitchy-black to reddish with patches of yellow pubescence varying also in intensity. Much of the pubescence is rubbed off a few days after emergence. It is common in England, but only known from two or three localities in Scotland and Ireland. The larvae are found boring into various hardwood timbers where fungal decay is present. From such situations attack may occur in adjoining softwoods. Pollard willow trees are very often attacked, but the species assumes great economic importance when oak timbers usually of large dimensions, and

Fig. 35. *Ochina ptinoides* (Marsham). Length: 3·2 mm.

because of their size, containing some unsound wood have been used in the construction of many of our historic buildings.

Xestobium rufovillosum cannot fly and could not be expected to walk far, and therefore it is virtually certain that, when found indoors, it has been infesting the timber since it was originally hewn and used for constructional purposes. There has been much laboratory work carried out on this species and its biology is fairly well known. A point of special interest, however, is the rapid crawling habit of the young newly hatched larvae

Fig. 36. *Xestobium rufovillosum* (Degeer). Length: 7 mm.

when they are seeking entry into the wood. This habit is soon lost when tunnelling commences. The biology of this species is dealt with in detail in Chapter Four.

Ernobius

The genus *Ernobius* has given rise to a number of difficulties, mainly due to the proportions of the different segments of the antennae which have led to the occurrence of well-established varieties. The pronotum is as wide as the elytra and the tibiae and tarsi are long and slender. The eyes are large. Pubescence on elytra is evenly distributed and there is fine puncturation. Johnson (1965) has investigated the species raised by Dr. David Sharp which have been the cause of most of the confusion within this genus. As a result the genus now stands with five British species.

Ernobius nigrinus (Sturm 1837). The body of this species is black and the tarsi are reddish. The antennae have the third segment distinctly longer than the second, the fifth much longer than the sixth, and the eighth distinctly transverse. It is rare.

Ernobius pini (Sturm 1837) (=*oblitus* Sharp 1916). In his key to the Sharpian species, Johnson does not use the antennal segments which are variable and have caused so much difficulty. He separates *pini* from *mollis* by the C-shaped punctures on the pronotum and by the characteristic aedeagus, among other things; and from *mulsantianus* by the anterior tibiae being straight in the female and turned outwards in the male, and also by the aedeagus. In length the beetle is 2·0–3·3 mm.

Ernobius mollis (Linnaeus 1758) (Fig. 37). This species has been variously known as *Anobium molle*, *Liozoum consimile* and *Ernobius mulsantianus*. The punctures on the disc of the pronotum are deep and circular. The front tibia is turned slightly outwards, more so in the male than the female. The average length is 3·5–4·5 mm, but the recorded size range is from 2·8–6·2 mm.

There is one named aberration:

ab. *reversus* Sharp 1916 in which the aedeagus is a mirror image of the typical form.

The biology of *Ernobius mollis* is given in greater detail later in this chapter (see page 90).

Ernobius mulsantianus Sharp 1916. The punctures on the disc of the pronotum in this species are circular and shallow and the anterior tibiae are curved inwards, more so in the male than the female; also the aedeagus is characteristic.

Ernobius angusticollis (Ratzeburg 1847) (=*parvicollis* Mulsant 1863). The sides of the pronotum are strongly angled. The tarsi are shorter than in the other species. The beetle is black in colour with the elytra and tarsi reddish. Segments 6–8 of the antennae are slightly elongate and of equal length. The beetle is 3–4 mm in length and very rare, but found in Surrey.

ANOBIINAE

The subfamily ANOBIINAE consists of four genera: *Stegobium* Motschulsky (1860) (the *Sitodrepa* of Thompson 1863), *Gastrallus* du Val (1860), *Anobium* Fabricius (1775) and *Dendrobium*. The genus *Dendrobium* Mulsant was dropped by Kloet and Hincks and the single species *denti-*

colle (Creutzer 1796) included in *Anobium* but the present author has reverted to the use of the original genus.

Stegobium

There is a single species, *paniceum* (Linnaeus 1758), (Fig. 38) which is a small beetle only 2–3·5 mm in length and reddish in colour. The pronotum

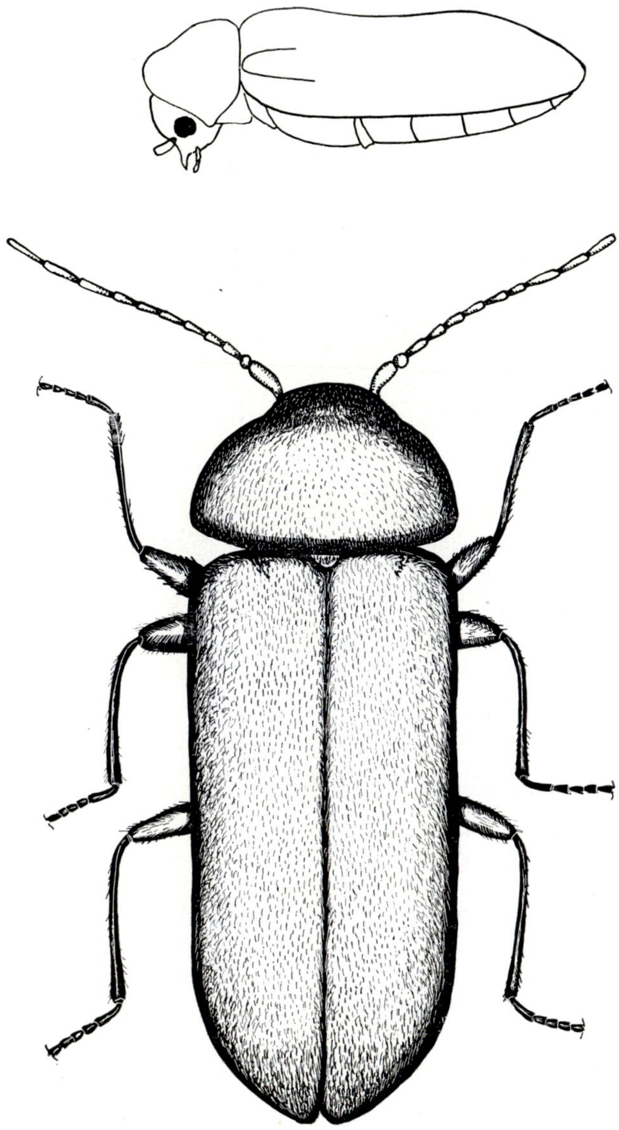

Fig. 37. *Ernobius mollis* (L.). Length: 4·5 mm.

Fig. 38. *Stegobium paniceum* (L.). Length: 2·5 mm.

is as wide as the elytra and completely masks the head when seen from above, and all margins are convex. The puncturation is confused, but long pubescence occurs in longitudinal rows on elytra. Joy omits England from the distribution, but it is today a very common pest in certain stored foods such as dried soup powders and various packeted farinaceous materials.

Gastrallus

Gastrallus has a single British species, *immarginatus* Müller 1821 originally recorded in error as *laevigatus* (Olivier 1790).

Anobium

The genus *Anobium* contains two British species—*punctatum* (Degeer 1774) and *fulvicorne* Sturm 1837. *A. striatum* Olivier 1790 nec Fabricius 1787 and *domesticum* (Geoffroy in Fourcroy 1785) are the chief synonyms of *punctatum* and have been widely used.

Anobium punctatum (Figs. 11 and 39) is variable in size from 2·5 to 4·5 mm. It is variable also in colour and may occur from a light reddish yellow through dark chocolate-brown to a pitchy-red. The legs are usually reddish-brown and the last three joints of the antenna are said to be longer in the male than in the female. The pronotum is approximately as wide as the elytra (if the small tubercles at the posterior angles are taken into account). The disc is raised into a crest and there are two foveae. Pronotum and elytra are covered with a very fine pubescence and there are longitudinal rows of fine sculpture on the elytra. The biology of this species has been dealt with at the beginning of this chapter.

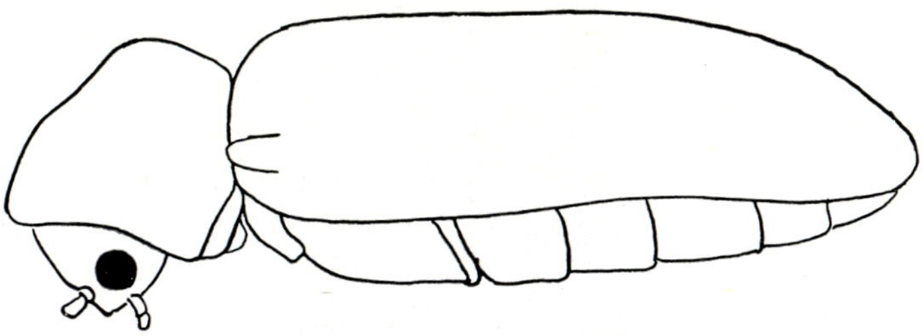

Fig. 39. *Anobium punctatum* (Degeer), lateral view.

Anobium fulvicorne Sturm. This insect measures up to 4 mm in length and is dull black in colour but the antennae, tibiae and tarsi are reddish. The antennae are rather shorter than in *punctatum* and the elytra are rather less rounded at the apex. The tarsi are more distinctly thickened towards the apex. Cox considers this species to be rather common.

Dendrobium

The genus *Dendrobium* contains one species, *denticolle* (Creutzer 1796) (Fig. 40). In length it varies from 4·5 to 5·5 mm and is pitchy in colour and covered with short greyish pubescence, but in the hinder angles of the pronotum the pubescence is a little longer and yellowish. There are two foveae at the hind part of the pronotum and at the sides there are narrower wing-like extensions. The antennae are short, scarcely as long as the pronotum, and the last three segments are much longer than the remainder. This insect is very rare, found only in southern England as far north as Worcester, and is collected from old oaks and hawthorn.

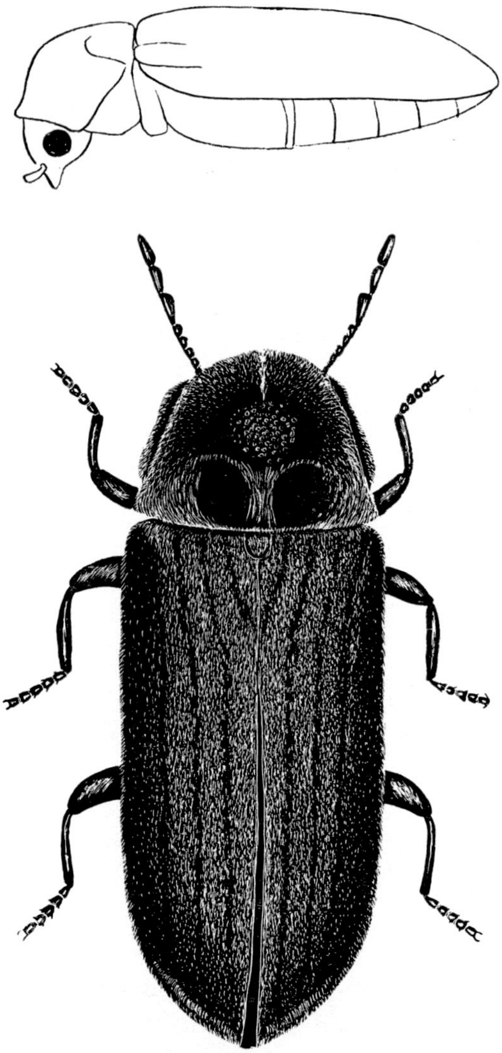

Fig. 40. *Dendrobium denticolle* (Creutzer). Length: 4·5 mm.

PTILININAE

The PTILININAE is made up of one genus *Ptilinus* Mueller (1764) containing two British species. These are elongate, cylindrical beetles with the pronotum almost globular and usually darker in colour than the more reddish elytra. The latter are decorated with punctures arranged in more or less irregular rows. The antennae of the male are strongly pectinate whilst those of the female are serrate.

Ptilinus pectinicornis (Fig. 41). This species varies from 3·3 mm to 5·5 mm in length, but during 15 years' acquaintance with these beetles, the author has never seen them as small as some specimens in the Joy collection. The large pectinate antennae of the male are quite extraordinary.

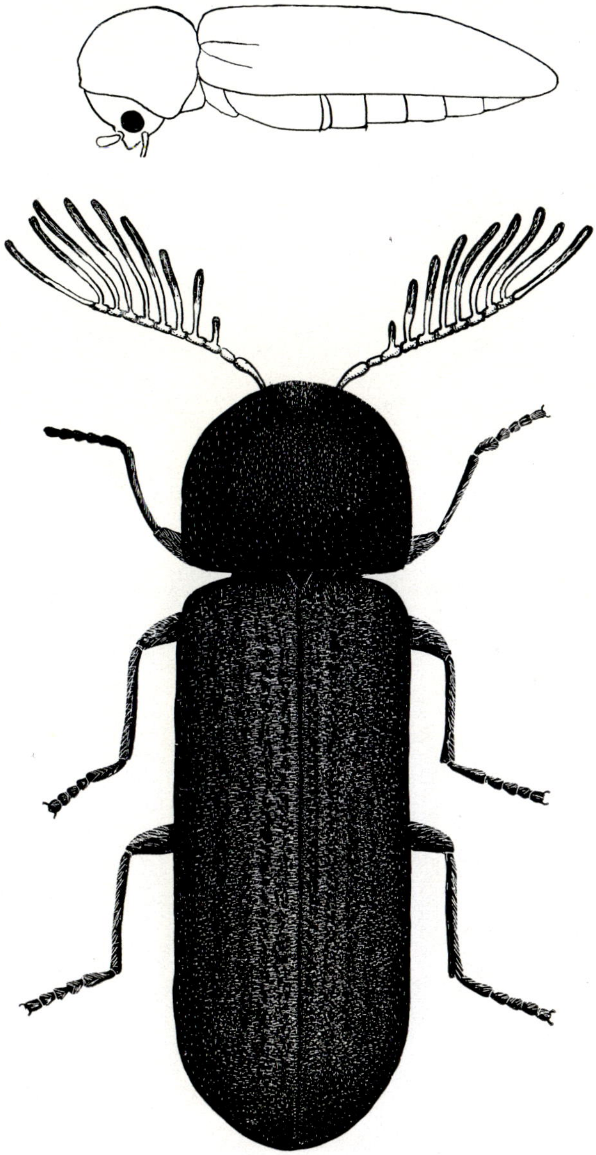

Fig. 41. *Ptilinus pectinicornis* (L.). Male. Length: 5 mm.

A description of the biology of this species as far as it is known, is given at the end of this chapter (see page 97).

Ptilinus fuscus Geoffroy in Fourcroy 1785, was placed in the British list in 1931 by Donisthorpe on the strength of the record by Taylor in 1910 of specimens bred out of the wooden base of a pair of scales at Sandown, Isle of Wight, under the name of *P. costatus* Gyll.

XYLETININAE

The XYLETININAE consists of two genera, *Xyletinus* Latreille (1810) and

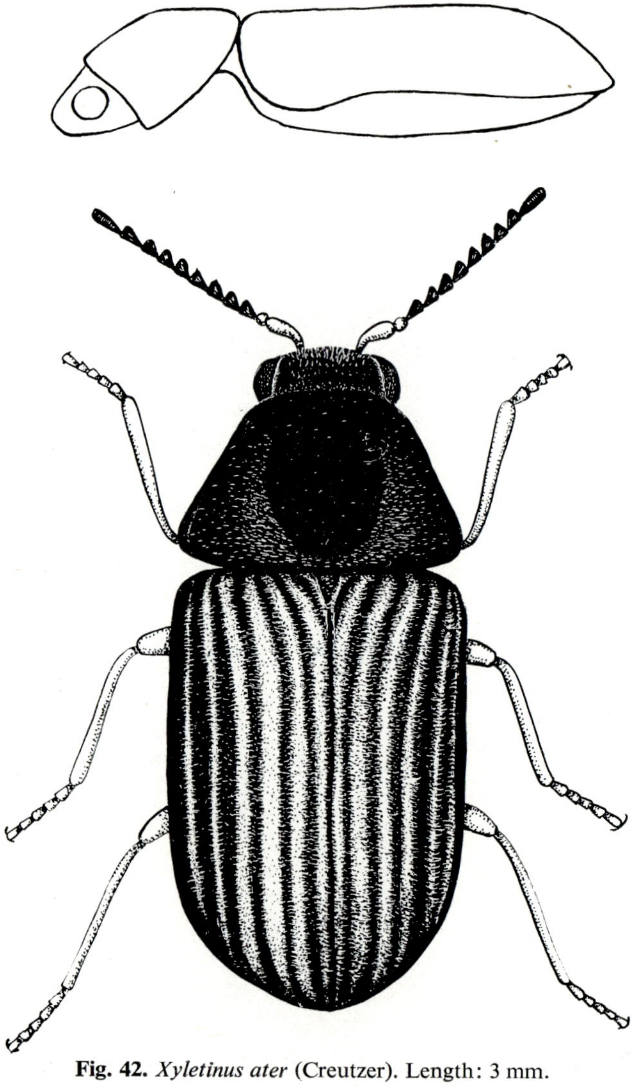

Fig. 42. *Xyletinus ater* (Creutzer). Length: 3 mm.

Lasioderma Stephens (1835) but this subfamily is not equivalent to the older family XYLETINIDAE which, in addition, contained the genera *Eledona*, *Haplocnemus* and *Ochina*.

Xyletinus ater (Creutzer 1796) (Fig. 42) are black beetles measuring from 2·5–4·0 mm in length and with the first three segments of the antennae and the legs reddish-yellow. It is rare, but found in England south of Nottingham in decaying wood.

Lasioderma serricorne (Fabricius 1792) (Fig. 43) is a smallish beetle, from 2·5 to 3·5 mm in length. It is shining red in colour and covered with extremely fine puncturation and very short pubescence. It is a common stored food-product insect and cosmopolitan in distribution. It shows a particular preference for tobacco and its products and infests the freshly dried leaf and all its subsequent stages of manufacture. It is universally

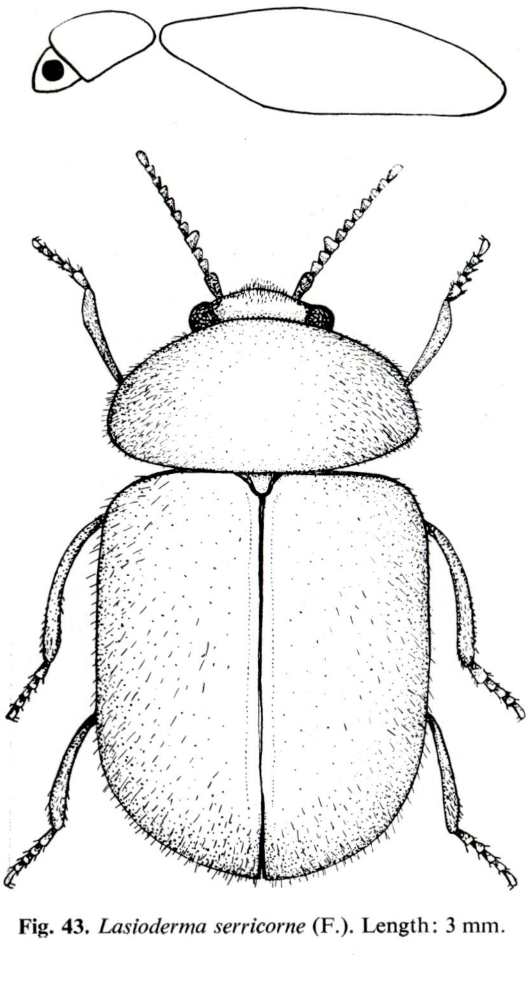

Fig. 43. *Lasioderma serricorne* (F.). Length: 3 mm.

known as the cigarette beetle. It has been found infesting a wide variety of dried vegetable substances and has been recorded from wood which was already infested with wood-boring larvae.

DORCATOMINAE

The subfamily DORCATOMINAE was formerly considered a separate family, the DORCATOMIDAE, following on from the BOSTRYCHIDAE but is now (it seems wrongly to the author) included in the ANOBIIDAE. There are three genera; *Dorcatoma* Herbst (1792), length 2·0–2·6 mm, *Caenocara* Thomson (1859), length 1·8–2·5 mm, and *Anitys* Thomson (1863), length 2·4–3 mm. All species of these genera are somewhat globular and may be distinguished as follows:

The elytra of *Dorcatoma* are elongate whilst those of *Caenocara* and *Anitys* are quadrate. The species of *Caenocara* are black whilst *Anitys rubens* (Hoff.), the single species of its genus, is red. In addition the eyes of *Caenocara* are prominent and the basal margin of the pronotum is sinuate at the sides. The eyes of *Anitys rubens* are not prominent and the basal margin of the pronotum is straight.

Five species, all found in dead trees, are contained in *Dorcatoma*. In *D. flavicornis* (Fabricius 1792) the eyes are small and the elytra dull, the puncturation resembles *D. chrysomelina* and the pubescence resembles *D. dresdensis*. This is a rare insect found south of Yorkshire. All the remaining species have larger eyes. *D. chrysomelina* Sturm 1837 is rare, but found in Lancashire and to the south. The elytra are dull and covered with very small punctures. *D. dresdensis* Herbst 1792 and *serra* Panzer 1796 are both less strongly punctured and the elytra are more shining. These two species may be differentiated by the pubescence which is recumbent in *dresdensis* and erect in *serra*. Both the two latter species are very rare and whilst *dresdensis* is found only in Berkshire and Norfolk, *serra* is found in S.E. England as far north as Berkshire. The two species of *Caenocara* are easier to identify if both are present! *C. bovistae* (Hoffmann 1803) (Fig. 44) is duller with stronger and more concentrated puncturation, the legs are pitchy and the length varies from 2 to 2·5 mm. It is rare and found in puff ball fungi in counties south of Lancashire.

Caenocara subglobosa Mulsant 1864 on the other hand is more shining and the puncturation is finer and much less concentrated. It is, on the whole, smaller than *bovistae*, being from 1·7 mm to 2·0 mm in length. It is very rare, found only in Norfolk in the fungus *Lycoperdon gemmatum*.

Anitys rubens (Hoffmann 1803) is rare, found in decayed oak in southern England as far north as Yorkshire.

The biology of *Ernobius mollis*

Unfortunately most accounts of this species have suffered from brevity. having usually appeared as an addendum to descriptions of the economically more important species of the ANOBIIDAE, such as *Anobium puncta-*

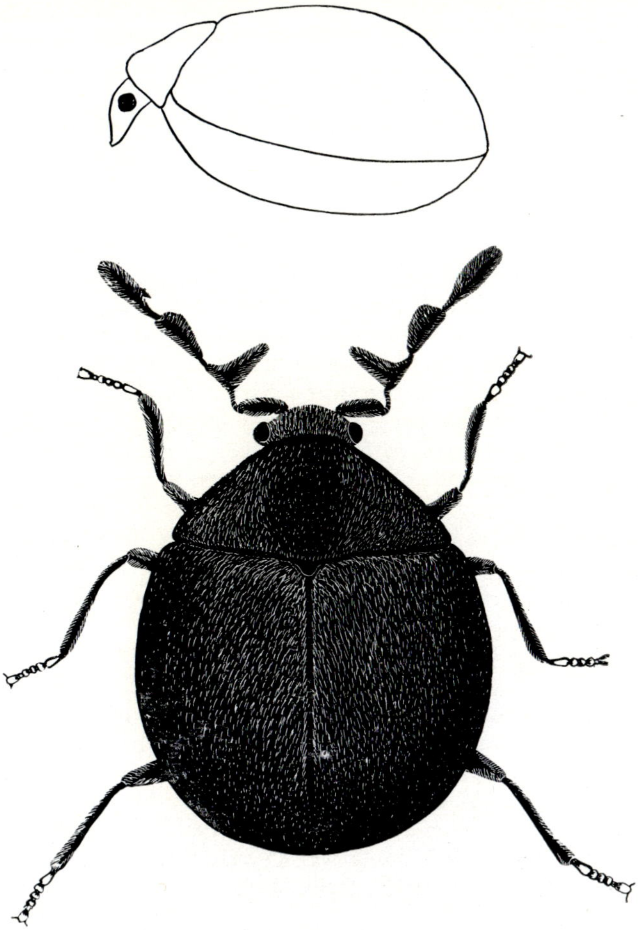

Fig. 44. *Caenocara bovistae* (Hoffmann). Length: 2·5 mm.

tum and *Xestobium rufovillosum*. Apart from the earlier descriptions of the adult, Munro (1915) and Parkin (1933) described the fully grown larva and gave keys for the separation of the larvae from those of the commoner wood-boring ANOBIIDAE. However, Miss Phyllis Gardiner, in 1953, gave a comprehensive account of the morphology and biology of this species.

World distribution

Ernobius mollis has almost a world-wide distribution. It would appear to be indigenous to the north temperate regions, being common in northern Europe, especially in Scandinavia. Simeone, however, considered this species to have been introduced into the North American continent during the last hundred years. He now gives the range as from Nova Scotia to South Central Ontario, and southwards to Texas and Florida. In the United Kingdom and Ireland it is encountered wherever its biological habitat occurs.

It has been introduced, probably from Europe, into the Southern Hemisphere, where its extent of damage appears to be greater than in its native territories. In South Africa, Tooke has given an account of the phenomenal spread of the species from its introduction, in or about 1935 in some packing-case timber, to 1949 when it was to be found everywhere in the Union of South Africa, where unbarked coniferous timber was used.

In New Zealand, where it has also been introduced, the larvae are able to tunnel throughout the wood of dead, standing pine trees, which have a diameter of up to 20 cm. In milled timber also, larvae are reported as being present at least 5 cm below the bark. In Australia, too, *Ernobius mollis* has, on occasion, been extensively present. This is due to the increased use of the timber of *Pinus radiata*. The adult beetles, although originating in the bark left on boards underneath floors, made their way through the boards and gnawed through the floor coverings.

This species has also been reported from New Caledonia and it would be surprising if it were not now established on the South American continent, although, so far, no records of such an occurrence have been traced.

Host trees

Ernobius mollis attacks only softwood trees—no hardwood trees having been recorded as supporting infestations. Gardiner gives the following list of recorded host species, but remarks that doubtless innumerable other coniferous trees are susceptible: common spruce (*Picea abies*), Austrian pine (*Pinus nigra* f. *austriaca*), Monterey pine (*P. radiata*). Scots pine (*P. sylvestris*), Cluster pine (*P. pinaster*), Canary Island pine (*P. canariensis*), Loblolly pine (*P. taeda*), Larch (*Larix decidua*), and Douglas fir (*Pseudotsuga taxifolia*).

The larvae normally consume only the bark, more generally that part of the bark lying closest to the outer sapwood, but the sapwood is scored also, with the tunnels lying half in the inner bark and half in the outer sapwood (Fig. 45). The species is thus a cambium feeder. Instances of feeding more generqlly in sapwood have already been given, but in addition the larvae are sometimes found in the vicinity of the brown pith area. Gardiner states that the larvae appear to like resin, since they are often found burrowing in sites adjacent to resin canals.

Life History

Adult emergence

The main emergence period is from late June to early August. Gardiner records the first emergence as 26th May and the latest as the first week in

October. Kemner, however, states that in Sweden pairing occurs in May and August and that the August adults are produced from the eggs laid in May. Gardiner is sure that in England there is only one generation each year, in fact the wide variation in larval size suggests that some individuals may require longer than one year to complete the life cycle. It is possible that late-emerging adults may successfully overwinter. Tooke says that

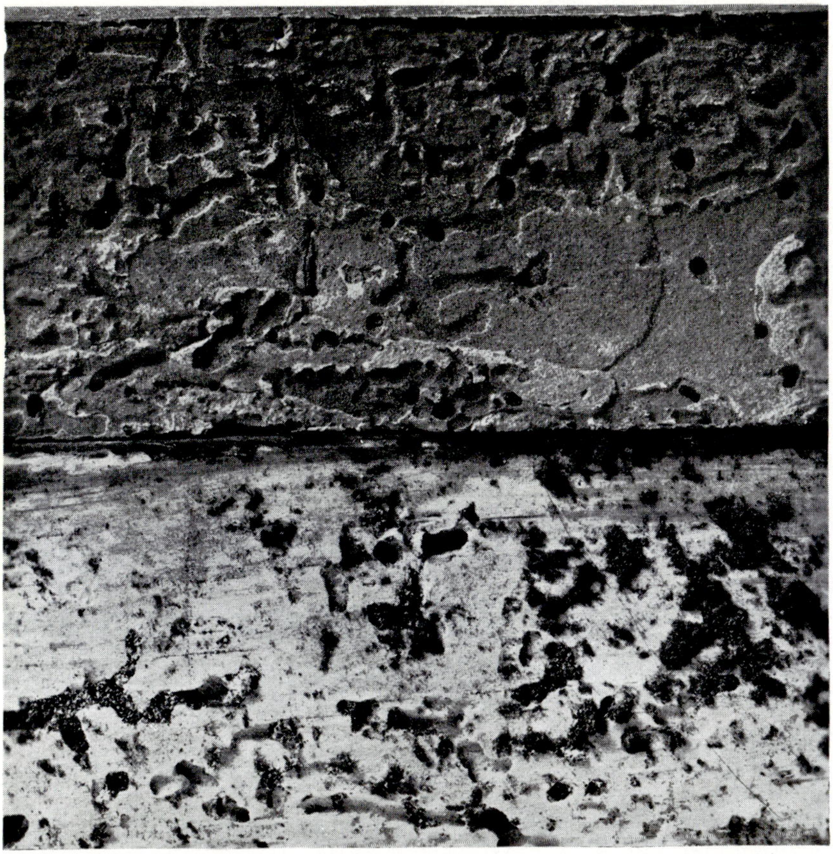

Fig. 45. Larval tunnels of *Ernobius mollis*: above, in underside of bark; below, in outer sapwood (some tunnels are filled with frass). Natural size.

in South Africa adults may be found all the year round in heated rooms. The adults emerge usually through the bark, constructing a circular flight hole about 2 mm in diameter. They will emerge through wood in cut timber and will gnaw through paper or cloth to gain their release.

There is always a period of 6 to 12 days which elapses after eclosion of the adult and this is partly taken up by a resting period and by the work of boring out through the bark. On emergence the reproductive organs of both sexes are fully mature and copulation takes place within a few hours, or, occasionally, as long as 12 days afterwards.

Twenty-four hours is the usual period for the beetles to remain in copula, but pairing for as long as six days is said not to be uncommon, and they often remain hidden during this time in bark crevices or old flight holes. The same beetles often pair several times during the laying of successive batches of eggs.

Egg laying

Eggs are laid only at night and this usually occurs one or two days after copulation. Gardiner relates that during her work on this species she found that females that had been feeding on spruce bark as larvae could not be induced to lay eggs on larch, although the transferred larvae would thrive in this timber.

When the female is seeking an egg-laying site, she runs over the bark, exploring it with her antennae, and when satisfied, the ovipositor is inserted and a crevice found with the terminal palps. In about half a minute the egg passes down the ovipositor with the micropyle hindmost and it is wedged in position. No adhesive material is present. Although the eggs are sometimes deformed by being tightly wedged they appear to hatch satisfactorily. According to Kemner the number of eggs laid is from 10 to 15, but Gardiner found that the maximum number of eggs laid by a single female was 66. The average number laid during one year's experiments, however, was 15, but the following year the average was 26. Dissection shows that the potential egg number per female is 120. The duration of the egg-laying period varied from one day (number of eggs 1 to 25) to nine days (35 eggs).

The length of adult life is up to one month in both sexes and neither sex has been observed to feed.

The length of the egg stage was found to vary between 10 days and 3 weeks, with an average of 13 to 14 days under ordinary laboratory conditions. The eggs become opaque immediately before hatching.

Larvae

When the egg hatches the larva eats part or all of the shell and the transfer of symbiotic organisms from the female parent may take place in this way. There is an extremely high viability rate, with not less than 84 per cent of the eggs hatching in the experimental work carried out by Gardiner. When first hatched the larva can walk fairly rapidly over the bark surface and may do so for one or more days, thus showing a similarity to the first stage of *Xestobium rufovillosum*. Some difficulty seems to be experienced by the larva in making the initial entry into the bark, but this is usually achieved eventually in the vicinity of knots and sawn surfaces of the bark. When once the larva has begun to bore, the body becomes curved. The actual ingress holes of the young larvae can be identified by the small piles of bark-coloured faecal pellets. Gardiner states that the first-stage larva makes a number of blind galleries radiating from the point of entry, about 1 mm below the surface and from 0·25–0·5 mm in width, and that the first stage lasts about one month. The total length of the gallery of the *Ernobius mollis* larva is surprisingly short. Tooke says that they rarely exceed 8–12 mm in length. They begin at a width of about 0·5 mm and increase to about 3·5 mm. The frass (Fig. 46) is characteristic of this insect and is loosely packed in the galleries, the pellets being bun-shaped. When the larva has been eating sapwood, the pellets are yellowish-white, but when the larva is boring in the bark, they are dark brown.

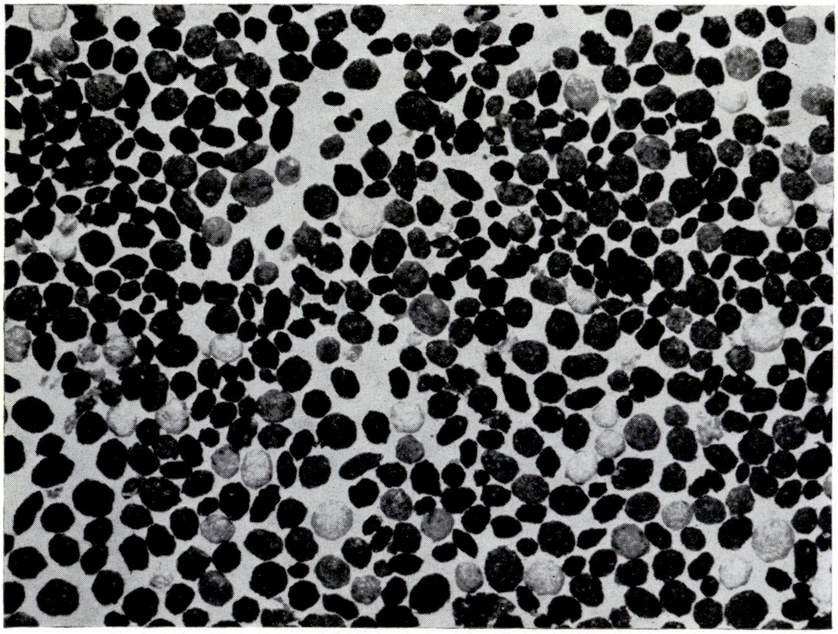

Fig. 46. Frass of *Ernobius mollis*. Magnified 10 times.

It has not been possible so far to ascertain the total number of instars, on account of the larva's habit of eating its cast skin. Gardiner was able to show, however, that larger larvae, hatched early in the summer are usually actively tunnelling during the winter months, whereas the smaller larvae—those hatched late in the season—are further retarded in their growth by undergoing a period of diapause. Moreover, it is considered that there is a genetical basis for the wide variation in size of larva and tendency to diapause.

With regard to nutrition, Parkin included this species in his work on the digestive enzymes of wood-boring beetle larvae and he found that the larvae of *Ernobius mollis* are able to break down the cell wall carbohydrates —cellulose and hemicelluloses. They are also able to utilize the protein (a proteinase being of general occurrence in the larval gut of wood-boring beetles), as well as the starch and sugars which may be present in the cell contents.

Description of larvae

Perris (1854) gave a general description of the larva, but later authors have disagreed with some of his data. Parkin's account (1933) of the fully grown larva is detailed and the figures enable it to be differentiated from the commoner wood-boring anobiids. Gardiner (1953) has described the newly hatched larva, and has given an account of the digestive system of the mature larva. Böving (1954) also gives a detailed account of the fully grown larva with several figures. Larvae are illustrated in Fig. 47.

Newly hatched larva

The newly hatched larva measures about 0·7–0·8 mm in length and the body is only slightly curved, the full curvature being acquired during the successive stages. In addition the thoracic region is only slightly swollen and there are two protruberant cushions, one on each side of the anus, which help the young larva to progress over the surface of the bark when it is seeking a convenient situation for making its ingress hole. Janisch (1923) described similar structures which he called prolegs, in the non-wood-boring anobiid larva of *Stegobium paniceum*. These organs are lost by the time the larva attains the second stage. Another feature of the early stage, apart from differences in the number and placing of the abdominal spinules, is the eye spots, or ocelli, which are more obvious than in the fully developed larva, but are nevertheless so inconspicuous that both Perris and Parkin considered that they were absent altogether.

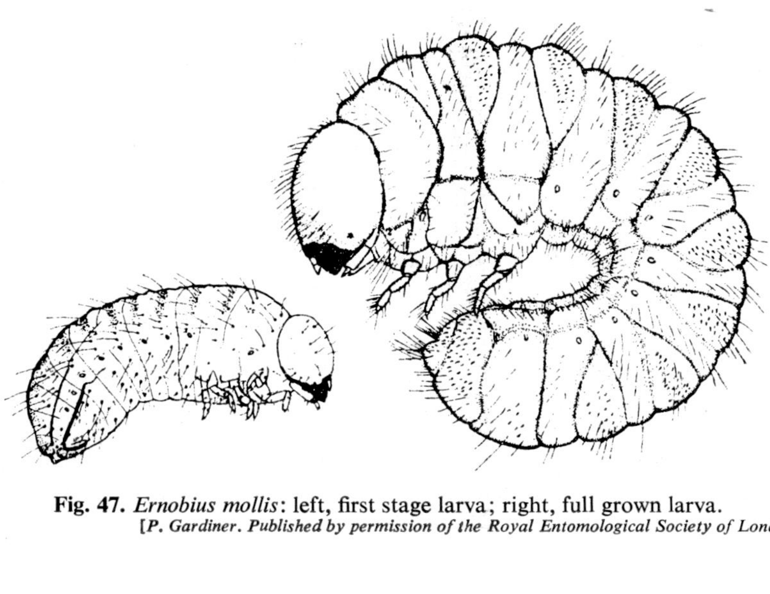

Fig. 47. *Ernobius mollis*: left, first stage larva; right, full grown larva.
[*P. Gardiner. Published by permission of the Royal Entomological Society of London.*]

Mature larva

In addition to the information which has already been given, Munro mentions the spinules of this species as being stouter and more marked, and the spiracles more easily discernible (they are reddish brown) than is the case in the other common wood-boring anobiids. Böving uses characters of the epipharyngeal setae and the number and situation of the prodorsal asperities (about forty-five on third thoracic segment and about thirty on the sixth abdominal segment) in order to distinguish this larva from all other larvae of ANOBIIDAE known to him.

Pupation

When the larva is ready to pupate it constructs a pupal chamber at the extreme end of the larval gallery which is usually sited between the inner face of the bark and the outer surface of the sapwood (it may be wholly in the bark or wholly in the sapwood). The larva plugs the entrance with a mass of faecal pellets and gnawed wood fragments. Within a few days the prepupal phase begins, recognizable by the larva assuming a linear position, losing mobility and developing swellings in the positions to be occupied by the wings. The shape of the adult abdomen appears also. This stage lasts for only about two days, when the final larval skin is cast. The pupae can be sexed quite easily, using the description and figures given by Gardiner, who states that the pupal stage lasts for 10 days or rather less. When the adult eclodes, the body is at first almost colourless, with the appendages yellow, but the cutting edge of the mandibles is deeply pigmented. Between 6 and 12 days are spent by the adult in the bark before it finally bites its way out. During this time the parts harden and become darker in colour, whilst the reproductive organs reach such a degree of maturation that copulation can take place immediately after the adults emerge from the bark.

The biology of *Ptilinus pectinicornis*

Cymorek has made a study of the mating behaviour and biology of this species. The unfertilized female sits on the wooden surface with the head bent forwards, but with the abdomen lifted up slantingly by the fully extended back legs. A small, conical, shining and damp gland becomes exposed on the anal segment. This position is taken up for relatively short periods of time, from a quarter to half a minute, alternating with periods at the normal resting position. Cymorek was able to prove experimentally that a sexual vapour is emitted from the gland, which excites the males in the vicinity and they rapidly fan their highly pectinate antennae by means of which they can ascertain the direction from which the vapour has travelled. The maximum distance covered by a male to find a female was 1 metre, but in the open it was found that in the case of the closely related species *Ptilinus fuscus*, a distance of 10 metres was covered in flight to reach a sexually displaying female.

When the male reaches the female, she often shams dead, but this reflex is broken by the male taking up a position on top of her and rapidly stroking her head and antennae. After the immission the male turns through 180° so that both animals hang on to one another, abdomen to abdomen. Between 5 and 15 minutes afterwards copulation is broken by the male turning again towards the female. It has sometimes been stated that this species will pair inside the wood, but from his detailed observations Cymorek considered that achievement of copulation in the flight hole was impossible, although a pair already copulating could probably withdraw into a flight hole. Possibly such a withdrawal has prompted this assertion.

The brood chamber

The females of *Ptilinus pectinicornis* bore breeding passages into the timber in order to lay their eggs, and two types of attack are recognizable as follows:

The primary attack

This consists of a short bore-hole made by the female after copulation. It is made at right angles to the direction of the grain and is usually bored in wood that has not previously been attacked. The female prefers to bore into vertical surfaces and selects a crack or crevice as the initial boring site. If a bore-hole is made on a wooden surface cut across the grain, the tunnel turns off at right angles a short distance from the surface so that the boring principle of across the grain is maintained. In hornbeam, *Carpinus betulus*, the boring of a brood chamber of 6–8 mm was accomplished in two to three days, whilst in poplar (*Populus* spp.) a depth of 8–12 mm was bored in 24–36 hours.

The secondary attack

This requires the presence of a pupal chamber and flight hole. In this case the breeding passage is arranged as a cross passage, or as an angled extension from the pupal chamber. Very often the females use their own pupal chambers for the construction of such secondary brooding galleries. In the latter case, the female only leaves the wood to mate, then returns, head first, into the flight hole which she herself has constructed. When the brooding gallery is under construction the female appears from time to time at the flight-hole entrance, throwing out the fragments of wood that result from her boring activities. The females are normally active only from dusk to dawn, but when occupied in brood gallery construction they are active by day also.

The preference for using old flight holes to initiate the construction of a brood gallery explains why timber attacked by this species is often very rapidly destroyed. Only a small proportion of females leave the old breeding place to initiate a primary attack. The female of *Ptilinus pectinicornis* will lay her eggs also in the flight holes of *Anobium punctatum*.

Egg laying

The extremely long, thin-shelled, glassy and pointed eggs are inserted into the lumen of wood vessels and are never found in the brood gallery itself. The eggs have flexible shells so that when deposited in the vessel lumen they are stretched and accommodate themselves to the dimensions of the latter. The length of eggs laid in poplar measured 1·5 mm in length and 0·075 mm in width and were situated 0·5 to 1·0 mm from the inside edge of the brood gallery. The females do not leave the brood gallery, even after the eggs have been laid, and Cymorek states that they often remain in death with the relatively massive pronotum plugging up the entrance to the brood gallery for the protection of the brood.

The egg-laying pattern is thus seen to be similar to that of LYCTIDAE and some BOSTRYCHIDAE and is not in the least like that of such ANOBIIDAE as *Anobium punctatum*, *Xestobium rufovillosum* and *Ernobius mollis*, so that grave doubt must be cast on the true relationship of *Ptilinus pectinicornis* with the ANOBIIDAE.

Timbers attacked

In Germany the vast majority of attacks by this species, out of doors, is made in beech, *Fagus sylvatica*; standing and felled trunks and branches are the usual sites of attack. In Turkey this species is recorded from *F. orientalis*. The present author has observed attacks by *Ptilinus pectinicornis* out of doors in elder, *Sambucus nigra* (composite attack with *Anobium punctatum*). This occurred at Leatherhead, Surrey, in a living tree where a patch of sapwood had been removed, exposing the heartwood. In addition, ash has been seen to be severely attacked in a cut trunk used for an upright for an outhouse. This happened at Bletchingley, in Surrey. In addition to the timbers already mentioned, maple, oak, alder, plane, sycamore and elm have been quoted by various authors. It has also been found in common plywood, together with *A. punctatum*.

Several authors have reported the occurrence of *Ptilinus pectinicornis* attack in softwoods, but it has been denied by others. Cymorek was able to get larvae to feed well on the sapwood of conifers, including spruce, but nevertheless he considers that such attack could not possibly be initiated naturally, on account of the very narrow cell lumens of the softwood, unless such wood was joined with a susceptible hardwood. In his experiments he could not get females to lay eggs in the brood galleries which they excavated, some of which were only 1–2 mm in length, although one was unusually long at 15 mm. Cymorek considers that this species ranks second to *Anobium punctatum* (amongst the ANOBIIDAE) as the cause of damage to dry wood in Europe, but in the United Kingdom it must be placed amongst the species of much lesser importance.

Nutrition

Parkin has described how the mycetomes in the larva of *Ptilinus pectini-*

cornis have migrated forward from the anterior end of the midgut to form a pair of lobed organs dorsoventral to the front of the gizzard. A pair of slender ducts, however, connects them with the midgut. Parkin was making an analogy with similar organs in the larvae of the anobiids *Stegobium paniceum* and *Lasioderma serricorne*, although it is now known that the biology of *Ptilinus pectinicornis* is not typically anobiid, but approaches that of lyctids. Parkin considers that there is strong morphological evidence for concluding that the yeast-like symbionts secrete a fluid containing vitamins of the B group, which are required for the nutrition of the larva, but which are unobtainable in sufficient quantities in the wood in which the larvae tunnel. Parkin goes on to say that when the larva pupates the mycetomes look little different from those of several other species of anobiid larvae.

REFERENCES

ALTSON, A. M. 1922. Beetles Damaging Seasoned Timber. *Timber Trades J.*, 15th April–13th May, 1922.

ANON. 1951. *Furniture Beetles.* British Museum (Natural History) Economic Series No. 11, pp. 1–34.

ANON. 1952. *Anobium punctatum* (de Geer). (Holzschädlingstafeln.) *Holz Roh-u. Werkst.* **10:** 119–20.

ANON. 1955. *Forest Products Research Report for* 1954. H.M.S.O. London.

ANON. 1955. The Common Furniture Beetle. Pamphlet No. 13. *Anobium punctatum* de Geer. New South Wales Forestry Commission. Division of Wood Technology, Sydney.

ANON. N. D. *Timber Preservation.* British Wood Preserving Association jointly with Timber Development Association.

BAKER, E. W. and WHARTON, G. W. 1952. *An Introduction to Acarology.* New York.

BECKER, G. 1938. *Naturwissenschaften* **26**: 462–3.

BECKER, G. 1938. The Testing of the Insecticidal Effects of Wood Preservatives using *A. punctatum* as the Test Insect. *Arb. Physiol. angew. Ent. Berl.* **5**: 357–60.

BECKER, G. 1940. Beobachtungen über Schädlichkeit, Frass und Entwicklungsdauer von *Anobium punctatum* de Geer ('*Totenuhr*'). *Z. PflKrankh.* **50** (3–4): 159–72.

BECKER, G. 1942. Ökologische und Physiologische Untersuchungen über die holzzerstörenden Larven von *Anobium punctatum* Deg. *Z. Morph. Ökol. Tiere* **39**: 98–151.

BECKER, G. 1943. Zur Ökologie und Physiologie holzzerstörender Käfer. *Z. angew. Ent.* **30**: 104–118.

BECKER, G. 1949. Beiträge zur Ökologies der Hausbockkäfer-Larven. *Z. angew. Ent.* **31**: 135–74.

BECKER, G. 1953. Untersuchungen über die Eignung von Kontaktinsektizeiden für den Holzschutz. *Holz Roh-u. Werkst.* **11**: 45–50.

BECKER, G. and WEBER, W. 1952. *Theocolax formiciformis* Westwood (HYM. CHALCID) ein Anobienparasit. *Z. Parasitenkunde* **15**: 339–56.

BEESON, C. F. C. 1941. *The Ecology and Control of the Forest Insects of India and the Neighbouring Countries.* Forest Research Institute, Dehra Dun, India.

BJERKANDER, C. 1783. Beskrifning pa en Hallon-Mask. *K. svenska Vetensk. Akad. Handl.* **4**: 246.

Bletchly, J. D. 1952. A summary of Some Recent Work on the Factors affecting Egg-laying and Hatching in *Anobium punctatum* de Geer. (COLEOPTERA-ANOBIIDAE) *Trans. Ninth Int. Congr. Ent.* **1**: 728–34.

Bletchly, J. D. 1953. A Note on the Prepupal Stage and the External Sex Characters of the Pupae of *Anobium punctatum* (COLEOPTERA, ANOBIIDAE). *Bull. ent. Res.* **43**: 587–90.

Bletchly, J. D. 1957. The Biological Work of the Forest Products Research Laboratory, Princes Risborough. III. The Work of the Entomology Section, with particular reference to the Common Furniture Beetle *Anobium punctatum* Deg. *Proc. Linn. Soc. Lond.* **168**: 111–15.

Bletchly, J. D. 1959. The Influence of Soft Rot on the Susceptibility of Beech to Attack by the Common Furniture Beetle (*Anobium punctatum* Deg.) *British Wood Preserving Association Convention Record*, p. 5.

Bletchly, J. D. and Farmer, R. H. 1959. Some Investigations into the Susceptibility of Corsican and Scots Pines and of European Oak to attack by the Common Furniture Beetle, *Anobium punctatum* Deg. (Col., Anobiidae). *J. Inst. Wood Sci.* **3**: 2–20.

Bollard, E. G. 1953. Nitrogen Metabolism of Apple Trees. *Nature, Lond.* **171**: 571.

Bollard, E. G. 1957a. Composition of the Nitrogen Fraction of Apple Tracheal Sap. *Aust. J. biol. Sci.* **10** (3): 279–87.

Bollard, E. G. 1957b. Nitrogenous Compounds in Tracheal Sap of Woody Members of the Family Rosacae. *Aust. J. biol. Sci.* **10** (3): 288–91.

Bollard, E. G. 1957c. Translocation of Organic Nitrogen in the Xylem. *Aust. J. biol. Sci.* **10** (3): 292–301.

Bose, S. R. and Sardar, N. S. 1937. *Proc. R. Soc. (B.)* **123**: 193–213.

Bouche, P. F. 1834. *Naturgesch der Insekten*, p. 187. Berlin.

Böving, A. G. 1927. The Larva of *Nevermannia dorcatomoides* Fisher, with Comments on the Classification of the ANOBIIDAE according to their Larvae. *Proc. Ent. Soc. Wash.* **29**: 51–62.

Böving, A. G. 1954. Mature Larvae of the Beetles—Family ANOBIIDAE. *Dan. Biol. Medd.* **22**, No. 2: 3–177, 279–98.

Brownlee, K. A. 1949. *Industrial Experimentation.* H.M.S.O. London.

Buchner, P. 1927. Wood Nourishment and Symbiosis. *Int. Congr. Zool.* **10** (Budapest).

Buck, F. D. 1950. *Grynobius excavatus* Kugelann and *G. kiesenwetteri* Edwards (COL. ANOBIIDAE) one species. *Ent. mon. Mag.* **86**: 127–8.

Buddeberg, C. D. 1885. Beitrag zur Biologie einheimisher Käferarten Jahrb Nassau. *Ver. Naturk.* **38**: 81–3.

Busvine, J. R. 1966. *Insects and Hygiene.* 2nd. Ed. Methuen. London.

Buxton, P. A. 1931. The Measurement and Control of Atmospheric Humidity in relation to Entomological Problems. *Bull. ent. Res.* **22**: 431–47.

Campbell, W. G. 1929. *Biochem. J.* **23**: 1290–3.

Campbell, W. G. 1935. *Nature, Lond.* **136**: 299.

Campbell, W. G. and Bryant, S. A. 1940. A Chemical Study of the Bearing of Decay by *Phellinus cryptarum* Karst, and other Fungi on the Destruction of Wood by the Death-Watch Beetle (*Xestobium rufovillosum* de Geer). *Biochem. J.* **34** (10–11): 1404–14.

Cann, F. R. 1932. Laboratory Tests of Insecticides for use against Wood-boring Insects. *Ann. appl. Biol.* **19**: 291–4.

Cann, F. R. 1938. An uncommon type of Furniture Beetle. *The Builder* **154**: 654–5.

CASIMIR, J. M. 1955. The Common Furniture Beetle. *Australian Forestry* **19**: 26–34.

CASIMIR, J. M. 1958. *Ernobius mo* Linn., An Introduced Bark Beetle. *Technical Notes* **11** (3): 24–7, September 1 8, Issued by the Division of Wood Technology, Forestry Commission of N.S.W

CLARKE, S. H. 1938. *Nature, Lond.* **12**: 899.

COCKAYNE, A. H. 1919. The Wood-borer and its Control. *N.Z. J. Agric.* **18**: 198–99.

COLEBROOKE, E. M. 1937. The Aural Detection of the Larvae of Insects in Timber. *J. Sc. Instrum.* **14**: 119.

DENNE, W., SPILLER, D. and KELSEY, J. M. 1944. Research on *Anobium punctatum* de Geer. The Flight Period at Auckland. *N.Z. J. Sci. Tech.* **26** (Ser. A.): 152–4.

DIÉGUEZ, J. M. 1960. *La carcoma de la madera y metodos para combatirla.* (*Anóbidos Dañinos en mueblas y construcciones*). Servicio de Plagas Forestales, Ministerio de Agricultura, Madrid.

DOMINIK, J. 1959. Remarks upon Insects Destroying Timber Structure on Water. *Sylwan* **103** (4): 49–53. (*Biol. Abs.* 12237, **36**: 1016, 1961.)

D.S.I.R. 1955. *Forest Products Research*, 1954. H.M.S.O. London.

D.S.I.R. 1958. *Forest Products Research*, 1957. H.M.S.O. London.

ERICHSON, W. F. 1841. Zur Systematischen Kenntnis der Coleopteren-Larven. I. Beitrag. Die Larven der Coleopteren. *Wiegmann Arch. Naturgesch.* **7** (1): 98–9.

ERICHSON, W. F. 1898. *Naturgeschichte der Insekten Deutschlands. Coleoptera*, p. 124.

EVANS, J. W. 1943. *Insect Pests and their Control.* Dept. of Agriculture, Tasmania.

FALCK, R. 1930a. *Zellulose-Chem.* **11**: 89–91.

FALCK, R. 1930b. Scheindestruktion des Holzes durch die Larven von *Anobium. Zellulose-Chem.* **11**: 128–9.

FISHER, R. C. 1932. *Report of the Forest Products Research Board for the Year* 1930, p. 20. H.M.S.O. London.

FISHER, R. C. 1935. *Nature, Lond.* **135**: 102.

FISHER, R. C. 1937. Studies of the Biology of the Death-Watch Beetle, *Xestobium rufovillosum* de Geer. A Summary of Past Work and a Brief Account of the Development Stages. *Ann. appl. Biol.* **24**: 600–13.

FISHER, R. C. 1941. Studies of the Biology of the Death-Watch Beetle, *Xestobium rufovillosum* de Geer. IV. The Effect of Type and Extent of Fungal Decay in Timber upon the Rate of Development of the Insect. *Ann. appl. Biol.* **28**: 244.

FISHER, R. C. 1942. Furniture Beetles and other Wood Borers. *Ann. appl. Biol.* **20**: 313–16.

FISHER, R. C. 1949. Household Timber Insects. *The Sanitarian*, Nov. 1949.

FISHER, R. C. 1952. Some Aspects of the Biology of Timber Insects. *Science Progress*, No. 158: 213–32.

FISHER, R. C. 1958. Current Problems in Woodworm Control. *Ann. appl. Biol.* **46**: 111–17.

FISHER, R. C. 1959. Developments in the Control of Timber Insects. *Chem. and Ind.*

FISHER, R. C. and TASKER, H. S. 1940. The Detection of Wood-boring Insects by means of X-rays. *Ann. appl. Biol.* **27**: 92–100.

FRAENKEL, G. 1936. *Nature, Lond.* **137**: 237–8.

GAHAN, C. J. 1925. *Furniture Beetles.* Brit. Mus. (Nat. Hist.) Econ. Ser. No. 11.

GARDINER, P. 1953. The Morphology and Biology of *Ernobius mollis* L. (COLEOPTERA-ANOBIIDAE). *Trans. R. ent. Soc. Lond.* **104**: 1–24.

GARREN, K. H. 1938. *Phytopathology* **28**: 839–45.

GOSSWALD, K. 1939. *NachrBl. dt. PflSchutzdienst, Berl.* **19**: 17–19.

GOURLAY, E. S. 1930. Preliminary Host-list of the Entomophagous Insects in New Zealand. *N.Z. Dept. Sci. Ind. Res. Bull.* **22**: 6.

HEITZ, E. 1927. *Z. Morph. Ökol. Tiere* **7**: 279–305.

HERFS, A. 1926. Ökologische Untersuchungen an *Pediculoides ventricosus* (Newp.). *Berl. Zoologica* **74**. Stuttgart.

HICKIN, N. E. 1949. The Common Furniture Beetle, *Anobium punctatum* de Geer (COL. ANOBIIDAE). *Ent. mon. Mag.* **85**: 213–14.

HICKIN, N. E. 1953. The Common Furniture Beetle, *Anobium punctatum* de Geer (COL. ANOBIIDAE). *Entomologist* **86**: 216.

HICKIN, N. E. 1959. The President's Address. The British ANOBIIDAE. *Proc. S. Lond. ent. nat. Hist. Soc.* 1958.

HICKIN, N. E. 1960. The Problem of Wood Damage by *Anobium punctatum* de Geer in the United Kingdom, with some Notes on the Current Commercial Practice of Control. *Trans. XI Int. Congr. Ent.* Vienna.

HICKIN, N. E. 1972. *The Woodworm Problem.* 2nd Edn. Hutchinson. London.

HOPF, H. S. 1937. Protein Digestion of Wood-boring Insects. *Nature, Lond.* **139**: 286–7.

HOPF, H. S. 1938. *Ann. appl. Biol.* **25**: 390–405.

HORN, O. 1930. *Gesamm. Abh., Kohle* **10**: 23–8.

HOULBERT, C. 1903. *Les Insects ennemis des Livres*, pp. 25–80. Paris.

JANISCH, E. 1923. Zur Bekämpfungsbiologie des Brotkäfers, *Sitodrepa panicea. Arb. bio. Reichsanst.* **12**: 243–84.

JOHNSON, C. 1965. Taxonomic Notes on Coleoptera; No. 2—Sharp's species of the genus *Ernobius* (Anobiidae). *Entomologist* **100**: 175–180.

JONES, S. R. and RITCHIE, S. D. 1937. *Tech. Studies.* **5**: 179–81.

JOY, N. H. 1932. *A Practical Handbook of British Beetles.* Witherby. London.

JUDEICH, K. F. and NITSCHE, J. 1895. Lehrbuch der Mitteleuropäischen. *Forstinsektenkunde* **1**: 341–7.

KELSEY, J. M., SPILLER, D., and DENNE, R. W. 1945. Biology of *Anobium punctatum.* Progress Report. *N.Z. J. Sci. Tech.* (*B*) **27**: 59–68.

KELSEY, J. M. 1946. A Note on rearing *Anobium punctatum* de Geer. *N.Z. J. Sci. Tech.* (*B*) **27**: 329.

KELSEY, J. M. 1946. Tests with Timber Preservatives in New Zealand. *N.Z. J. Sci. Tech.* (*B*) **27**: 446–57.

KELSEY, J. M. 1947. Insects attacking Milled Timber. Poles and Posts in New Zealand. *N.Z. J. Sci. Tech.* (*B*) **28**: 65–100.

KELSEY, J. M. 1949. A Note on the Life-cycle of *Anobium punctatum* de Geer. *N.Z. J. Sci. Tech.* (*B*) **20**: 211–14.

KELSEY, J. M. 1958. Symbosis and *Anobium punctatum* de Geer. *Proc. R. ent. Soc. Lond.* (*A*) **33**: 21–4.

KEMNER, N. A. 1915. De ekonomiskt viktiga vedgnagende Anobierna. *Medd. Cent. Anst. Forsoks.* Jordbr. No. 108 (Ent. Avd. No. 9).

KERRICH, G. 1960. Private Communication.

KIRK, H. B. 1905. On the Occurrence of Starch and Glucose in Timber. *Trans. N.Z Inst.* **37**: 379–80.

KLOET, G. S. and HINCKS, W. D. 1945. *A Check List of British Insects.* Stockport.

KOCH, A. 1934. *Prakt. Mikroscopie* **9**: 3–12.

LINSCOTT, D. 1967. Investigations into the Susceptibility of Timber to Attack by *Anobium punctatum* (Deg.). M.Sc. Thesis, Lond. Univ.

LINSCOTT, D. 1971. The Susceptibility of Imported and Newly-seasoned Home-grown Hardwoods to *Anobium punctatum* (Deg.). *J. inst. wd. Sci.* No. 28: 36–39.

LYNGNES, R. 1956. A Useful Chalcidid, *Trigonoderus tristis* Walk. *Z. angew Ent.* **39**: 368–75.

MANSOUR, K. and MANSOUR-BEK, J. J. 1934a. *J. exp. Biol.* **11**: 243–56.

MANSOUR, K. and MANSOUR-BEK, J. J. 1934b. The Digestion of Wood by Insects and the supposed Role of Micro-organisms. *Biol. Rev.* **9**: 363–82.

MANSOUR, K. and MANSOUR-BEK, J. J. 1937. *Enzymologia* **4**: 1–6.

MASON, C. G. W. 1952. Some Notes in the Incidence and Control of Wood-boring Insects in New Zealand. *N.Z. Sci. Rev.* **10**: 103–6.

MER, E. 1893. *C.R. Acad. Sci., Paris.* **117**: 694–6.

MER, E. 1903. *Mem. Soc. natn. agric. Fr.* **140**: 233–47.

METCALFE, M. E. 1932. The Structure and Development of the Reproductive System in the COLEOPTERA with Notes on its Homologies. *Q. J. Micr. Sci.* **75**: 49–129.

MILLER, D. 1925. Forest and Timber Insects in New Zealand. *N.Z. State Forest Service Bull.* No. 2, p. 50.

MÜLLER, W. 1934. Untersuchungen über die Symbiose von Tieren mit Pilzen und Bakterien. 3 Mitt, über die Pilzsymbiose holzfressender. Insektenlarven. *Arch. Mikrobiol.* **5**: 84–107.

MUNRO, J. W. 1915. The Larvae of the Furniture Beetles. *Proc. R. phys. Soc. Edinb.* **19**: 220–36.

NIXON, G. E. J. 1943. A Revision of the SPATHINAE of the Old World (HYMENOPTERA, BRACONIDAE). *Trans. R. ent. Soc. Lond.* **93**: 173–456.

NORMAN, A. G. 1936. *Biochem. J.* **30**: 1135–7.

O'DWYER, M. H. 1937. *Biochem. J.* **31**: 254–7.

O'DWYER, M. H. 1939. *Biochem. J.* **33**: 713–7.

O'DWYER, M. H. 1940. *Biochem. J.* **34**: 149–52.

OLIVIER, M. 1790. *Encyclopedie Methodique, Histoire Naturelle Insectes*, Vol. V. Paris.

OUDEMANS, A. C. 1937. *Tijdschr. Ent.* **80**.

PARKIN, E. A. 1933. The Larvae of some Wood-boring ANOBIIDAE. *Bull. ent. Res.* **24**: 33–68.

PARKIN, E. A. 1936. *Ann. appl. Biol.* **23**: 369–400.

PARKIN, E. A. 1940. The Digestive Enzymes of some Wood-boring Beetle Larvae. *J. Exp. Biol.* **17**: 364–77.

PARKIN, E. A. 1952. Symbiosis in *Ptilinus pectinicornis* L. *Nature, Lond.* **170**: 847.

PERRIS, E. 1854. Histoire des Insects du Pin Maritime. *Ann. Soc. Ent. Fr.* **2**: 622–40.

PERRIS, E. 1878. *Larves des Coleopteres*, 228–49. Paris.

PIC, M. 1912. In Junk and Schenkling's *Coleopterorum Catalogus*, Part 41.

PICARD, F. 1919. Les Insectes du Figuier. *Ann. Epiphyties* **4**: 46–7.

POCHON, J. 1939. *C.R. Acad. Sci., Paris* **208**: 1684–6.

PRELL, H. 1927. On a Hibernating Cage for Insects. *J. econ. Ent.* **20**: 830–34.

RATZEBURG, J. T. 1839. *Die Forst-Insekten,* **1.** *Die Käfer*, 41. Berlin.

REIKHARDT, A. N., KARAKULIN, B. P., and ISACHENKO, V. B. 1930. *Pests of Timber and their Control.* 60 pp. Moscow, Gosudarstv, sel'skokhoz. Izd. (In *R.A.E.* **18**: 188.)

RIPPER, W. 1930. *Z. vergl. Physiol.* **13**: 314–33.

ROUGHLEY, T. C. and WELCH, M. B. 1929 Wood-borers damaging Timber in Australia. *Bull. Technol. Mus., Sydney.* No. 8, p. 13.

ROUZET, M. H. 1849. Histoire des Metamorphoses des *Anobium abietis* et *striatum Ann. Soc. Ent. Fr.* **7**: 305–13.

RUPERTSBERGER, M. 1880. *Biologie der Käfer Europas,* 174–5.

RUPERTSBERGER, M. 1894. *Die Biologische Literatur über die Käfer Europas von* 1880, pp. 178–80.

SAALUS, U. 1923. Die Fichtenkäfers Finnlands. *Ann. Acad. Sci. Fenn.* (A) **22**: 206–15, 701.

SAVORY, J. G. 1955. The Role of Microfungi in the Decomposition of Wood. *B.W.P.A. Conv. Rec.* pp. 3–33.

SCHANDERL, H. 1942. Über die Assimilation des Elementaren Stickstoffs der Luft durch die Hefesymbionten *Rhagium inquisitor* L. *Z. Morph. Ökol. Tiere* **38**: 526–33.

SCHUCH, K. 1937. Experimentelle Untersuchungen über den Nahrungswert von Kiefern- und Fichtenholz für die Larven des Hausbockkäfers (*Hylotrupes bajulus* L.). *Z. Pfl. Krankh.* **47**: 572–82.

SEILLIERE, G. 1905. *C.R. Soc. biol., Paris* **57**: 940–1.

SHARP, D. 1901. *Insects.* Pt. II, p. 247. Cambridge Nat. Hist. Vol. VI.

SHARP, D. 1916. Additional British species of *Ernobius. Ent. mon. Mag.* **52**: 178–179; 219.

SMITH, J. H. and FORBES, A. C. 1944. How to defeat Wood-borers, *N.Z. J. Agric.* **68** (2): 83–8.

SPEIGHT, R. 1918. New Zealand Timbers and the Borer. A Note on the Susceptibility of New Zealand Timbers to the Attack of the Borer, *Anobium domesticum. N.Z. J. Sci. Tech.* **1**: 142–4.

SPILLER, D. 1948. A Note on the Susceptibility of Heat-treated Wood to attack by the Common House-borer, *Anobium punctatum* de Geer. *N.Z. J. Sci. Tech.* (*B*) **30**: 162–3.

SPILLER, D. 1948. Toxicity of Pentachlorophenol to the Common House-borer, *Anobium punctatum* de Geer. I, Residual, Contact and Ovicidal Action. *N.Z. J Sci. Tech.* (*B*) **30**: 142–53.

SPILLER, D. 1951a. Note on Susceptibility of *Pinus radiata* D. Timber to attack by the Common House-borer, *Anobium punctatum* de Geer. *N.Z. J. Sci. Tech.* (*B*) **32**: 38–9.

SPILLER, D. 1951b. Digestion of Alpha-cellulose by Larvae of *Anobium punctatum* de Geer. *Nature, Lond.* **168**: 209–10.

SPILLER, D. 1952. A Study of Control of Infestations of the Common House-borer, *Anobium punctatum* de Geer. *N.Z. J. Sci. Tech.* (*B*) **33**: 447–459.

SPILLER, D. 1960. Artificial Egg-laying Sites for the Common House-borer, *Anobium punctatum* de Geer. *N.Z. Entomologist* **2**: 19–23.

STEPHENS, J. F. 1839. *A Manual of British Coleoptera.* London.

STRAND, A. 1951. On the species of *Grynobius* Thoms. (COL. ANOBIIDAE). *Norsk. Ent. Tidskr.* **8**: 224–7.

SWAN, D. C. 1934. The Hay Itch Mite, *Pediculoides ventricosus* (Newp.) (PEDICULOIDIDAE, ACARINA) in South Australia. *J. Agric. South Aust.* **37**: 1289–99.

SWINGLE, H. S. 1925. Digestive Enzymes of an Insect. *Ohio J. Sci.* **25**: 209–18.

TASCHENBERG, E. L. 1879. *Praktische Insektenkunde* **2**: 77.

TAYLOR, J. M. 1968. Prevention of insect attack in plywood. Part 2: *Anobium* attack in birch plywood. *Wood* **33** (3): 43–44.

THOMSON, G. M. 1922. *The Naturalisation of Animals and Plants in New Zealand*, 292–3. Cambridge University Press.

TOOKE, F. G. C. 1949. Beetles Injurious to Timber in South Africa—A study of Their Biology, Prevention and Control. *Science Bulletin* No. 293—Dept. of Agric. (Ent. Ser. No. 28), Union of South Africa.

TOTH, L. 1952. The Role of Nitrogen-active Micro-Organisms in the Nitrogen Metabolism of Insects. *Tijdschr. Ent.* **95**: 43–59.

TRÄGARDH, I. 1938. Untersuchungen über die Verbreitung und das Auftreten der holzzerstörenden Insekten in öffenlichen Gebäuden in Schweden. *Z. Pfl. Krankh.* **48**: 295–302.

TRÄGARDH, I. 1938. Survey of the Wood-Destroying Insects in Public Buildings in Sweden. *Bull. ent. Res.* **29** (1): 57–62.

TRÄGARDH, I. 1947. Final Survey of the Wood-Destroying Insects in Public Buildings in Sweden. *Bull. ent. Res.* **38** (1): 117–22.

UVAROV, B. P. 1928. Insect Nutrition and Metabolism. A Summary of the Literature. *Trans. ent. Soc. Lond.* **76**: 255–343.

VAN EMDEN, F. I. 1925. Bericht über die entomologische Überwachung der Speicher und Kulturen der Firma Caesar und Loretz A.G. im Jahre, 1923. *Jber. Caesar Loretz* 167–232. Halle, 1924.

VERHOEFF, K. W. 1923. Beiträge zur Kenntnis der Coleopteren-Larven mit besonderer Berücksichtigung der Clavicornia. *Arch. Naturgesch.* (*A*) **89**: 16.

WEBSTER, F. M. 1910. A predaceous and supposedly Beneficial Mite, *Pediculoides* becomes obnoxious to Man. *Ann. ent. Soc. Am.* **3** (1): 15–39.

WEISS, F. 1905. Nogle Vand-og Kvealstofbestemmelser i stammer af Fyr og Gran, Det. *Forstl. Forsoksv.* **1**: 113–16.

WESTWOOD, J. O. 1839. *An Introduction to the Modern Classification of Insects*, **1**: 269 *et seq.* London.

WIGGLESWORTH, V. B. 1927. *Biochem. J.* **21**: 797–811.

WILSON, S. E. 1933. *Ann. appl. Biol.* **20**: 661–90.

ZACHER, F. 1927. *Die Vorrats-Speicher-und Materialschädlinge und ihre Bekämpfung*, 94–9. Berlin.

CHAPTER FOUR

The Anobiidae. II

An account of the biology of *Xestobium rufovillosum*

Historical

It is seldom possible when writing of insects to go back almost three centuries for an account of an insect whose identity can be established without question. This is the case, however, with the Death Watch Beetle, whose tapping noise, produced by both sexes in the adult stage, was sufficiently distinctive to merit the attention of the seventeenth-century naturalists. The first writer on this subject is thought to have been Swammerdam, who wrote in his *History of Insects*, some time between 1637 and 1680, about the ticking noise made by the beetle he called *Scarabeus sonicephalus*. *Sonicephalus* can be freely translated as 'head noises', so that an accurate description of the mechanism by means of which the insect makes its tapping noise was then given. Two other early writers were Allen, in 1695, and Derham, in 1701, who also gave accounts of the tapping habits of this insect. Over a hundred years went by before any biological information about this insect was published, but then Bouché, in 1834, and Ratzeburg, in 1837, mentioned the occurrence of the life stages in decayed hardwoods. Westwood followed in 1839 and mentioned in his *Introduction to the Modern Classification of Insects* that he had reared the Death Watch Beetle from larvae found in rotten whitethorn wood. He stated that 'it remained more than three years before it arrived at the perfect state'.

Little else of value was recorded about the Death Watch Beetle during the remainder of the nineteenth century, with the exception of the statement of Taschenberg, in 1879, that it had an annual life cycle. As will be shown later, by reference to the work of Fisher, this is true only under exceptionally favourable conditions. In the early twentieth century, however, an event occurred which was to provide a great impetus to biological studies on this insect. In 1914, Sir Frank Baines made a report on the condition of the roof timbers of Westminster Hall. Sir Frank, who was then Director of Works, H.M. Office of Works, pointed out the serious condition of the roof timber in this great Norman monument. Furthermore, in 1924, at a public meeting of the Royal Society of Arts, with Sir Aston Webb in the chair, he stated that the attack of the Death Watch Beetle appeared to be so prevalent that there was probably hardly a medieval structure in this country which was not, or had not been, infested by it at some time. Immediately following the Westminster Hall report, Kemner in Sweden (1915) gave a full account of the insect, the most comprehensive that had been produced up to that time.

Sir Frank Baines stimulated H. Maxwell Lefroy, Professor of Entomology at the Imperial College of Science and Technology in London, to carry out an investigation into the habits of the Death Watch Beetle and into methods of exterminating it. Lefroy had spent some years in India with the resounding title of Imperial Entomologist and had published important contributions to economic entomology. With his characteristic energy Lefroy spent a great part of the rest of his life devising chemical means of destroying Death Watch Beetles, and it is fair to say that in Lefroy's laboratory at Imperial College, the modern industry of *in situ* wood preservation came into being. Lefroy gave a history of the Death Watch Beetle in a paper in 1924, which provoked a most exhilarating discussion. Lefroy mentioned that the insect in the wild is attacked by another beetle which hunts it down in its galleries and which, unfortunately, is comparatively rare and has apparently been found only in trees, not in buildings. We are not told whether this was *Opilo mollis* or *Korynetes caeruleus*. The former species of clerid would fit the case, but it is strange that the latter species was not recorded from the many Death Watch infestations which Lefroy visited and investigated. Kimmins, who was Lefroy's assistant and worked on the biology of Death Watch, read a paper to the South London Entomological Society in 1933 on this subject, and he makes no mention of clerid predators, although he does instance difficulty in laboratory rearing of Death Watch caused by the parasitic mite *Pediculoides* (=*Pyemotes*) *ventricosus*. If the species *Korynetes caeruleus* was indeed rare, or absent, indoors up to that time, then a considerable change has taken place in recent years. There seems to be no reason why these two competent observers should have overlooked it, and therefore it seems most probable that, as Death Watch infestations in this country have remained more or less static, or are, if anything, perhaps less extensive now than in former years, the great increase of *K. caeruleus* is due to the great increase of *Anobium punctatum*, its alternative prey.

The writer has seen a 1 lb. jam jar half full of *Korynetes caeruleus* beetles, collected over a few years from a Death Watch Beetle infestation in an Essex house.

Meanwhile interest in the nutrition of insects had been growing, especially concerning the way in which the larvae of wood-boring insects obtain nutrients, apparently from cellulose, and Breitsprecher, Heitz, Campbell, Ripper, Müller, and Norman included the Death Watch Beetle amongst the species they studied. Campbell, working at the Forest Products Research Laboratory, Princes Risborough, showed, in 1929, that when the Death Watch larva is feeding on oak heartwood, the frass has a lower cellulose content that the uneaten wood. Breitsprecher and Heitz demonstrated the presence of micro-organisms, yeast-like in their biology, in the gut of larvae, but Ripper who showed that cellulase was present in the larval gut, did not consider this enzyme was due to the presence of micro-organisms.

Parkin (1940), whose work has already been referred to in connection with *Anobium punctatum*, found that the larvae of *Xestobium rufovillosum* contained enzymes enabling them not only to break down the carbohydrates of the cell wall, namely cellulose and hemicelluloses but to utilize protein and in addition, when they are present, starch and sugars in the cell contents.

An account of the Death Watch Beetle by Altson was published in 1922 in the *Timber Trades Journal*, and the British Museum pamphlet on furniture beetles, published in 1925 in the economic series by Gahan, included an account of the insect. There followed in 1927 the second great impetus to the study of Death Watch biology when an Entomology Section was established at the Forest Products Research Laboratory at Princes Risborough. This was brought about, in the main, by the efforts of Professor J. W. Munro of the Imperial College of Science and Technology, who had previously been entomologist to the Forestry Commission, and from his work there he knew well the importance of wood-boring insects. The actual decision to found the Entomology Section, however, was probably taken as a result of widerspread *Lyctus* attack in newly manufactured furniture. An enquiry was initiated at the Imperial Forestry Institute, Oxford, under the direction of Professor Munro, and afterwards the work was transferred to Princes Risborough. R. C. Fisher was appointed to the new Entomology Section, in which he served until his death in 1959. During his years at the Princes Risborough Laboratory he and his staff made a large number of contributions to our knowledge of the biology of wood-boring insects. These are contained in original papers, reviews, and annual reports, but the monument to Fisher's work is probably his series of papers on the biology of the Death Watch Beetle (1937, 1938, 1940, and 1941) and the account of this insect which follows is largely taken from these publications.

Classification

The genus *Xestobium* was erected by Motschulsky in 1845 and it is contained within the subfamily ERNOBIINAE, together with the British genera *Ochina* and *Ernobius*. The specific name *rufovillosum* is due to Degeer in 1774 but *tessellatum* de Villers 1789 has been widely used. In addition to the latter, Pic, in the catalogue of Junk and Schenkling, has listed the following synonyms: *faber* Thunb., *fatidictum* Blumenb., *fuscum* L., *pertinax* Herbst, *pulsator* Schall., *pulsatorium* Scriba, *rubiginosum* Müll., and *squalidum* Lec. Of these however, *fuscum* of Linnaeus is most probably *Lyctus linearis* of Goeze and in addition there is some confusion concerning the name *pertinax*.

Distribution

Xestobium rufovillosum occurs throughout Europe, including Corsica. In Africa it is recorded from Algeria, and in America from the United States, although Simeone in a recent survey of anobiid beetles in the north-eastern states says that it does not appear to have achieved prominence there and that its scattered distribution, as in the case of *Anobium punctatum*, is probably indicative of its spread by commercial means. Simeone did not think that he had enough collections of this species to generalize about its approximate range. Craighead in 1950 listed *Xestobium*

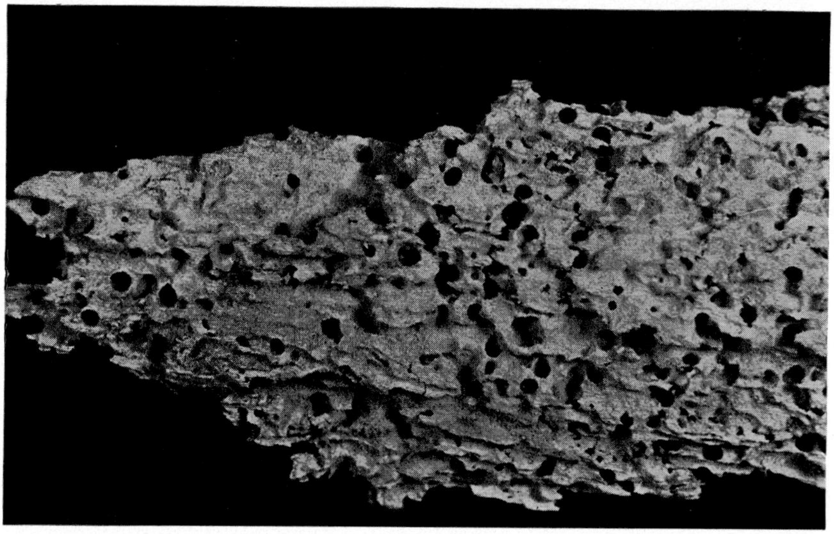

Fig. 48. Appearance of internal damage to oak by Death Watch.

rufovillosum as destructive in the New England states, but Simeone did not find one attack in 720 building structures examined in New York State, although he found it in seventeen collections of wood-feeding anobiids deposited in the larger museums in north-eastern United States and in adjacent Canada.

Death Watch Beetle is also recorded from the Pacific Island of New Caledonia (it has already been mentioned that *Anobium punctatum* is recorded from there also—quite outside the expected range of both these temperate anobiids). Although *Anobium punctatum* occurs in both Australia and New Zealand, to which countries it seems certain it was carried in the old furniture of the early European settlers, *Xestobium rufovillosum* does not occur in the Australasian region. It is extremely rare for it to be found in furniture, so that its chances of being transported in this way are remote.

Fisher considers that, as a timber pest, this species is of greater economic importance in England than elsewhere in Europe, yet in England it is on the fringe of its distribution. It is extremely rare in Scotland and has been found only twice in Ireland (in each case in situations where timber from England had most probably been used in building construction—in Leinster House, Dublin, and in a mansion in Waterford). Commenting on the Death Watch situation in England, Fisher thought this was due to the large number of buildings in which oak, the principal timber attacked, has been used. The occurrence of this species, as revealed by surveys of properties, is discussed in a later chapter.

The natural habitat

When Lefroy carried out his work on Death Watch Beetle, he was handicapped by lack of insect material. Fisher likewise had initial difficulties until he found a natural habitat from which he was able to obtain the insect in abundance. This habitat was old, decayed, willow trees on the banks of a stream in Oxfordshire, and subsequently, two of the trees were transplanted to the grounds of his laboratory. The insect is found out of doors in the decaying branches, or in parts of the trunk, of a number of hardwood trees, but the two most commonly infested species are oak and willow. In addition, it has been recorded in ash, beech, hornbeam, poplar, sweet chestnut, whitethorn, etc. Ratzeburg (1837), however, records it in yew, and Hopkins (1893) in black spruce in Virginia, which is not to be expected from our present knowledge.

Host timber indoors

Indoors, Death Watch is most usually found in oak, but it has been recorded in elm, chestnut, alder, and walnut, amongst the hardwoods (Figs. 48, 49 and 50) and it is not altogether unusual for coniferous timbers to be attacked if they abut infested hardwood. Fisher states that there is no case on record of an attack of Death Watch Beetle in a building constructed entirely of coniferous timber, which tends to confirm the opinion that attacks originate only in hardwoods.

Description of adult beetle

Xestobium rufovillosum is the largest of the British ANOBIIDAE, the length varying from about 5 to 7 mm (Figs. 51 and 52). The females are in general slightly larger than the males but this cannot be relied upon for separating the sexes. Fisher did not find Taschenberg's observation that the ninth antennal segment in the male is shorter than the seventh and eighth together, but not in the female, to be accurate. Fisher found no external, morphological feature of the beetles which could be used for sex differentiation which was a great handicap to his subsequent experimental work.

Fowler, in his *Coleoptera of the British Isles*, Vol. IV, gives the following description: 'A large and conspicuous species, convex, subcylindrical, of a dark chocolate brown colour, thickly tessellated with patches of yellowish pubescence, which gives the upper surface a variegated appearance; the upper surface is finely and thickly scabrous; head rather deeply sunk in thorax; thorax about as broad as elytra, contracted in front and posterior margin produced, central furrow obsolete, scutellum semi-circular; elytra parallel-sided, with traces of raised lines; legs ferruginous, stout and robust, tarsi short and broad, with fifth joint transverse.'

Fig. 49. Death Watch attack in a wooden boss showing little evidence from the front

Harris (1959) has shown how the sexes of the beetle may be separated by means of secondary characters—by no means an easy task. Reliance is placed on a series of differences in the last visible sternite (Fig. 53). In the male the posterior margin is slightly emarginate while the average ratio of median length to basal width is 1:3·6 the limits being 1:3·4 to 1:3·8. The surface contour is convex both laterally and antero-posteriorly and the sculpture is less rugose than the female and is slightly shining. In the female the posterior margin of the last visible sternite is continuously curved though occasionally flattening out. The average ratio of median

Fig. 50. Death Watch attack in the same wooden boss showing heavy attack at back adjacent to damp wall.

length to basal width is 1:2·9 the limits of the ratio being from 1:2·3 to 1:3·2 so that there does not appear to be any overlapping in this character. The surface contour is convex laterally but not or scarcely so antero-posteriorly and it is rugose and dull.

X. rufovillosum may be distinguished from all our other ANOBIIDAE by the absence of striae on the elytra, and from *Ernobius* by its stouter tibiae and the broader fifth joint of the tarsus. Generally, this species can easily be identified by the widely flanged prothorax. The yellow pubescence is easily rubbed off, the wing cases of the insect often thus becoming more reddish and shining.

Adult emergence

The earliest appearance of beetles in an infested building is recorded as occurring in mid-March and the latest in early June, but the more usual emergence occurs in the latter part of April and at the beginning of May. The earliest date mentioned by Fisher, from personal observation, was 7th April and by Kimmins, 24th March. Linscott records unusually late emergence of Death Watch from a large oak timber kept in an unheated wood store in 1961. Seventeen live beetles were found on 17th July, a further eleven on 19th July and on the 21st an egg was laid. On 24th July a further three emerged and the last one emerged on 1st August. Lefroy suggests that in certain cases where the damage to the wood is severe, the adults may not emerge to the outside of the wood, but pairing and subsequent egg-laying could take place within the wood, the insects never

Fig. 51. Adult *Xestobium rufovillosum.*

being visible to a human observer. In recent years there have been instances of Death Watch infested cavities in large structural timbers where the extent of the attack could not possibly have been assessed from outside.

Adult metamorphosis

Although the reader may think that the life cycle of this insect is being described in reverse order, it is in fact convenient, before discussing other features of the adult, to begin with the eclosion. The actual date of this has given rise to some misunderstanding, in that the information given by Munro and Lefroy is not acceptable to Fisher. Munro (1928) states that pupation occurs in the spring, but the adults rarely come out of the timber until autumn, or even the following spring, and Lefroy (1924) indicated that pupae may be found in June and September, and that the adults emerge during May and June of the following year. Another view, that of Gahan and Laing (1932) and Kimmins (1933–4) is that the larvae pupate in late summer or early autumn, but the adults do not emerge from the timber until the following spring. Fisher carried out a number of observations on the insect out of doors, and in infested timber buildings, and in general confirmed that the latter view was correct. Out of doors he found that the earliest date at which pupae were found was 18th July, and over several years he found pupae in decayed willow trees during July, August, and occasionally September, the length of the pupal stage being three to four weeks. This finding was in general agreement with Kimmins, who states that the pupal stage lasts about three weeks 'under laboratory conditions'. The late autumn and winter is then spent in the adult stage, but the reproductive organs are not developed. These mature slowly in the female and it is not until April that there are two or three well-developed eggs in each ovariole. In the males the testicular follicles are at first elongate and cylindrical, but by the end of September, or in early October, the follicles are no longer cylindrical and thereafter become pear-shaped as they pass their contents into the *vasa deferentia* and seminal vesicles, which become distended. Between December and April the follicles further diminish in size until they are difficult to see. At emergence the males are usually sexually mature.

Fisher found a number of exceptions to the periods given above at which pupae could be found when he examined the timber from infested buildings. The most usual variation was the finding of pupae during the

Fig. 52. Four Death Watch Beetles and flight holes on fungally decayed beam.

winter months, when adults only would have been expected. He thought that the higher temperatures experienced indoors were usually responsible for some of the abnormal developments.

The males emerge first and the females are ready for copulation on emergence, although their reproductive organs are not quite so advanced as those of the males. The exit holes are circular and 2·1–3·1 mm (average, 2·6 mm) in diameter.

Adult activities

Fisher considered that the general activity of the insect was mainly influenced by temperature. Out of doors, on warm, sunny days, he

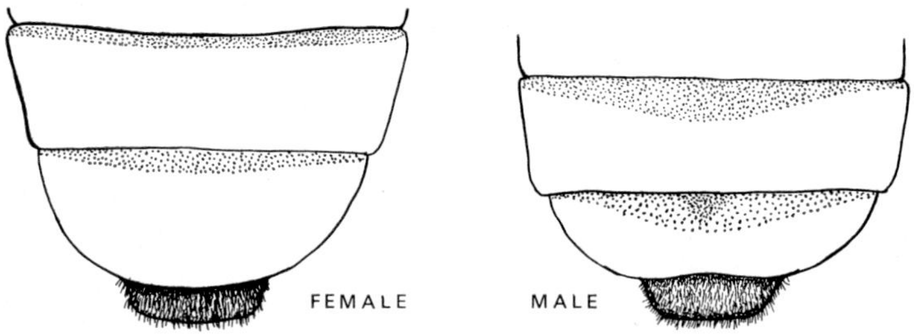

Fig. 53. Ventral view of posterior abdominal segments of adult Death Watch.

observed the adults walking over the surface of attacked trees and tapping frequently. They were most active from 4 to 6 p.m. Kimmins, however, suggests that they are most active at night. Fisher found that in the laboratory at 14° C they moved about slowly, but their activity increased greatly at temperatures of 17–20° C, tapping, pairing, and egg-laying being observed.

Tapping

The tapping noise is made by both sexes knocking the frons of the head against the wood on which they are standing. To the unaided ear, the number of taps is apparently seven or eight. However, a tape recording played back at quarter speed clearly reveals that the beetle normally taps 12 times. Occasionally there are 13, and rarely 11 taps. Each set of 12 taps takes about 1·25 seconds, and when in 'full cry', the interval between each set is 1·25–2 seconds. Tapping is probably a sex call and the observer can imitate it tolerably well with a pencil, at least to the extent of stimulating surrounding beetles to begin tapping themselves! Fisher thought that the males tap more frequently than do the females. Tapping may continue after pairing. From time to time, however, there is misunderstanding about the identity of insects making ticking or tapping noises. Gahan (1918) has written at length on the confusion that has existed over two centuries between the ticking of the booklouse and the tapping of the the Death Watch beetle. He also mentions the various anobiid beetles, including *Anobium punctatum*, which have been described by continental authors as making tapping noises. In Germany the whole family is known as the Death Watch beetles. The earliest known reference in English to the ticking noise made by beetles occurs in the work by John Wilkins, Dean of Ripon, in his book, *An essay towards a Real Character and a Philosophical Language*. It is dated 1668 and contains the description of a beetle, *Scarabeus domesticus*—frequent about houses—making a noise

Fig. 54. Larva of Death Watch Beetle leaving its egg.

like the minute ticking of a watch, by striking the bottom of his breast against his belly!

Flight and distribution

Very seldom does the Death Watch beetle fly. The only reference to actual flight that can be found is that of Kimmins, who reported that Lefroy was once watching some beetles when one of them suddenly spread out its wings and flew rapidly away. The wings, however, are well developed and the insect will use them momentarily to regain balance if it slips when walking, or if it falls on to its back. Fisher thought that it was questionable whether the beetles ever flew intentionally. Of recent years much has been learnt from the dissection of the wing muscles of beetles and perhaps this research will throw some light on why records of flight are so scanty.

The beetles walk rapidly and as the length of adult life may often be prolonged, it is conceivable that an adult could reach another tree with decaying parts if that tree was in close proximity, e.g., a row of pollard willows along the bank of a stream. Such Death Watch-attacked decaying trees could be washed along a river by flood and deposited many miles from their original site. It might be of interest to ascertain the resistance of the larvae to a few days in waterlogged wood.

When consideration is given to means of dispersal from one building to another, however, difficulties become apparent. It is inconceivable that the insect could, by walking, find its way from one building containing some decaying hardwood structure to a similar building. Such buildings are not common, they are almost always old properties and isolated from each other. If the building is heated then wood containing emerging adults

Fig. 55. Death Watch larva and bun-shaped pellets in frass. A flight hole is also shown.

could be introduced into the building as firewood. Against this argument is the fact that many churches containing a Death Watch attack were not heated until fairly recent times.

Although the insects are commonly found on church floors, where they have dropped from infested roof rafters, it would be most unusual for them to be found crawling up plain walls. The ease with which the insect feigns death, by pulling in its antennae and legs tightly against the ventral surface of the body (a similar position to that occupied in the pupal stage), when disturbed is probably the reason for the fall from the roof whilst the insects are still alive.

There seems, therefore, no explanation to account for their presence in many buildings other than their introduction as larvae in large timbers of which some part supported a fungal infection. Many of the very large and sometimes angled timbers would almost certainly be taken from trees which were past their prime. There may be some difficulty in visualizing the Death Watch attack in the roof of Westminster Hall as a direct result of unnoticed larvae in the oak of that 'miracle of Gothic architecture' which Richard II had erected in 1394. But when Sir Frank Baines examined the roof in 1913 he 'found that many of the principal rafters, purlins, and some of the main collar beams, were hollowed out to a thin shell'.

Pairing and oviposition

Although Fisher records that pairing took place most frequently in the late afternoon, it has been observed at all times of day and evening when conditions such as temperature favoured activity. Although it is only observed when both insects are on the surface of the wood, it is virtually certain that it takes place also in cavities inside large timbers. Such cavities are often a feature of Death Watch attack.

Fisher's account of the copulation is as follows: 'The male mounts upon the back of the female, and when coition has been established gradually falls over backwards and finally rests on its back, with legs and antennae drawn closely into the sides of the body. The female is capable of supporting the male in this position and was seen on occasion to crawl over the surface of the wood with the male thus attached. At laboratory temperature (18–20° C) the duration of copulation varied from $\frac{1}{2}$ to $1\frac{1}{4}$ hours.'

There are no records of egg-laying having been observed either in buildings or out of doors under natural conditions, but an account has been given of observations in the laboratory. The female moves slowly over the wood surface exploring cracks and crevices with her antennae, and then, by turning round, with the pair of small palps on the extremity of the ovipositor. When she has found a situation which suits her, the ovipositor is further extruded into the crevice and then, from five to seven seconds afterwards, a contraction of the abdomen occurs and the egg can be seen moving down the ovipositor. Occasionally eggs are laid at night, but most are laid during the day and early evening and Fisher thought that as the period of greatest activity of the beetle occurred during the afternoon, this was probably the period of main oviposition also.

The oviposition site, under laboratory conditions, is in cracks and crevices, roughened surfaces and amongst the broken wood fibres that usually occur on the transverse surface of a piece of wood. The eggs are also laid amongst the frass in old tunnels and in the vessels of oak, and are frequently difficult to observe. They are laid singly, in pairs, or in small groups, and there appears to be a cement-like substance present when the eggs are first laid so that they adhere to each other and to the wood. It is of interest to note, in view of the subsequent behaviour of the newly emerged larva, that oviposition is not confined to decayed wooden surfaces.

Kimmins found that egg-laying commences between 5 and 11 days after pairing and Fisher found that this period was 8 and 13 days when the beetles were left out of doors, but the latter author also mentions records of 36 days for this period. It seems obvious from the experimental data given by Fisher that this period is a very elastic one and depends on temperature.

The number of eggs laid varied from 39 to 201, but the more usual number is from 40 to 60. Fisher found that the egg-laying potential of the females differed on account of the variation in the number of ovarioles in

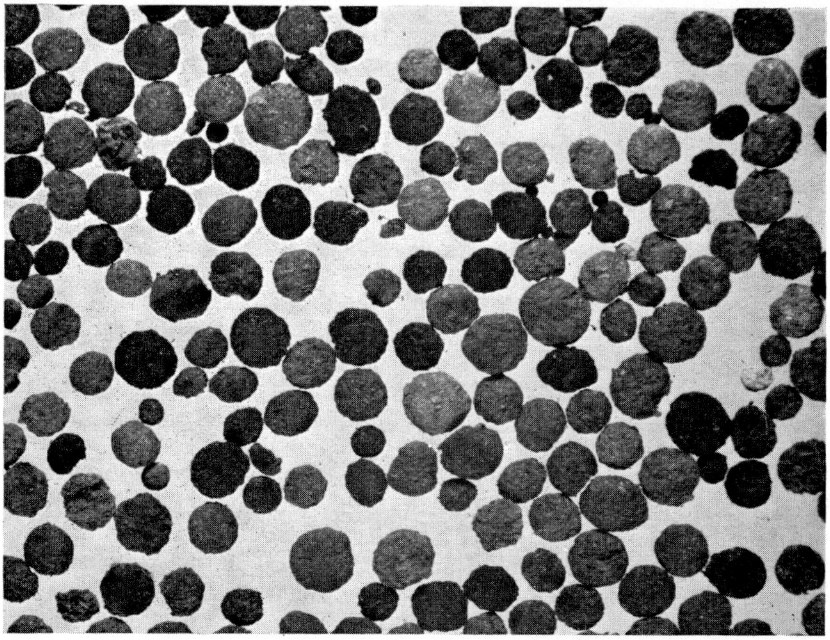

Fig. 56. Bun-shaped pellets in the frass of Death Watch larvae. Magnified 10 times.

each ovary. Temperature and humidity also have considerable effect on the numbers of eggs laid.

Length of adult life

The female may live as long as 10 weeks after pairing and the males up to 9 weeks. As the larvae pupate during July and August and the length of the pupal stage is only 3 weeks, the total length of duration of the adult stage is from 10 to 11 months, but only during April, May, and June is it reproductive.

Eggs

When first laid the egg is pearly-white and somewhat lemon-shaped but slightly more pointed at one end. In length it measures about 0·6–0·7 mm and at its greater width it measures 0·4–0·5 mm. The micropyle is visible at the anterior end as a circular translucent patch which later becomes more pronounced. The difference in appearance between the micropyle and the remainder of the shell of the egg is due to the reticulate appearance of the latter, the area around the micropyle being smooth. The egg becomes more opaque shortly before hatching, the sides collapse to some extent and the micropylar area becomes more evident.

Length of egg stage and viability

Out of doors the length of the egg stage is about five weeks but the period may be shorter if the temperature is high. Fisher found that 297 eggs had a mean incubation period of 35·7 days at an average temperature of 14·5° C, whilst 14 eggs had a mean incubation period of 25·2 days at an average temperature of 18·3° C. The average survival of eggs out of doors was found to be about 95 per cent.

Under laboratory conditions Fisher found that no hatching took place at a relative humidity of 23 per cent nor at 30° C. At 15° C the egg stage duration did not vary much, being 49·7 days at 41 per cent RH, 41·3 days at 53 per cent RH, 45·6 days at 75 per cent RH, and 44·9 days at 95 per cent RH. Again at 20° C there is little difference between the various relative humidities. The period is 23·6 days at 41 per cent RH, 21·3 days at 53 per cent RH, 20·0 days at 75 per cent RH, 20·7 days at 86 per cent RH, and 21·1 days at 95 per cent RH. The incubation is still less at 25° C, being 15·3 days at 41 per cent RH, 15·7 days at 53 per cent RH, 14·2 days at 75 per cent RH, 15·7 days at 86 per cent RH, and 12·5 days at 95 per cent RH.

Fisher kept a few eggs at 10° C and found that after eight weeks they had not hatched; but they hatched on subsequent exposure to 20° C. Under the above conditions of relative humidity and temperature (excepting 23 per cent RH, and 30° C) the viability varied from 61·3 per cent (20° C at 41 per cent RH), to 97·0 per cent (15° C at 75 per cent RH).

In considering the factors affecting oviposition, viability of the egg and length of incubation period in relation to the occurrence and spread of Death Watch in a building, Fisher came to the following general conclusions. Warm dry conditions during egg laying are less favourable for the spread of infestation than warm moist conditions. The latter may be localized in structural timbers due to water leakage of some sort. Central heating may produce conditions suitable for egg laying and hatching in roofing and other timbers by causing condensation of moisture, unless adequate ventilation is provided. Fisher also thought that he had evidence which showed that life cycles carried out at a temperature raised above that which the species normally experiences, caused a decrease in fecundity in later generations and he gave this as a possible explanation of a gradual cessation of attack in buildings. Against this however must be set the opposite sort of evidence in that in many cases Death Watch attack has continued in a limited and circumscribed area for centuries.

Larvae

The first instar larva which hatches from the egg (Fig. 54) walks actively over the surface before starting to tunnel into the wood. This contrasts with the larva of *Anobium punctatum*, which bores straight into the wood through the base of the egg. The difference of behaviour means that the eggs do not necessarily have to be laid where conditions are suitable for the larvae.

The full grown larva is typically anobiid—fleshy, soft and curved and of scarabaeoid eruciform type (Fig. 55). It has six small legs and is creamy white in colour. It is up to 11 mm in length. The thoracic region is swollen and the whole body is covered with long golden-yellow setae standing erect. The last three abdominal segments are also somewhat swollen but not to such a degree as the thoracic segments. The body is normally held hook-shaped. Areas of spinules are present on the prenotal folds of the metathorax and of the first eight segments of the abdomen. They are present also laterally and ventrally on the ninth and tenth abdominal segments. The frass contains large, bun-shaped faecal pellets (Fig. 56).

Pupae

The pupa again is typically anobiid, the only points worthy of comment are the thin soft cuticle, the movable abdomen and the complete absence of spines and pubescence. In colour it is at first milky-white but afterwards the parts darken progressively starting with the eyes.

The pupae can be sexed readily. Female pupae are from 7·5 to 8·5 mm in length and from 2·5 to 3·5 mm at the widest part of the abdomen, but sex can most readily be differentiated by the pair of prominent knob-like processes carried by the last two segments (Fig. 57). Each process consists of two segments of which the basal one is much the larger. There is also a slit-like furrow situated in the centre of the raised posteror part of the eighth sternite.

The male pupa is from 7 to 8 mm in length and is from 2·5 to 3·0 mm in breadth at the abdomen. A pair of small bilobed protuberances which are curved and wrinkled replace the prominent processes of the female. There are also two spherical tubercles which are of different sizes. The

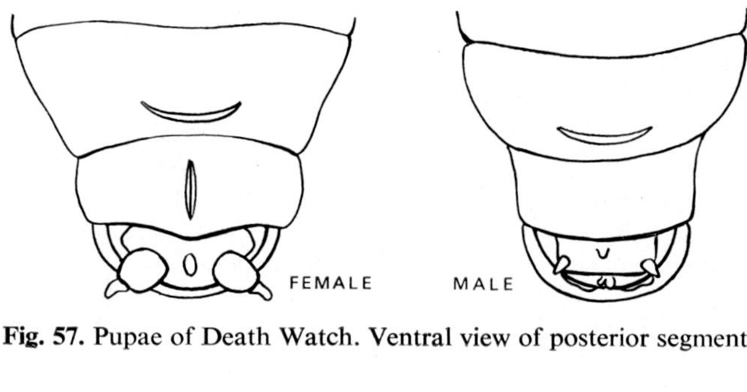

Fig. 57. Pupae of Death Watch. Ventral view of posterior segments.

slit-like depression present in the female on the eighth sternite is absent in the male.

Death Watch and fungal decay

The association of Death Watch attack with fungally decayed wood has already been noted but the relation of the latter to the rate of development of the Death Watch larva was studied in detail by Fisher and his colleagues at Princes Risborough. They first investigated the species of fungi causing decay of the hardwood trees where Death Watch attacks were occurring naturally out of doors. They thought it probable that the fungus *Polyporus sulphureus* (Bull.) Fr. caused decay of heartwood in old parkland oaks which had obviously supported an attack of Death Watch at some time even if the infestation was no longer active. Newly felled parkland oak of the brown colour characteristic of the attack of the beefsteak fungus *Fistulina hepatica* Fr. has been stated by Fisher to contain the larvae and beetles of Death Watch, and the stag-headed and hollow condition of oaks typically supporting a Death Watch infestation has been associated by Braid (1924) with attack by this same species of fungus. These observations are supported by the writer from the numerous examples of brown oak found attacked by Death Watch in buildings. This timber is often extremely hard, making the sawing of it very difficult but nevertheless it is bored by Death Watch larvae.

With regard to willow trees which Fisher found to be such a prolific source of Death Watch material, the following fungal species were identified: *Polyporus fumosus* (Pers.) Fr., *Polystictus versicolor* (L.) Fr., *Trametes suaveolens* (L.) Fr., and *Coniophora cerebella* Pers. In addition a heavy attack of Death Watch was found associated with *Xylaria hypoxylon* Grev. The larvae did not occur in the soft, most heavily decayed wood.

In buildings where severe Death Watch attack occurs it is invariably associated with decay to a greater or lesser extent (Fig. 58). Often the limitation of the decay is shown by the grouping of the flight holes of the beetle and indeed unsuspected pockets of decay may be identified by this means. However, although the situation of the decay is most often found and is usually most severe where the timbers are built-in, such as the extremities of tie-beams, wall plates, or indeed anywhere where the timber is not ventilated (as Lefroy pointed out), the particular species of fungus may be difficult to identify. Cartwright and Findlay (1936) considered that *Phellinus cryptarum* Karst. is probably able to cause a more rapid decay of oak than any other species and indeed Death Watch attack is often found either adjacent to the actively growing fungus or present in the rotted wood after the fungus has died out. This fungus requires very moist conditions and usually occurs indoors either in roofing timbers where a leakage has occurred and, due to a covering of plaster, insufficient ventilation is present, or in the centre of a large beam. Mangin and Patouillard described an attack of oak beams in the Palace of Versailles by *P. cryptarum* and a subsequent attack of Death Watch. A point of special interest concerning the decay of oak caused by this fungus is that the medullary rays are resistant and the subsequent Death Watch attack shows the medullary rays are also resistant to the insects.

Death Watch attack associated with the two commonly occurring fungi *Coniophora cerebella* Pers., the cellar fungus, and *Merulius lacrymans* (Wulf.) Schum., the Dry Rot fungus, has often been observed in churches and older buildings. Indeed these two fungi are the two most commonly

Fig. 58. Death Watch damage to centre of beam following the course of heartrot attack.

occurring species in such situations. The latter fungus however usually attacks softwoods which Death Watch only rarely infests, i.e. when it happens to be adjacent to decayed hardwood. *Coniophora cerebella* decays both hardwoods and softwoods. Ramsbottom (1937) has drawn attention to the great damage caused formerly to wooden vessels of the Royal Navy by *Merulius lacrymans*, but our knowledge of Death Watch attack in such vessels is confined to H.M.S. *Victory*, where there is evidence also of an attack by *Coniophora cerebella*. The fungus *Polyporus sulphureus* (Bull.) Fr., was associated also with the fungi mentioned above in decaying ship's timbers and indeed is not uncommon as causing oak joists to decay in buildings, the fungus originating from the standing tree. It is known that this fungus can lie dormant in timber converted from such trees but can revive and cause further damage if, subsequently, the timber absorbs a quantity of moisture. Furthermore in timbers of large dimensions, the inside dries very slowly and the fungus can decay fairly large cavities in the wood before the drying process puts an end to the fungal activity.

Fisher mentions *Poria vaillantii* (D.C.) Fr., as being rarely found in oak and elm attacked by Death Watch, although this species normally only attacks softwoods. *Fistulina hepatica* has been already commented upon as decaying oak subsequently attacked by Death Watch outdoors as well as indoors.

The mould *Paecilomyces varioti* Bairier has been recorded as being associated with the tunnels of Death Watch larvae which are characteristically tanned a bright golden colour (golden oak).

It can thus be concluded that Death Watch attack is associated with wood decay caused by a fairly wide variety of fungal species.

Length of life cycle of Death Watch Beetle

Fisher's classical work at Princes Risborough on this subject was extraordinarily difficult. It is not easy to assess accurately the extent of decay of a piece of wood being eaten by the larvae as the years go by, and furthermore some of the larvae took 10 years to complete the life cycle. In willow which was not decayed, the larvae either died soon after they had penetrated the outer surface or, in a few cases, they developed very slowly. Larvae only 3 mm in length were found in sound willow nine and a half years after they had emerged from the egg. In another case a larva completed its development in nine years owing to the presence only of traces of fungal decay.

In decayed willow the completion of the life cycle was much more rapid. In thoroughly decayed wood the life cycle was completed in 10–17 months, the range being probably due to the variation in the severity of the decay and the differences in the proportion of sapwood and heartwood present.

Again in oak sapwood where decay was slight the life cycle was prolonged. With severe decay by *Phellinus cryptarum* at 22–25° C and 80–90 per cent relative humidity the length of the life cycle varied from 14 months to something more than two years. With decay by *Polystictus versicolor* it took 18–19 months. Out of doors in oak attacked by *Phellinus cryptarum* the life cycle varied from three to seven years. There were a number of instances of young larvae endeavouring to establish themselves in sound oak but dying within a few weeks. In heartwood similarly, the larvae failed to establish themselves when it was sound, but when decayed by *P. cryptarum* the life cycle was as short as 16 months. Brown oak heartwood was attacked by Death Watch but development of larvae was exceedingly slow.

Fisher also obtained an annual life cycle in naturally decayed Banak (*Virola merendonis*) but in the case of two softwoods (Scots pine (*Pinus sylvestris*) decayed by *Coniophora cerebella* and Sitka spruce (*Picea sitchensis*) decayed by *Phellinus cryptarum*) although the decay was severe, the larvae made numerous tunnels but did not develop and even after one and a half to two and a half years were not much more than half grown.

In a long series of experiments Fisher compared the duration of the life cycle with the degree of decay in the case of oak sapwood decayed by *Phellinus cryptarum* (producing a white rot), willow decayed by *Polystictus versicolor* (also producing a white rot) and oak sapwood attacked by *Coniophora cerebella* (producing a brown rot). His method was to cut blocks of sound wood $6 \times 1\frac{1}{2} \times 1\frac{1}{2}$ inches which were then decayed by the

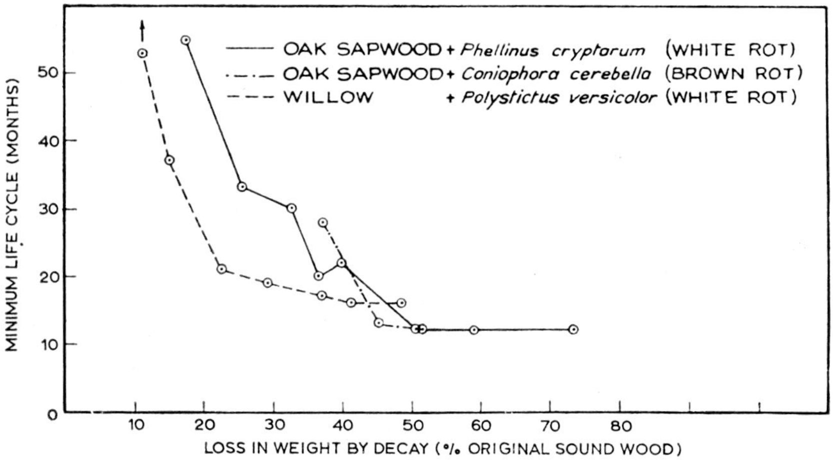

Fig. 59. Relation between extent of decay in oak and willow and duration of life cycle of *Xestobium rufovillosum*. (Fisher 1941).

particular species of fungus after initial sterilization. In this way varying degrees of decay were obtained and measured by the loss in weight expressed as a percentage of the original oven dry weight. About thirty eggs were laid on each block and groups of blocks were maintained at a temperature of 22–25° C and 80–90 per cent relative humidity which corresponded to an equivalent moisture content of 18–20 per cent. The minimum length of life cycle was taken as the period until the first exit hole appeared. The blocks were then cut up and all larvae and pupae—alive and dead—examined. There was sometimes considerable variation amongst the individual blocks in a group, although as far as possible groups of four samples were maintained as closely as possible in constant conditions. The first point to note from his results was that the young larvae were not able to penetrate into the undecayed control blocks and the attack therefore was never initiated. On the other hand in all cases where the wood was decayed to some degree, the life cycle was completed. The important result of Fisher's work was the establishment of the rule that the extent of the decay, within limits, determined the length of the life cycle and the rate of disintegration of the wood due to the larval boring. Where decay was very slight—giving only 18 per cent loss in weight—the life cycle was as long as 55 months. At the other end of the scale with severe decay amounting to as much as 73 per cent loss in weight, the complete life cycle was not only passed through in less than 12 months but due to the larval boring the wood was completely disintegrated. The relationship between the minimum length of life cycle and extent of decay is shown in Fig. 59 and the relation between the minimum length of life cycle and the density of the decayed oak and willow is shown in Fig. 60.

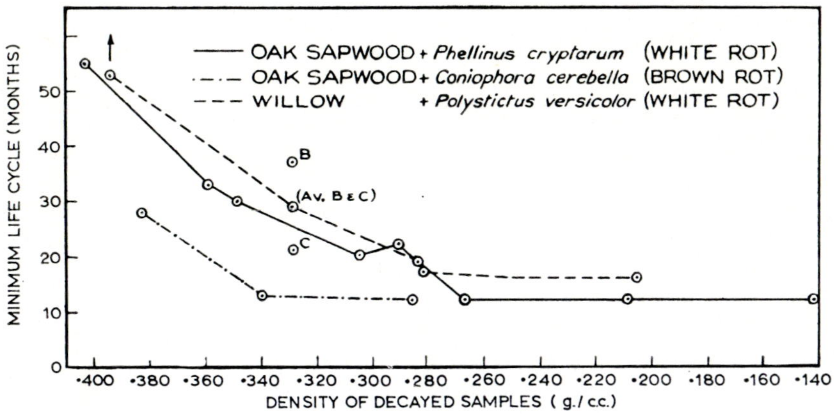

Fig. 60. Relation between density of decayed oak and willow and duration of life cycle of *Xestobium rufovillosum*. (Fisher 1941).

Campbell and Bryant (1940) studied the way in which *Phellinus cryptarum* decayed oak wood from the chemical aspect and they found that the major components were broken down simultaneously. They then compared the chemical composition of the decayed wood with that of the frass and came to the conclusion that the Death Watch larvae were utilizing components of the wood and not the fungal tissue. The suggestion was made, therefore, that the development of the larva was not dependent on the availability of food components resulting from fungal decay. The rather surprising conclusion was reached that fungal decay was necessary for purely physical reasons, i.e. the rate of conservation of energy in the metabolism of the larva reaches the required amount only when it is feeding on wood which has been 'softened'.

Fisher discussed the nutritional requirements of the Death Watch larva but made no mention of Staniland's (1930) note on the presence of yeast cells which were found to be invariably present in large numbers in the hepatic diverticula and closely connected parts of the alimentary canal of larvae and adults. In the adult females they are found in large numbers also in the spermathecae and vagina and doubtless are passed on to the eggs. Campbell and Bryant thought that the small nitrogen requirements of the insect could be met from the wood. It is apparent that Fisher hardly shared his colleagues' opinion on this, as he obviously wondered whether the fungal mycelium present in the wood could provide a rich source of nitrogen and he instances the work of Becker (1938) with the House Longhorn Beetle *Hylotrupes bajulus*. This author found that the weights of larvae of this species when maintained on pine sapwood containing peptone were ten to fifteen times as great as the weights of larvae in the controls. Fisher also thought that fungal decay could liberate accessory growth substances and, in addition, he discussed the possibility of the fungal decay bringing about a concentration of 'pre-existing ferment in the larval gut'.

REFERENCES

ALLEN, B. 1695. An account of the *Scarabaeus Galeatus Pulsator* or the death-watch, taken August, 1695. *Philos. Trans.* **245**: 376.

ALTSON, A. M. 1922. Beetles Damaging Seasoned Timber. *Timb. Trades J.* 15th April to 13th May.

ANON. 1935. The Common Furniture Bcetle. *Forest Products Research Leaflet* No. 8.

ANON. 1940. Beetles Injurious to Timber and Furniture. *Bull. For. Prod. Res.* No. 19. H.M.S.O.

BAINES, F. 1914. Report on roof of Westminster Hall. Cd. paper, **7436**.

BAUMBERGER, J. P. 1919. A Nutritional Study of Insects, with Special Reference to Micro-organisms and their Substrata. *J. exp. Biol.* **28.**

BECKER, G. 1938. Zur Ernahrungsphysiologie der Hausbokkäfer-Larven (*Hylotrupes*) *bajulus* L.). *Naturwissenchaften* **26**: 462.

BOUCHÉ, P. A. 1843. *Naturgesch der Insekten*, 187.

BRAID, K. W. 1924. Some Observations on *Fistulina hepatica*, and Hollow Staghead Oaks, *Trans. Brit. mycol. Soc.* **9**: 210.

BREITSPRECHER, E. 1928. Beiträge zur Kenntnis der Anobiidensymbiose. *Z. Morph. Okol. Tiere* **11.**

BUCHNER, P. 1921. Studien an intracellularen Symbionten. III. Die Symbiose der Anobiinen mit Hefepilzen. *Arch. Protistenk.* **42.**

CAMPBELL, W. G. 1929. The Chemical Aspect of the Destruction of Oakwood by Powder-post and Death-Watch Beetles, *Lyctus* spp. and *Xestobium* sp. *Biochem. J.* **23**: 1290–3.

CAMPBELL, W. G. 1930. The Chemistry of the White Rots of Wood. I. The effect on wood substance of *Polystictus versicolor* (Linn.) Fr. *Biochem. J.* **24**: 1235.

CAMPBELL, W. G. 1932. The Chemistry of the White Rots of Wood. III. The Effect on Wood Substance of *Ganoderma applanatum, Fomes fomentarius, Polyporus adustus, Pleurotus ostreatus, Armillaria mellea, Trametes pini,* and *Polystictus abietinus*. *Biochem. J.* **26**: 1829.

CAMPBELL, W. G. and BRYANT, S. A. 1940. A Chemical Study of the Bearing of Decay by *Phellinus cryptarum* Karst. and other Fungi on the Destruction of Wood by the Death-Watch Beetle, *Xestobium rufovillosum* de Geer. *Biochem. J.* **34**: 1404.

CARTWRIGHT, K. St.G. 1929. Notes on Fungus associated with *Sirex cyaneus*. *Ann. appl. Biol.* **16**: 182.

CARTWRIGHT, K. ST.G., CAMPBELL, W. G. and ARMSTRONG, F. H. 1936. The Effect of Progressive Decay by *Polyporus hispidus* Fr. on the Strength of English Ash (*Fraxinus excelsior* L.). *Proc. Roy. Soc.* (*B*) **120**: 76.

CARTWRIGHT, K. ST.G. and FINDLAY, W. P. K. 1936. *The Principal Rots of English Oak*. London, H.M.S.O.

DERHAM, Wm. 1701. *Philos. Trans.* **271**: 832.

ESCHERICH, K. 1900. Über das regelmässige Vorkommen von Sprosspilzen in dem Darmepithel eines Käfers. *Biol. Zbl.* **20.**

FISHER, R. C. 1937. Recent Work on Wood-destroying Insects. *J.R. Soc. Arts.* **85**: 407–25.

FISHER, R. C. 1937. Studies of the Biology of the Death-Watch Beetle *Xestobium rufovillosum*. I. A Summary of Past Work and a Brief Account of the Development Stages. *Ann. appl. Biol.* **24**: 600–13.

FISHER, R. C. 1938. Studies of the Biology of the Death-Watch Beetle, *Xestobium rufovillosum* de Geer. II. The Habits of the Adult with Special References to the Factors affecting Oviposition. *Ann. appl. Biol.* **25**: 155–80.

FISHER, R. C. 1940. Studies of the Biology of the Death-Watch Beetle *Xestobium rufovillosum* de Geer. III. Fungal Decay in Timber in Relation to the Occurrence and Rate of Development of the Insect. *Ann. appl. Biol.* **27**: 545.

FISHER, R. C. 1941. Studies of the Biology of the Death-Watch Beetle *Xestobium rufovillosum* de Geer. IV. The Effect of Type and Extent of Fungal Decay on the Rate of Development of the Insect. *Ann. appl. Biol.* **28**: 244–260.

FISHER, R. C. 1941. Building Reconstruction and the Death-Watch Beetle. *J. chart. Surv. Instn.* **20** (10): 645–50.

FISHER, R. C. and PARKIN, E. A. 1930. Presence of a Yeast in the Death-Watch Beetle, *Xestobium rufovillosum*. *Nature* **125**: 892.

FOWLER, W. W. 1890. *Coleoptera of the British Isles*. **4**: 191.

GAHAN, C. J. 1918. The Death-Watch: Notes and Observations. *Entom.* **51,** 661: 121–5.

GAHAN, C. J. 1925. Furniture beetles. Brit. Museum Pamph. Econ. Ser. **2.** 3rd ed. (1932), by F. Laing.

GRAHAM, S. A. 1925. The Felled Tree Trunk as an Ecological Unit. *Ecology* **6**: 397.

HARRIS, E. C. 1959. The Secondary Sex Characters of Death-Watch Beetle. *Ent. mon.. Mag.* **95**: 208–9.

HEITZ, E. 1927. Über intrazellulare Symbiose bei holzfressenden Käferlarven. *Z. Morph. Ökol. Tiere* **7**: 279–305.

HOPKINS, A. D. 1893. Catalogue of West Virginia Forest and Shade Tree Insects *Bull. W. Va. agric. Exp. Sta.* **32**: 189.

JOY, N. H. 1932. *A Practical Handbook of British Beetles*, **1**: 456.

JUDEICH, F. and NITSCHE, H. 1895. *Lehrbuch der Mitteleuropäischen Forstinsektenkunde* **1**: 341–7.

KARAWIEW, W. 1899. Über Anatomie und Metamorphose des Darmkanals der Larve von *Anobium paniceum*. *Biol. Zbl.* **19.**

KEMNER, N. A. 1915. De ekonomiskt viktiga vedgnagande Anobierna. *Medd. Cent. Anst. Försöksv. Jordbr., Stockh.*, No. 108.

KIMMINS, D. E. 1933–34. Notes on the Life History of the Death-Watch Beetle. *Proc. S. Lond. ent. nat. Hist. Soc.* 133–7.

KOFOID, CHARLES A., *et al.* 1934. *Termites and Termite Control*. 2nd ed. 105–16. University of California.

LEFROY, H. M. 1924. The Treatment of the Death-Watch Beetle in Timber Roofs. *J.R. Soc. Arts* **52**: 260–6.

LINSCOTT, D. 1962. *Entomologist* **95**: 58–9.

MANSOUR, K. and MANSOUR-BEK, J. J. 1934. On the digestion of wood by insects. *J. exp. Biol.* **11**: 243–56.

MANSOUR, K. and MANSOUR-BEK, J. J. 1934. The digestion of wood by insects and the supposed role of micro-organisms. *Biol. Rev.* **11**: 363–82.

MANGIN, L. and PATOUILLARD, N. 1922. Sur la destruction des charpentes au chateau de Versailles par le *Phellinus cryptarum* Karst. *C.R. Acad. Sci. Paris* **175**: 389.

MAXWELL-LEFROY, H. 1924. The Treatment of the Death-Watch Beetle in Timber Roofs. *J. of Royal Soc. Arts.* **72**: 260–70.

MCKENNY HUGHES, A. W. 1954. Insect Infestation of Churches. *J. Royal Inst. Br. Arch*: 2–4.

METCALFE, M. E. 1932. The Structure and Development of the Reproductive system in the COLEOPTERA, with notes on its homologies. *Quart. J. micr. Sci.* **75**: 49–129.

MILLER, D. 1925. Forest and Timber Insects in New Zealand. *Bull. N.Z. For. Serv.* **2**.

MULLER, W. 1934. Untersuchungen über die Symbiose von Tieren mit Pilzen und Bakterien. III. Über die Pilzsymbiose holzfressender Insektenlarven. *Arch. Mikrobiol.* **5**: 84–147.

MUNRO, J. W. 1928. Beetles injurious to timber. *For. Comm. Bull.*, No. 9. H.M.S.O.

NORMAN, A. G. 1936. The Destruction of Oak by the Death-Watch Beetle. *Biochem. J.* **30**: 1135–7.

PARKIN, E. A. 1933. The Larvae of some Wood-boring ANOBIIDAE. *Bull. ent. Res.* **24**: 31–68.

PEARSE, A. S., PATTERSON, N. T., *et al.* 1936. The ecology of *Passalus cornutus* F., a beetle which lives in rotting logs. *Ecol. Monogr.* **6**: 455.

PIC, M. 1912. In Junk and Schenkling's *Coleopterorum Catalogus*, Pt. XLI.

RAMSBOTTOM, J. 1937. Dry Rot in Ships. *Essex Nat.* **25**: 231–67.

RATZEBURG, J. T. 1837. *Die Forstinsekten* **1**: 46.

RIPPER, W. 1930. Zur Frage des Zelluloseabbaus bei der Holzverdauung xylophager Insektenlarven. *Z. vergl. Physiol.* **8**: 314–33.

SAVELY, H. E. 1939. Ecology of Certain Animals in Dead Pine and Oak Logs. *Ecol. Monogr.* **9**: 323.

SCHEFFER, T. C. 1936. Progressive effects of *Polyporus versicolor* on the physical and chemical properties of red gum sapwood. *Tech. Bull. U.S. Dept. Agric.* **527**.

SHARP, D. 1922. *Insects*. Pt. II: 247. Camb. Nat. Hist.

SHARP, D. and MUIR, F. 1912. The Comparative Anatomy of the Male Genital Tube in Coleoptera. *Trans. ent. Soc. Lond.* **3**: 477–642.

STANILAND, L. N. 1930. Presence of yeast in the Death-Watch Beetle. *Nature, Lond.* **125**: 635.

STEINHAUS, E. A. 1940. The Microbiology of Insects with Special Reference to the Biological Relationships between Bacteria and Insects. *Bact. Rev.* **4**: 17.

SWAMMERDAM, J. 1672–80. *History of Insects*. Translated by Thoms. Flloyd. rev. by J. Hill (1758), **1**: 125.

TASCHENBERG, E. L. 1879. *Praktische Insektenkunde*, **2**: 77.

TILLYARD, R. J. 1926. *The Insects of Australia and New Zealand.*

UVAROV, B. P. 1929. Insect Nutrition and Metabolism; a Summary of the Literature. *Trans. ent. Soc. Lond.* **76**: 255–343.

WESTWOOD, J. O. 1839–40. *Introduction to the Modern Classification of Insects.* **1**: 268.

WILLIAMSON, H. S. 1923. The origin of 'golden oak'. *Ann. Bot., Lond.* **37**: 433.

WILSON, R. E. 1921. Humidity Control by means of Sulphuric Acid Solutions with critical Compilation of Vapour Pressure Data. *Industr. Engng. Chem.* **13**: 326–31.

CHAPTER FIVE

Powder Post Beetles

The families LYCTIDAE and BOSTRYCHIDAE

The LYCTIDAE, together with the BOSTRYCHIDAE, are known as powder post beetles on account of the larval habit of burrowing into the sapwood of susceptible hardwoods until nothing is left but a fine powder. This may take several generations but a thin veneer of the wood on the outside is left unsupported from the inside. This uneaten veneer may break and become detached. The outer veneer does, of course, show the flight holes made by the mature beetles.

LYCTIDAE

World distribution

The family LYCTIDAE consists of twelve genera and altogether sixty-three species have so far been described. Each of the world faunal regions has its own indigenous lyctid genera with which is found a lyctid fauna originally derived from other regions, but now introduced and established through commerce in timber.

The British species

We are in an immediate difficulty in determining which species of LYCTIDAE can be considered British. Fowler, in 1891, stated that *Lyctus brunneus* Stephens was very rare in Britain and that *Lyctus linearis* (Goeze) was common, but in 1923 Altson considered that the reverse was then the case and that only these two species had been recorded in the British Isles.

A study of the N. H. Joy collection of British Coleoptera has been made by the present author; this revealed only these two species. Munro, in his 1915 paper, in which he refers to the LYCTIDAE as Furniture Beetles, includes *Lyctus brunneus* and *L. canaliculatus* Fabricius (this is *linearis* Goeze). Since the time of Altson, however, a number of Lyctids have been recorded as being found in the British Isles. Kloet and Hincks list four of these as 'established introductions'. It now seems virtually certain that the only truly indigenous British species is *L. linearis*; *L. brunneus* being an introduction from North America sometime around the beginning of the nineteenth century.

The established introductions of Kloet and Hincks are as follows: *Lyctus sinensis* Lesne 1911; *L. planicollis* Leconte 1854; *L. cavicollis*

Leconte 1866 and *Trogoxylon parallelopipedum* (Melsheimer 1846). Kloet and Hincks include the last named species in the genus *Lyctus*.

The synonomy of the Linnaean species *Silpha fusca* 1767 may never be resolved. Some workers have considered this to be the species *linearis* (Goeze 1777) and it was accordingly given the name of *Lyctus fuscus* (Seidl 1875). This is the name given by Kloet and Hincks, but by Joy (in his collection) specimens are named *linearis* and this appears to be the current American practice—see Gerberg. The LYCTIDAE are, however, placed by Joy in the CLAVICORNIA immediately preceding the LATHRIDIIDAE. Joy states that in addition to *Lyctus brunneus* Stephens, and *L. linearis* (Goeze) two additional species have sometimes been found in imported timber. He does not name them but says that the elytra are without striae. It is not proposed to give here any further synonomy—it has been given adequate treatment elsewhere—except to mention the name *canaliculatus* Fabricius 1792 as further synonomy for *linearis*. The name will be found in some collections.

Biology

Most work on the biology of the British lyctids has been carried out on *Lyctus brunneus*, the other species having had a smaller share of attention from research workers. The account which follows refers to *L. brunneus* unless otherwise stated.

When the adult beetles emerge from their flight holes, which they make by biting through the outer skin of the wood, they are at once active and sexually mature. Copulation usually takes place at dusk, or night, immediately following the emergence, although on occasion it takes place during the day. Females are much more common than males and the latter each fertilize several females. The females live about six weeks but the males only for about two or three weeks, although Parkin thought there was little difference between the sexes in this respect.

Lyctus brunneus flies well and is attracted to light. When an infestation occurs indoors the adult beetles will often be seen crawling around windows.

Oviposition commences within two or three days of copulation and normally takes place at night. The suitability of the wood for the nutritional requirements of the subsequent larvae is somehow perceived by the beetle. Development does not take place—indeed eggs are not laid—in sapwood containing less than about 3 per cent of starch in the xylem parenchyma.

The female beetles make a close examination of the surface, and appear to 'bite' the wood transversely. This action gives rise to very small grooves which have been called 'feeding marks' by Fisher (1928), 'tasting marks' by Parkin (1936) and 'mandibular grooves' by Rosel (1969). Neither of these terms is satisfactory, however, for it has been shown that no material is ingested by the beetle; there is no evidence to support the view that the beetle uses this method to perceive the starch content of the wood (grooves are not made on starch-free timbers), and the grooves are in the wood, not the mandibles as Rosel's term implies.

Altson (1923a), Gay, Bletchly and most recently Rosel all believe that the grooves are made to break the surface of the wood to facilitate oviposition. The actual oviposition site is invariably the lumen of a xylem vessel, so that the making of grooves to open-up pores is more important when end grain sites are unavailable or are minimal.

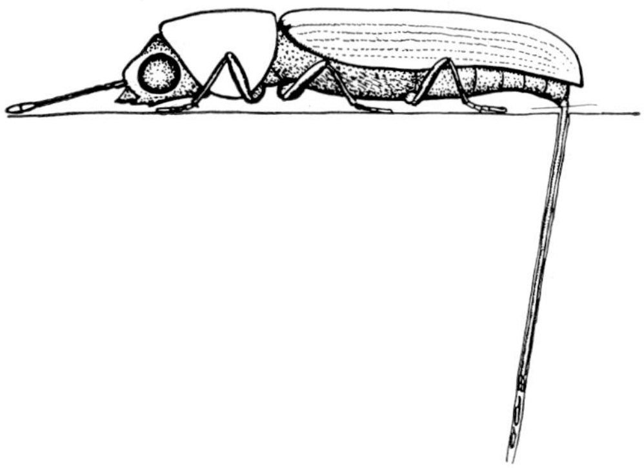

Fig. 61. Female *Lyctus* laying eggs, diagrammatic.

The ovipositor is of great length, being almost as long as the length of the body and, indeed, on occason it may be stretched even longer. It is very flexible and at the tip are two-segmented pygidial palps. Before the ovipositor is inserted into the lumen the actual position for insertion is explored by the palps. The female can rest in almost any convenient position when laying as the ovipositor can be held directly backwards in the same line as the abdomen or held at right angles to the abdomen underneath the body (Fig. 61). A depth of 7·75 mm into a vessel has been reached and instances are known of as many as eight eggs in a single vessel. More usually, however, one, two or three eggs are laid per vessel. Altson records that vessels exposed by the formation of the gallery leading from the pupal chamber to the flight-holes are sometimes utilized for egg laying.

Obviously in such a method of egg-laying the diameter of the lumen of the vessel in the timber is critical, as the ovipositor must be able to penetrate it. The following measurements have been recorded.

	Average diameter of ovipositor	Minimum diameter of ovipositor
brunneus	0·078 mm	0·056 mm
linearis	0·083 mm	—
planicollis	0·076 mm	—

The eggs of *brunneus* are long and cylindrical and slightly tapering to the posterior end. In colour they are whitish and translucent and the ends are hemispherical. There are some longitudinal creases and a long thread is borne by the anterior end.

Parkin considered that the average length was 1 mm and Altson gave measurements of 0·8 mm to 1·25 mm in length and 0·15 mm to 0·175 mm in width. Parkin, who worked with *linearis*, *planicollis* and *sinensis* in addition to *brunneus*, stated that the eggs of these species were similar to *brunneus*. This has been confirmed by Rosel who also studied several species.

Incubation periods vary with temperature from 19–20 days at 15° C to 14–15 days at 20° C and 7–8 days at 26° C. There is little difference between the species.

The development of the embryo within the egg has been described by Altson (1923a) (Fig. 62) and Rosel (1969). After a few days the posterior half of the egg becomes a little more transparent than the anterior half. Later the larva becomes faintly visible at the posterior end, whilst a quantity of yolk granules and fat body, termed the 'residual yolk mass', lies in the anterior half. The full outline of the embryo only becomes apparent on the last day of development. The larva then feeds on the residual yolk mass for three or four days.

Development to this point is similar in all the species studied by Rosel. Eclosion from the egg, however, varies with the species. The egg of

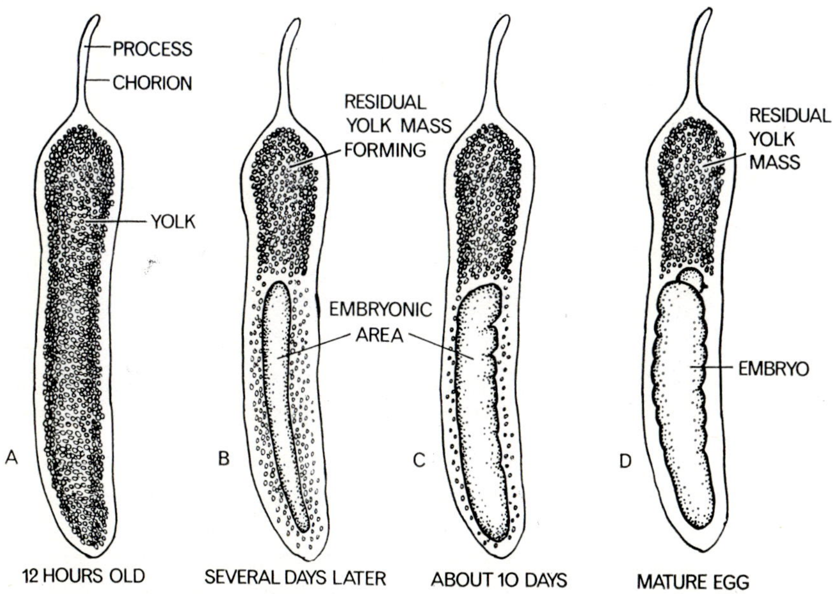

Fig. 62. Development of *Lyctus* egg, diagrammatic (after Altson).

L. brunneus is first broken at a point about one-third of the egg-length from the anterior end and in close proximity to the position of the head of the embryo. The posterior end may be broken at a later stage, but the larva always leaves by the anterior end. The larvae of *L. planicollis*, *L. africanus* Lesne and *Minthea rugicollis* (Walker) steadily move forwards eating the residual yolk mass. They fill the posterior end of the egg shell with excreta and emerge through the anterior end. On the other hand, larvae of *L. discedens* Blackburn consume the residual yolk mass without fracturing the anterior end of the shell. When fully fed they emerge backwards from the posterior end of the egg.

These findings are contrary to descriptions given by Altson and others, and are explained by Rosel as due to the method of study. Previous workers had opened a vessel and observed the egg without a support on the open side. Rosel corrected this by covering the exposed egg with a glass cover slip.

The young larvae, when they hatch, are about 0·65 mm long and 0·23 mm wide (at the thorax, the widest part) and they are creamy-white in colour. The body is not curved in this instar but the size of the spiracle on the 8th segment is about six times the size of those on segments 1 to 7 —a feature which is characteristic of all larvae of LYCTIDAE. At the second instar the body of the larva is curved (Fig. 63) as in the more mature larvae of other lyctids, and also the larvae of related families such as ANOBIIDAE, BOSTRYCHIDAE, and PTINIDAE.

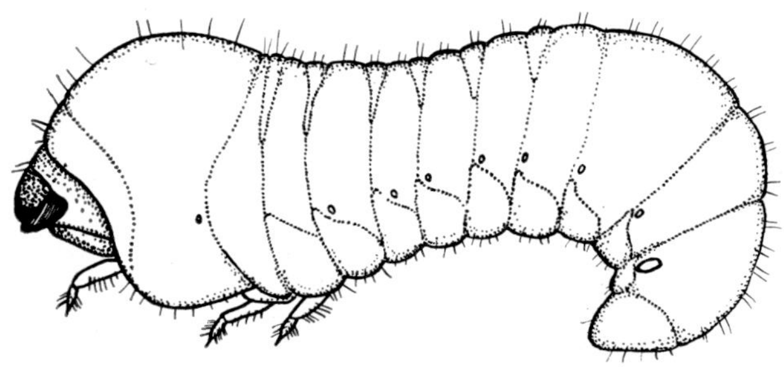

Fig. 63. Larva of *Lyctus*, lateral view.

At the time of 'hatching', or rupture of the chorion, the larva is too small to proceed along the vessel as it is not able to effect the necessary 'purchase', but the consumption of the remainder of the residual yolk mass increases its girth until it is able to do this. When this point has been reached ecdysis takes place and the body becomes arched which enables the larvae to crawl along the vessel before it pierces the wall and tunnels into the tissue.

A point of interest concerning the young larva is that the compound eyes, although present in all instars, are more pronounced in the first and second instars. The suggestion has been put forward that eyes are only necessary at this time. In the first instar, when backing out of the ruptured chorion the larva must travel towards the residual yolk mass in order to consume it before being able to traverse the vessel in which it finds itself. Within a few millimetres of the cut face of the vessel—at the point of entrance of the mother's ovipositor—light would be discernible so that in this instar the young larva is positively phototropic. At the second instar, however, when the larva must cut a way out of the vessel away from the light, the insect is obviously negatively phototropic.

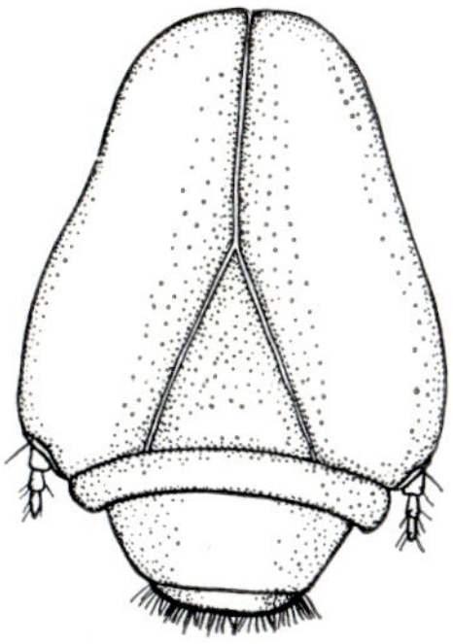

Fig. 64. Head of larva of *Lyctus* from the front.

In the early stages the larva excavates its tunnel with the grain but later the direction appears to be quite haphazard and so the grub will cut through the tunnels of other larvae or its own earlier tunnel. In badly infested wood the larvae tunnel through mostly faecal matter in order to find wood tissue which has not been eaten.

Lyctus larvae are unable to digest cellulose or hemicellulose but subsist on cell contents. Work carried out by Wilson, Tooke, Parkin, and others has shown that starch is the main foodstuff utilized by the larva, together with certain sugars, disaccharides, a polysaccharide, and some protein. Cellulose pentosans and lignin appear to remain unchanged after passing through the larval gut. In addition to starch and protein Parkin concluded that an unidentifiable water soluble substance is also a nutritional requirement. He detected in the larval gut the ferments amylase, invertase, maltase, lactase and proteinase, but in starch-rich wood the *Lyctus* bore-dust often contains much undigested starch.

A moisture content of the wood between 8 and 30 per cent is necessary for attack by *brunneus* but for *planicollis* the range is 6 to 32 per cent. In general the higher moisture content favours larval development.

When fully grown the larva may vary in size within wide limits, but 5 mm appears to be about the maximum length. It is strongly arched and the thorax is enlarged. The antennae are three-segmented and the labial palpi consist of one segment. The legs consist of three segments, the distal segments being paddle-shaped. When the larva has finished feeding it bores towards the outer surface and just underneath constructs a pupal chamber.

The pupae of lyctids are exarate—the adult appendages being free but covered by the thin pupal cuticle (Fig. 65). The duration of the pupal stage is recorded as being between 12 and 30 days but it seems reasonable to assume that, in the shorter life cycles that have been reported, the pupal stage lasts for much less than this.

The pupa is at first white then creamish and a few days before metamorphosis it darkens considerably in colour. After metamorphosis and presumably after the lapse of a few days for hardening of the adult parts, the insect bites a tunnel to the ouside surface of the wood. When a small

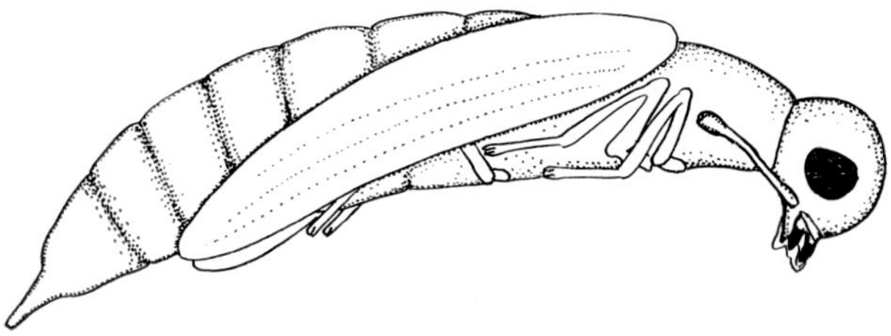

Fig. 65. Pupa of *Lyctus brunneus* from the side.

connection has been made to the outside the bore-dust is pushed through the aperture and on top of horizontal surfaces the small piles of usually cream-coloured powder are characteristic. When the flight hole is just large enough the adult struggles through. The flight hole of *Lyctus brunneus* is round with a clean-cut edge and 1·0–2·0 mm (average, 1·4 mm) in diameter. It is not unusual for materials which overlie the wood surface to be penetrated by the emerging adult. Such materials as paper, leather, softwood, heartwood and asbestos, lead, silver and plaster have been reported as being damaged in this way.

The length of the complete life cycle varies widely according to temperature, moisture content of timber, etc. *Trogoxylon parallelopipedum* has been recorded as completing its life cycle in 76 days. In Britain, in the case of *Lyctus brunneus* the probable usual length of life cycle is one year, both out of doors and in most situations indoors, but sometimes as in exceptionally warm conditions or in centrally heated premises there are two life cycles annually. In laboratory conditions of 25° C and 75 per cent relative humidity a complete life cycle is undergone in from 10 to 12 weeks.

The *maximum* length of life cycle is often of importance in disputes concerning *Lyctus* infestations. Gerberg states that under adverse conditions the life cycle may take from two and a half to four years or longer.

Host timbers

Lyctus beetles only infest hardwood timbers, and are limited to those with pores large enough to take the insect's ovipositor and those with a high enough starch content. The following timbers are known to be susceptible to attack by various species of *Lyctus*:

Acacia melanoxylon	Australian blackwood
Acer pseudoplatanus	Sycamore
Albizia lebbek	Kokko
Aucoumea klaineana	Gaboon
Beilschmiedia bancroftii	Yellow walnut
Canarium euphyllum	Dhup
Carapa guianensis	Crabwood
Carpinus betulus	Hornbeam
Castanea sativa	Sweet chestnut
Castanospermum australe	Black bean
Celtis soyauxii	African celtis (Itako)
Chlorophora excelsa	Iroko
Dalbergia latifolia	Indian redwood
Dipterocarpus spp.	Keruing, Gurjun, Eng
Dryobalanops aromatica	Borneo camphorwood (Kapur)
Dyera costulata	Jelutong
Entandrophragma cylindricum	Sapele
Eucalyptus gigantea	Tasmanian oak
obliqua	Tasmanian oak
regnans	Tasmanian oak
maculata	Spotted gum
other spp.	Ironbark
Fraxinus excelsior	Ash
Gossweilerodendron balsamiferum	Agba
Guarea thompsonii (probably susceptible)	Black guarea
Juglans regia	Walnut
Khaya spp.	African mahogany
Lovoa klaineana	African walnut
Mitragyna ciliata	Abura
Pterocarpus angolensis	Muninga
indicus	Amboyna
Quercus cerris	Turkey oak
ilex	Evergreen oak
petraea	Sessile oak
robur	Pedunculate oak
Robinia pseudoacacia	Robinia
Salix alba var. *caerulea*	Cricket bat willow
fragilis	Crack willow
other spp.	Willows
Schizomeria ovata	White birch
Shorea spp. and *Parashorea* spp.	Seraya, Meranti
Sloanea woollsii	Yellow carabeen

Tectona grandis	Teak
Terminalia bialata	White chuglam
ivorensis	Idigbo
procera	White bombway
superba	Afara
tomentosa	Indian laurel
Triplochiton scleroxylon	Obeche
Ulmus glabra	Wych elm
hollandica	Dutch elm
procera	Common elm

Plywood

Hardwood plywood is now frequently imported into the U.K., and most is susceptible to attack by *Lyctus* beetles. The most common timber is *Triplochiton*, used on its own or in conjunction with *Aucoumea*, *Juglans* or *Khaya*. Glue is usually urea-formaldehyde or phenol-formaldehyde, which larvae rarely penetrate. Thus the larvae normally remain in the veneer in which the eggs were laid.

Some protection is afforded by the incorporation of lindane or dieldrin in the glue during manufacture. The effect of lindane appears to be limited to just under four years, whilst dieldrin will give protection for at least four years when the plywood is stored at 25°C and 75 per cent relative humidity (Taylor).

Parasites and Predators

A summary of the recorded insects predacious or parasitic on lyctid beetles has been given by Gerberg. Amongst the HYMENOPTERA, species of BRACONIDAE, PTEROMALIDAE, BETHYLIDAE and CHALCIDIDAE have been recorded as parasitizing species of LYCTIDAE. In the COLEOPTERA several species of CLERIDAE are known to be predacious on LYCTIDAE, amongst which *Tarsostenus univittatus* (Rossi) (Fig. 66) appears to be commonest. Gerberg does not consider that parasites and predators are a factor in the artificial control of Lyctid beetles.

External morphology

Lyctid beetles are small, none of the British species exceeding 7 mm in length, and the colour range passes from reddish-chestnut through brown to black. They are devoid of marks. In shape they are elongated and flattened dorso-ventrally. The head is distinctly visible from above (in ANOBIIDAE the head is not visible from above) and slopes downwards at an angle. The vertex is rugose or punctate. The eyes are prominent, hemispherical but with the posterior margin flattened. The antennae are claviform, of eleven segments and inserted just anterior to the eyes; the club consists of two segments. The bidentate mandibles are heavily sclerotized, broad and robust at the base and there is a brush of setae on the outer face.

Fig. 66. *Tarsostenus univittatus* (Rossi). Length: 3·8 mm.

The labial palpi are three-segmented, the maxillary palpi four-segmented, the lacinia fringed and the galea is two-segmented.

The pronotum is roughly quadrangular, convex but with a median depression (median fovea). The curvature of the anterior margin varies in the different species as well as the shape of the anterior angles. The posterior angles are usually toothed. Light-coloured pubescence is present which may vary in texture.

The elytra are long and slender, and in the genus *Lyctus* there is a series of longitudinal ridges bearing fine hairs, and between the ridges rows of more or less regularly disposed punctations. The wings conform to the general pattern displayed by the BOSTRYCHOIDEA but an anal lobe is present which is characteristic of the LYCTIDAE within this super-family. The legs are slender and in the prothoracic legs the tibiae bear one or two short stout spines distally. There are five tarsal segments but the first in each case is very small.

Neither the male nor the female genitalia are sufficiently differentiated to serve as means of identification, as occurs in many other groups of insects.

Key for the identification of the British species of Lyctidae

1.	Pubescence on elytra confused. Femur of metathoracic legs enlarged. Anterior angles of pronotum prominent and acute . .	*Trogoxylon parallelopipedum*
–	Pubescence on elytra confined to longitudinal ridges. Femur of metathoracic legs not enlarged. Anterior angles of pronotum not prominent and not acute	2
2.	Y-shaped depression on pronotal disc. 5 rows of punctuations between elytral suture and first ridge	*Lyctus brunneus*
–	Depression on pronotal disc not Y-shaped. Fewer than five rows of punctations between elytral suture and first ridge	3
3.	Punctations on median third of elytra not in distinct rows. Reticulate punctation on pronotum	*Lyctus sinensis*
–	Punctations on median third of elytra in fairly distinct rows	4
4.	Punctations on elytra in single rows between ridges except between elytral suture and first ridge	*Lyctus linearis*
–	Punctations on elytra in double rows between ridges	5
5.	10th antennal segment wider than long . .	*Lyctus planicollis*
–	10th antennal segment not wider than long . .	*Lyctus cavicollis*

Notes on the species

It is not proposed to give a complete description of each of the species here. Such descriptions have already been adequately treated notably by Gerberg who has dealt in great detail. Characters of value, however, in identifying the species listed above are given.

Lyctus brunneus Stephens, 1830 (Fig. 67)

This is by far the commonest lyctid in Britain and can now be considered cosmopolitan, being found in all faunal regions of the world. It is thought to have been indigenous to North America. It is extremely variable in size.

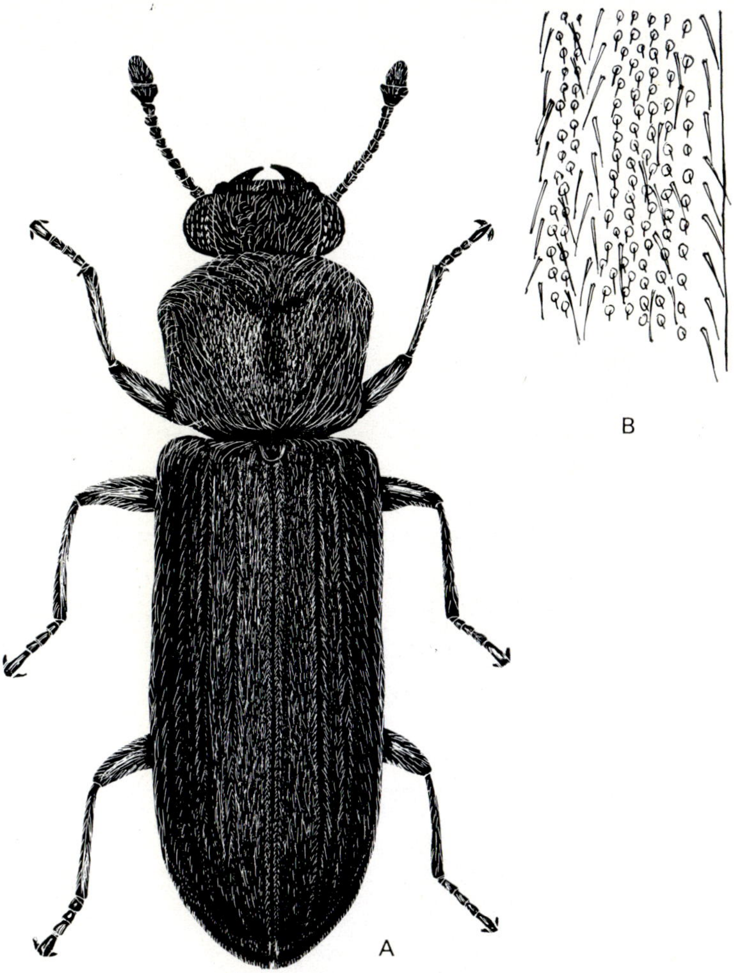

Fig. 67. *Lyctus brunneus* Steph. A. Adult; length, 5·5 mm. **B.** Part of left elytron from elytral suture.

A number bred by the writer from a wooden plate set with mother of pearl (Arabian in origin) were only 2 mm in length but a length of 7 mm has been recorded. There is also great variation in colour from very dark brown to reddish, the pronotum usually appearing darker than the elytra.

The antennae are about the same length as the pronotum with the basal two segments enlarged. The last but one segment of each antenna is shorter but wider than the last segment. The pronotum is wider at the front than behind, the anterior angles being rounded but distinct. The front of the pronotum is as wide as the abdomen and, except for a depression in the disc (median fovea), covered with fine hair. The fovea possesses two lateral arms which diverge towards the front. The whole of the pronotum is heavily punctated, giving the laterial margins the appearance of fine denticulation. There is a pronounced tibial spur.

The elytra have a series of longitudinal parallel ridges, being divided by rows of punctations. Between the elytral suture and the first ridge are five rows of punctations, between the next ridges are two rows, but thereafter only single rows of punctations separate the ridges. On the ridges and between the rows of punctations are longitudinal series of fine light-coloured hairs.

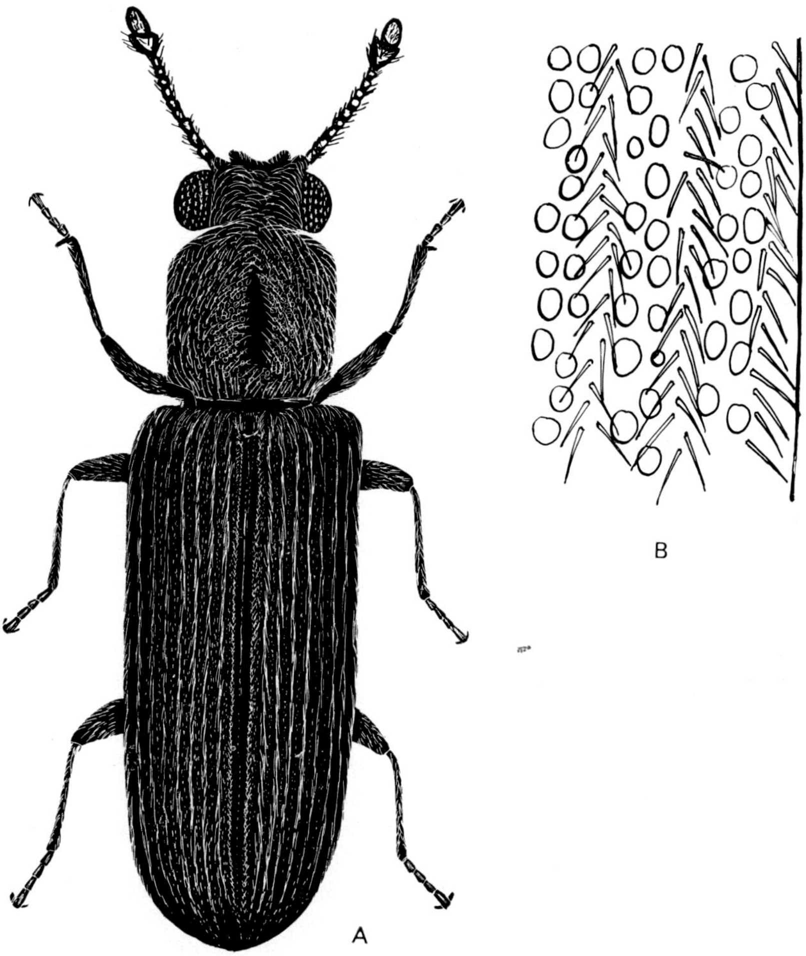

Fig. 68. *Lyctus cavicollis* Leconte. A. Adult; length, 5·0 mm. B. Part of left elytron from elytral suture.

The female can be distinguished from the male by possessing a pencil of hairs on the fifth sternite, which is absent in the male.

Lyctus cavicollis Leconte 1866 (Fig. 68)

This is a North American species which can be easily overlooked by reason of its resemblance to *linearis* (Goeze) and to *planicollis* Leconte. The characters by which it can be differentiated from these species are as follows:

The pronotum is rectangular and has the lateral margins almost parallel (converging slightly posteriorly) and sharply angled posteriorly. The anterior angles, although rounded, are distinct.

The punctations of the pronotum are separate and the whole of the pronotum, except the median fovea, is covered by fine long hairs.

The elytra possesses a series of longitudinal ridges on each of which a single series of fine hairs arises. Between ridges there is a double row of punctations.

It is possible to separate the sexes by examination of the fifth abdominal sternite. In the female it is triangular and on each side of the median line there is a triangular patch of hairs. In the male, the fifth sternite is rounded and hairs of the apical fringe are not arranged in triangular patches. The size varies from 2·5 to 5 mm.

Lyctus linearis (Goeze 1777) (Fig. 69)

Reference has already been made to the possibility that this is the *Silpha fusca* of Linnaeus 1767. Since 1777, when Goeze described this insect, it has received many synonyms which are listed by Gerberg. This native species was at one time the commonest of our lyctids, but for some years *Lyctus brunneus* has preponderated.

The frontal lobes are broad and prominently elevated which is an important character in differentiating this species from *Lyctus cavicollis*. The antennae are longer than the pronotum and the basal two segments are enlarged.

The roughly quadrangular pronotum has rounded anterior angles and the posterior angles are toothed. The lateral margins converge posteriorly. The disc is rugose with a deep median fovea which is circular and edged with long coarse hairs.

Elytra are about three times as long as wide and the longitudinal ridges, with a single row of fine hairs, are separated by a single row of large ovoid punctations except between the first pair of ridges (counting from the median line) where the punctations vary between one and two series. The elytra are rust-red, but the punctations are black and the legs light brown.

The sexes can be differentiated by the female having the fifth sternite more slender and tapering whilst in the male it is broadly and flatly rounded.

The length varies between 2·5 and 5 mm.

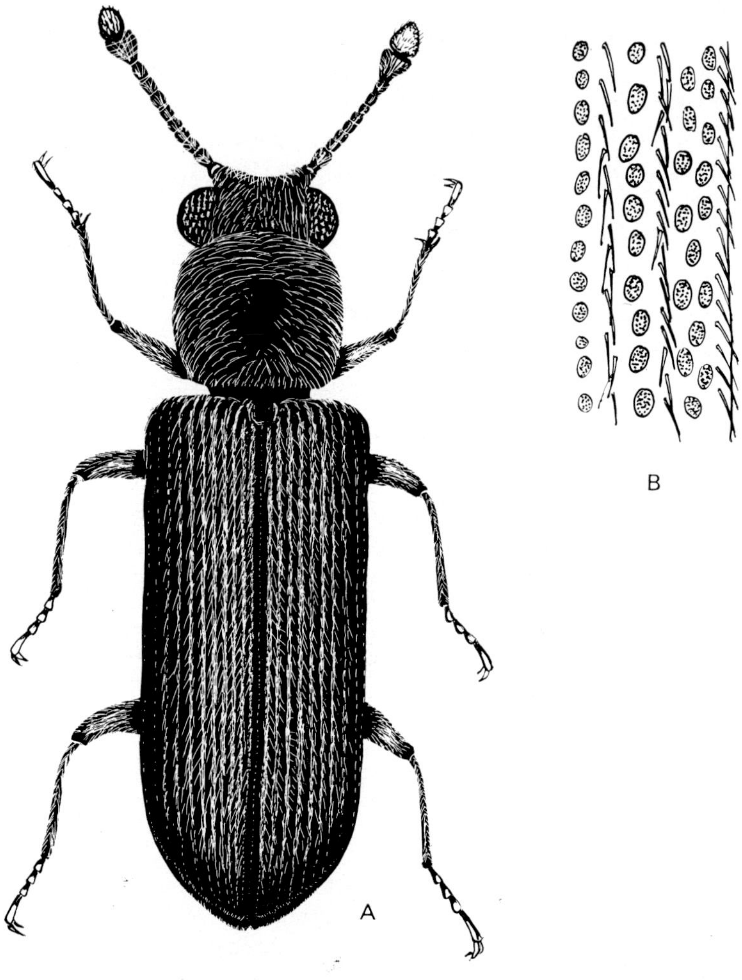

Fig. 69. *Lyctus linearis* (Goeze). A. Adult; length, 3·7 mm. B. Part of left elytron from elytral suture.

Lyctus planicollis Leconte 1854 (Fig. 70)

There is some doubt as to the correct name of this insect. It would appear that *carbonarius* Waltl 1832, named from Mexican material, should have priority but Gerberg retains *panicollis*.

The species is North American in origin where it is widespread, and is often reported from oak, hickory, and ash.

Characters which distinguish this species are: pronotum wider than long and the punctations distinctly separated. It can only with difficulty be differentiated from *cavicollis* and when series are bred the characters

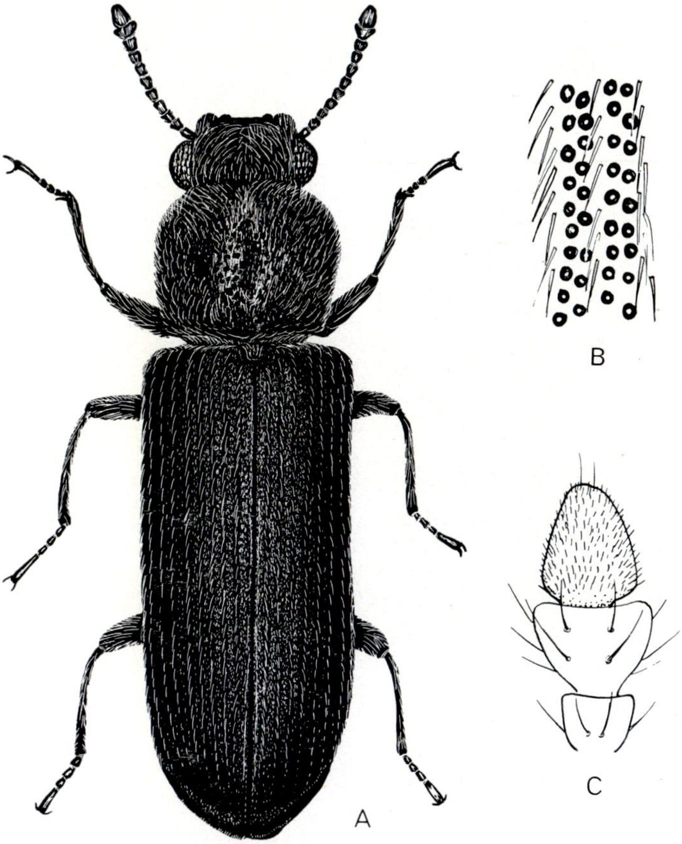

Fig. 70. *Lyctus planicollis* Leconte. A. Adult; length, 5·2 mm. B. Part of left elytron from elytral suture. C. Last three antennal segments.

often overlap. The tenth antennal segment is wider than it is long, whereas in *Lyctus cavicollis* this is not so. The length is not so variable as in some species, being 4 to 6 mm. The colour is black—shining where the covering of fine light-coloured hairs is rubbed. Antennae are as long as the pronotum and the legs are reddish.

The females possess a fringe of long setae on either side of the median line of the fifth sternite. The males have a rounded fifth sternite, the apical margin being fringed with setae.

Lyctus sinensis Lesne 1911 (Fig. 71)

This is an Asiatic species, Japan being a usual source of infestation. The pronotum is less in width than the elytra and the lateral margins converge posteriorly. It has reticulate punctations and the fine light-coloured hairs are thicker and longer on the pronotum and the head than on the elytra.

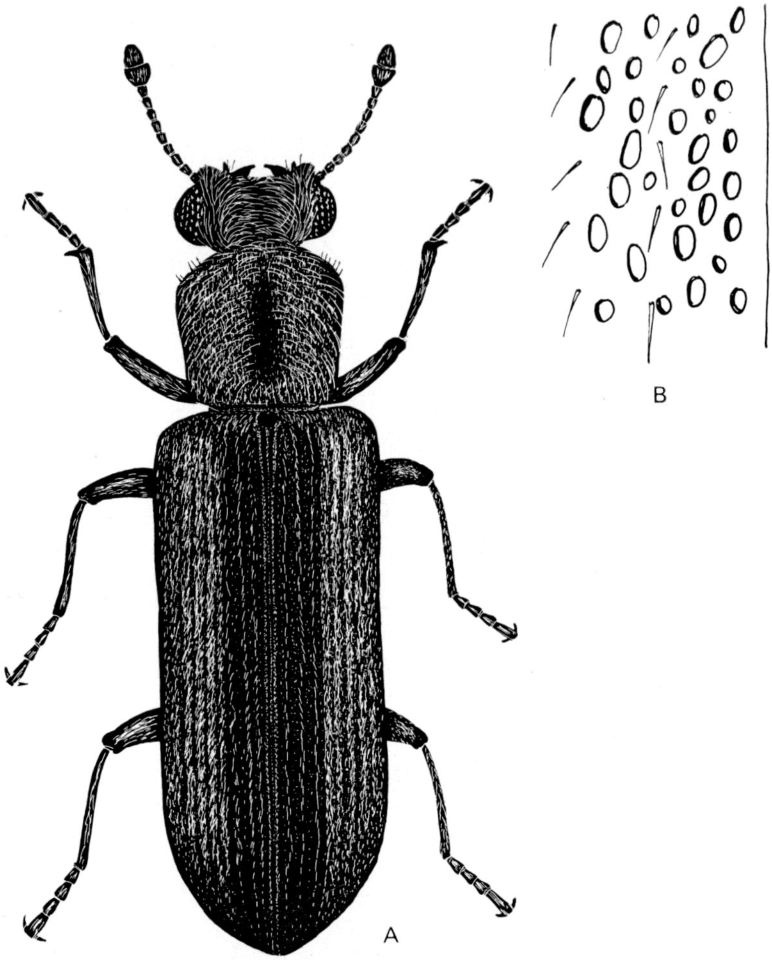

Fig. 71. *Lyctus sinensis* Lesne. A. Adult; length, 5·3 mm. B. Part of left elytron from elytral suture.

The lateral two-thirds of each elytron is lighter in colour than the median third. In the latter area, longitudinal ridges are absent, hairs are sparse and punctations are irregular.

Trogoxylon parallelopipedum (Melsheimer 1846) (Fig. 72)

This species is listed by Kloet and Hincks as being in the genus *Lyctus*. Presumably this was on account of the criticism of Leconte in his erection of the genus *Trogoxylon* (1861) and designating *Xylotrogus parallelopipedus* Melsh. 1846 as the type species. Arnett and Casey both considered that there was no justification for the erection of this genus and reverted to

Lyctus. Lesne, in 1921, however, was able to show that *Trogoxylon* was at a more advanced stage of evolutionary development than *Lyctus* and that there were connecting links.

The original distinction made by Leconte concerns the character of the

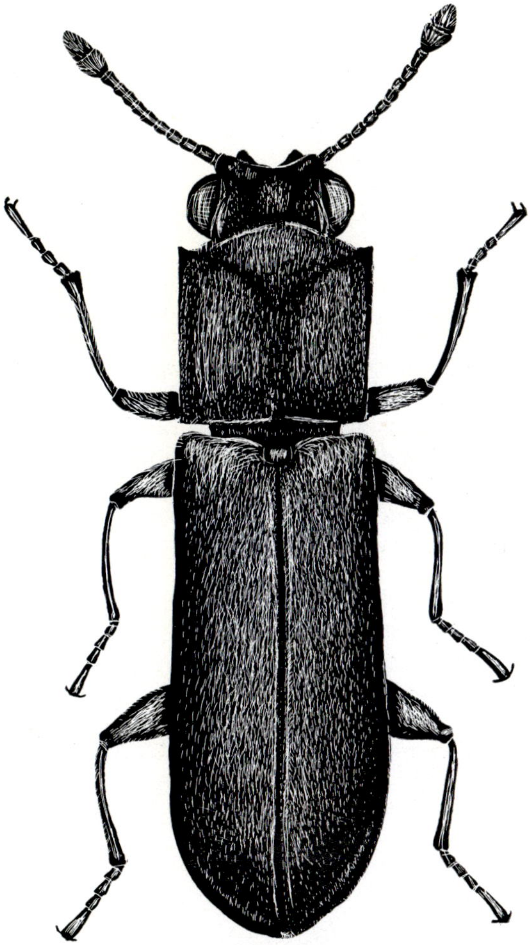

Fig. 72. *Trogoxylon parallelopipedum* (Melsheimer). Length: 3·8 mm.

anterior tibia which in his new genus had the outer-apical angle not prolonged, but Gerberg considers that the separation of this genus from *Lyctus* can be justified on the characters of confused pubescence of the elytra and the enlargement of the femur in the metathoracic leg.

Trogoxylon parallelopipedum can be readily differentiated from other lyctids occurring in Britain by the anterior angles of the pronotum being very prominent and acute, by the postclypeus being rounded and depressed and by the abrupt expansion of the antennal club. The antennae are slightly shorter than the pronotum and the latter is slightly wider than long. The base of the pronotum is approximately equal in width to the elytra and there is a shallow Y-shaped depression in the disc. The anterior margin of the pronotum is strongly convex and the posterior angles are prominent and acute. The prothoracic coxae are widely separated (this was thought to be a character of the genus by Casey). The elytra are about twice as long as wide and widen slightly posteriorly.

The length varies from 2·5 to 4·25 mm. This species is widely distributed throughout North America from whence it no doubt entered British ports.

Economic importance of the Lyctidae

In the limits set by nutritional and mechanical requirements, LYCTIDAE are polyphagous and the majority of commercial hardwoods are liable to attack. Those with plentiful vessels of adequate diameter are more likely to be attacked by *Lyctus* than where such vessels are not so plentiful, provided, of course, that the nutritional factor is present.

In most countries of the world *Lyctus* beetles are of economic significance and they are found infesting a wide range of wooden structures, furniture, structural timber, joinery, sports goods and many specialized wooden articles. In Great Britain, *Lyctus* damage was most widespread soon after the First World War. The entomological work at the Forest Products Research Laboratory, Princes Risborough, was commenced in 1927 due to the concern by sections of the Furniture Manufacturing Industry over *Lyctus* infestations. *Lyctus* damage again appeared to increase during the later stages of the last war and brought about many losses, causing embarassment to the hardwood timber industry and many furniture manufacturing and distributing concerns. Since that time, the growing use of tropical hardwoods, many of which are highly susceptible to *Lyctus*, has resulted in new infusions of the cosmopolitan *Lyctus* species into these islands. Although the *Lyctus* problem in Great Britain is small compared with that of *Anobium punctatum*, it is nevertheless substantial.

BOSTRYCHIDAE

Although the BOSTRYCHIDAE are joined with the LYCTIDAE to form the Powder post beetles, they are easily separated as in table on page 151.

Bostrychid damage in buildings in the United Kingdom is negligible, although it assumes importance in tropical countries. A large number of species is imported in parcels of timber, manufactured or semi-manufactured articles of wood and bamboo, and in packing case timber.

The family has a world-wide distribution, many species now being found throughout the warmer parts of the world. There are about 550 known species which vary from extremely small to moderate in size.

LYCTIDAE	BOSTRYCHIDAE
Antennae seldom serrate; club of two segments.	Antennae often serrate; club of three segments.
Beetle is dorsc-ventrally compressed.	Beetle is cylindrical.
Head is not hooded by prothorax.	Head is hooded by prothorax.
Front of pronotum and hind end of elytra simple, not ornamented with spines and protuberances.	Front of pronotum and hind end of elytra often ornamented with spines and protuberances.
Posterior end of body never truncated.	Posterior end of body often truncated.

British Bostrychidae

Four genera each containing one species are listed by Kloet and Hincks as British. A number of other species are introduced from time to time in timber, but these usually do not establish themselves.

Dinoderus minutus (Fabricius 1775) (Fig. 73)

This is a small insect 2·5–3·5 mm in length, pitchy-red in colour with the prothorax usually rather darker than the elytra. Beeson considers that this species is a pest only of bamboo and although on occasions the larvae are found in timber of various species and sometimes even in other materials, they will have originated in bamboo.

Joy lists also the species *ocellaris* Steph. as British and mentions it as being found chiefly in imported fruit, but Kloet and Hincks only list it as an importation. It is therefore somewhat strange that *minutus*, a bamboo-feeding species, is not also considered an imported beetle. The two species

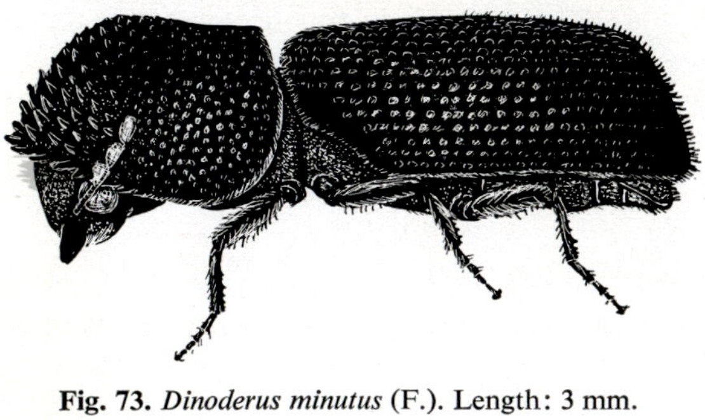

Fig. 73. *Dinoderus minutus* (F.). Length: 3 mm.

may be separated by the character of the raised hairs of the frontal region being short and few in *minutus* but long and thick in *ocellaris. Dinoderus bifoveolatus* Wollaston is another species of this genus listed as an importation.

In the East, from ancient times, it has been the belief that in order to minimize the attack by bamboo borers, the bamboo must be cut when the moon is waning. It is believed that when the moon is waxing the sap is being drawn up, whereas when it is waning the sap is drawn down. Beeson (1937) and Kirkpatrick and Simmonds (1958) have proved experimentally that there is no substance in this belief, although Beeson did show that there is a relation between the moisture content of the canes and the moon phase.

See also page 158.

Rhizopertha dominica (Fabricius 1792) (Fig. 74)

This bostrychid is a cosmopolitan pest of a wide variety of starch-rich grains, tubers, seeds and their ground or semi-manufactured products. The following have been recorded as being damaged by this species: wheat, barley, rice, maize, millet, sorghum, juar, dried potatoes, manioc

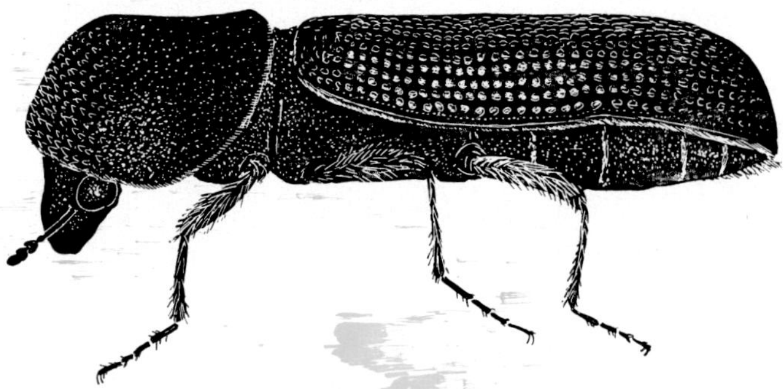

Fig. 74. *Rhizopertha dominica* (F.). Length: 2·7 mm

roots, waternuts and biscuits. In addition, a number of timbers have been listed by Beeson (1941) which this bostrychid is under suspicion of attacking, although in some cases the beetles were probably only hiding in the bark or sheltering in old larval galleries made by other borers.

The life cycle at 30° C and 30 per cent RH in wheat and maize has been observed to vary between 30 and 40 days, but the adult beetles have been maintained alive for 10 months.

Stephanopachys substriatus (Paykull 1800)

This is given in Kloet and Hincks' list of British insects but little information can be found, except that the related species *himalayanus* Lesne bores rambling galleries in the thick outer bark of *Pinus longifolia* from India.

Stephanapachys rugosus Ol. has been reported as being found in Pennsylvania homes where it was attacking the bark-covered joists.

Bostrychus capucinus (Linnaeus 1758) (Fig. 75)

With a body length of 9–14 mm, this beetle is much larger than the other British bostrychids. The elytra and tarsi are red and the thorax is dull black. The beetle lives in rotten wood and is quite rare in Britain.

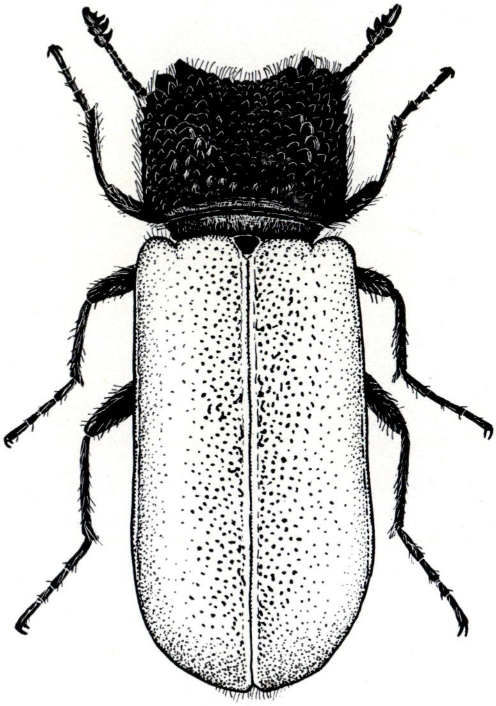

Fig. 75. *Bostrychus capucinus* (L.). Length: 10 mm.

Host trees

Bostrychids are usually found in the sapwood of hardwoods although some species, e.g. of the genus *Stephanopachys* are found in conifers, and some widely spread genera, e.g. *Dinoderus* are well known for their serious attack on bamboo canes.

In the forest, bostrychid attack usually commences in the dying or sickly trees but during timber extraction from the forest the newly felled trees are attacked. The insects are most usually polyphagous; it being unusual for a species to be restricted to a single genus of host plant.

Some BOSTRYCHIDAE are, in some regions, serious pests of tropical plantation crops such as rubber, coffee, and cocoa.

Bostrychid damage

Most countries of the world have some pest species of BOSTRYCHIDAE, but as already mentioned the greater number is found in warmer countries. Damage usually occurs at the time of timber extraction so that the main initial infestation takes place in timber yards and mills. Boocock (1960) considers that *Lyctus* and bostrychid beetles are undoubtedly the most serious pests with which saw-millers, plywood manufacturers, and other timber-using industries in hot countries have to cope. Brandsen says that bostrychids attack seasoning timber while it is at a much higher moisture content than is suitable for *Lyctus*. Consequently, where timber is attacked by bostrychids, the sapwood is often well powdered before *Lyctus* attack occurs. He also states that larvae working in the wood will continue to grow and feed as the wood dries out and may be found still active in seasoned timber. Such timber, however, will not be reinfested.

Larvae of Bostrychidae

The larvae have a curved body (Fig. 76) and when taken out of the wood remain C-shaped. They are creamy-white in colour. The thoracic segments are somewhat enlarged as are the anal segments. The narrowest part of the body occurs about the fifth abdominal segment.

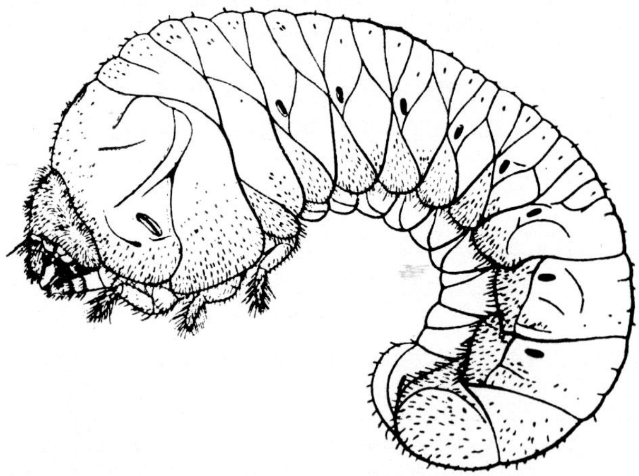

Fig. 76. Larva of *Heterobostrychus aequalis* Waterh. Length: 11 mm. (After Beeson).

The legs are well developed and the mandibles are gouge-shaped. The spiracles are annular but the eighth abdominal spiracle, which in the larvae of the LYCTIDAE is much larger than the preceding, is of similar size to those on the other segments.

Identification of bostrychid damage

In general terms, damage by bostrychid beetles is rather similar to that caused by *Lyctus*. Usually, however, the galleries made by the larvae of the former are larger, more circular in cross-section and, at least at an early stage in the attack, run mainly with the grain so that a transverse section of the sapwood shows a number of circular galleries of different sizes, which are all packed with bore-dust.

Nutritional requirements

Beeson (1941) has discussed the nutrition of *Dinoderus* in connection with the bamboo-cutting season. He says that it is generally known that bamboos soaked in water for considerable periods are practically immune to attack by *Dinoderus*, and it has been assumed that soaking leaches out certain water-soluble substances, possibly sugars, that are essential for the nutrition of the larva. Beeson then goes on to say that there is no satisfactory evidence for these beliefs, but that it is probable that (*a*) starch is an essential food substance, (*b*) glucose is not essential, (*c*) a soluble substance, possibly a disaccharide, is essential, and that (*d*) neither of the two essentials alone is sufficient. Beeson, however, does not mention the nitrogen requirements of the insect.

Parkin (1940), who included larvae of *Heterobostrychus brunneus* and *Bostrychoplites cornutus* Ol. in his investigations into the enzymes of wood-boring beetle larvae, came to the conclusion that larvae of BOSTRYCHIDAE fed exclusively upon cell contents—starch, sugars and protein—but they differed from LYCTIDAE in that they apparently were unable to utilize any portion of hemicellulose B.

Notes on some species of world wide importance

Heterobostrychus bruneus Murr.

This is a very common African species and an important pest of sawn timber throughout that continent. It often occurs in Britain in imported African timbers. In South Africa it is known as the Boxwood Borer. Both hardwoods and softwoods are attacked and Boocock states that heavy infestations in barked logs may reduce the whole of the sapwood to a mass of powder, held together only by a slender network of wood. He also states that in Ghana this beetle is a common pest of building timbers, especially in roofs and floors. The insect varies from 7–13 mm in length and is pitchy-black in colour and covered with fine reddish-brown hairs. In South Africa the life cycle takes about one year but, as is common in many bostrychids, a number of the brood takes two or even three years to complete development.

The species *aequalis* Waterh. is the commonest of the larger bostrychids in India and Beeson gives a long list of timber species which it attacks. He says also that this beetle is a regular pest in sawmills and in the factories of wood-using industries in India. It occurs in packing cases, boxes, plywood chests and panels, and in sapwood in furniture and fittings in buildings. The larval tunnel may reach a length of 38 cm and diameter of 6 mm.

The larva itself grows to about 15 mm. Beeson relates that if a tree is felled and not barked, nor the branches severed, the bark remains green for several months and during this time the starch in the sapwood is depleted. In this way the wood is rendered unsuitable for *Heterobostrychus* attack.

Bostrychopsis parallela Lesne (Fig. 77).

This species is most usually known as a borer of larger sizes of bamboo, but is also reported in a number of timbers. It has an annual life cycle, which is sometimes prolonged to two or three years, although a six-year life cycle has once been recorded in bamboo. The length of the adult varies from 9 to 14 mm and even though the larva is about 10 mm in length when mature, it has been recorded in the middle core of duali plywood in the Philippines. It occurs in Britain from time to time and the specimen figured was taken by Dr. P. Becker in the gardens of the Royal Horticultural Society at Wisley, Surrey. The species *jesuita* F. is the largest occurring in Australia. It may be up to 20 mm in length and the larvae bore tunnels up to 6 mm in diameter. It infests eucalyptus, wattles and orchard trees.

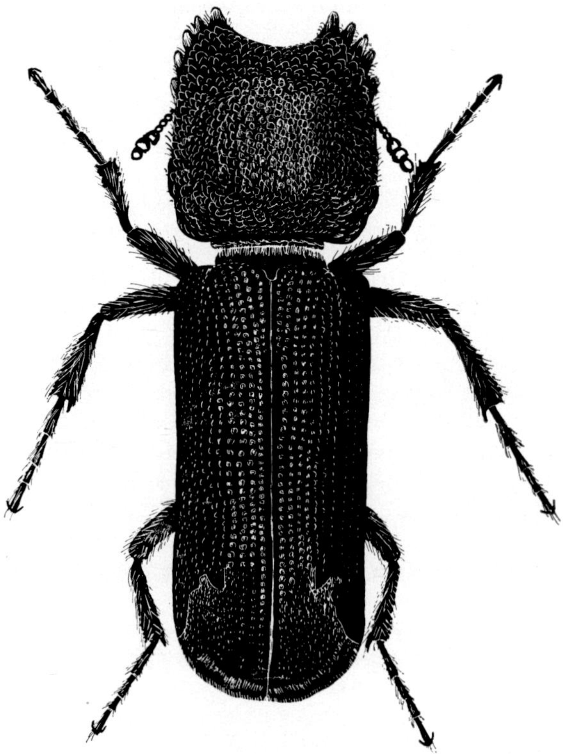

Fig. 77. *Bostrychopsis parallela* **Lesne. Length: 12 mm.**

Sinoxylon anale Lesne (Fig. 78)

This species is the commonest bostrychid in India. As a pest in the timber-using industries it is of the greatest economic importance and it has been carried to many parts of the world, including New Zealand and Australia, in timber. In Britain it sometimes emerges in great numbers from wood work originating from the East. It is rather a squat beetle about 4 mm in length with a pair of conspicuous pointed processes emerging from the posterior truncated part of the elytra.

Although primarily a pest of trees of the family LEGUMINOSAE, it has been recorded from a large number of timber species and is one of those capable of reducing the entire sapwood to dust. It is of interest to note that

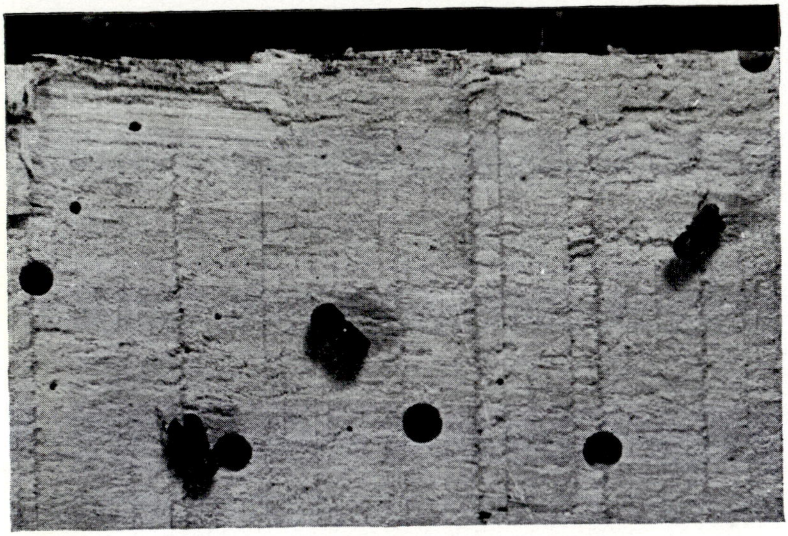

Fig. 78. *Sinoxylon anale* Lesne, adults and exit holes.

the male and female remain in the oviposition gallery long after the hatching of the eggs—guarding them against the attacks of predators. Many bostrychids die in the oviposition chamber and thus seal it effectively from predators; both ends of the adult beetle—the front of the pronotum and the hinder end of the elytra—are often heavily armoured with spine-like processes. The adult beetles of this species sometimes tunnel into the young shoots of living saplings and deform or kill them. Beeson says that in India the minimum life cycle is around three months and the maximum is over four years. There are numerous other species in the genus *Sinoxylon*.

Xylion cylindricus MacLeay and *X. collaris* Er.

These are common in New South Wales where they attack hardwoods, especially spotted gum (*Eucalyptus maculata*) used for pit-props and mining timbers. The second-named species has a reddish-yellow prothorax and shining black elytra.

Xylobiops basilare Say.

The red-shouldered hickory borer is found in the eastern United States. It is about 5 mm in length and is black with the basal third of the elytra dull reddish or yellow and with the posterior of the elytra concave and beset with three spine-like tubercles on each side. It is a pest of hardwood trees.

Polycaon stouti Lec.

This bostrychid is a native of the western United States and is remarkable in the adult being very unbostrychid-like in appearance. The head is inserted into the prothorax forwards and not beneath, as is more usual. The prothorax is, therefore, not hoodlike. It is black in colour and 20 mm in length. Although it breeds under natural conditions in *Eucalyptus*, maple, oak, fruit trees, madrona, manzanita and sycamore (Essig 1926), it is of considerable economic importance as it attacks hardwoods in lumber yards and buildings, as well as furniture.

The Bamboo Borer or 'Ghoon'.

Several species of bostrychids, but mainly of the genus *Dinoderus*, are of great economic importance throughout the regions of the world where bamboo is grown or used. In the main, three species of the genus are involved, *brevis, minutus* (F.) and *ocellaris* Steph., and they originated in Asia.

The adult beetles gain entrance to the cut-bamboo where the external rind has been cut (such as where side branches have been trimmed off), exposed cut ends, inside the end internodes of the hollow stems or where the nodes have been smoothed. The external rind is tough and seldom penetrated by the adults for gaining entrance except where two bamboo stems are pressed together. In this way the beetle can obtain sufficient leverage to gnaw its way in. All the bore-dust is ejected and pairing takes place inside the gallery which has been constructed. Later arrivals use tunnels which have been made earlier and eggs are laid in exposed pores in the wall of the pairing gallery. In India (Dehra Dun) there are three generations, or under the most favourable conditions four generations, annually, the first two of which usually taking 11 or 12 weeks. Plank (1948), however, found that in Puerto Rico the eggs hatch in 3 to 7 days, the larval period lasts for an average of 41 days (during which time the larva moults four times), the pupal stage lasts 4 days but after eclosion the adult remains within the pupal cell for 3 days. The average life cycle from egg to emerged adult was 51 days. The larvae bore their tunnels parallel to the long axis of the cane and more usually occur just beneath the inner wall, thus avoiding the outer zone where the fibrovascular tissue is more concentrated and the starch-bearing parenchyma is scarce or absent. The tunnels are filled with the tightly packed frass and whereas in *Lyctus* the frass consists of minute particles of a floury consistency, the frass of *Dinoderus* is rather coarser and the cellular structure can still be made out. Beeson presumes that the larva tears off small particles of tissues, tastes them, and rejects the horny fibres, as relatively long fragments occur in the frass. The palatable parenchyma is chewed into fragments much smaller than could be gripped by the tearing or chiselling action of the mandibles. This chewed material is presumed to have passed through the alimentary canal and digested without visible effect except that the starch grains have been removed. The larval frass shows no indication of being formed into faecal pellets.

When the adult bites its way out of the pupal chamber either it may bore directly out through the external rind or it may bore some considerable distance and perhaps find its way out through another exit or ingress hole. Generations follow each other until all the woody tissue is reduced to dust. The larvae of the subsequent generations make their way through the accumulated frass searching for unattacked morsels of wood. The outer and inner skins of the cane, however, remain intact until they are so fragile that they rupture under a slight stress. Beeson records five years as the longest period of infestation of a bamboo and a foot length of a bamboo, an inch or two in diameter, having 800 holes!

The Short-circuit Beetle or Lead Cable Borer.

A peculiar type of damage caused by certain BOSTRYCHIDAE can be conveniently described here. The beetles concerned cause great economic loss on account of their habit of boring into lead-covered cable. In California the species *Scobicia declivis* Lec., which is a reddish-brown cylindrical beetle up to 6 mm in length, bores holes 2–5 mm in diameter into the lead sheathing of aerial telephone cables. The holes are bored adjacent to the cable supporting rings where the beetles can obtain the necessary leverage for gnawing the lead. Burke *et al.* (1922) describe the damage and confusion resulting from a large number of short circuits occurring in telephone cables during the first rain after the emergence period of the beetle. The same species has also been recorded by Rivers (1886) as severely damaging wine casks. The former authors have given an account of the life cycle and they state that the female beetle bores directly into the wood where she lays her eggs. These hatch in 21 days and thereafter the larvae tunnel in the wood for about nine months. The pupal stage lasts for only about 14 days but, after eclosion, the beetle remains in the pupal chamber beneath the surface of the wood for a month, before finally gnawing a flight hole to the exterior.

Other species have somewhat similar lead-boring and cable-boring habits in other parts of the world. It is of interest to note that most records of insects attacking or boring into metals concern adults biting their way out from the pupal chamber but in the case of these species they cause damage by biting their way in, ostensibly to make the oviposition gallery.

REFERENCES

LYCTIDAE

ALTSON, A. M. 1922. On the Young Larvae of *Lyctus brunneus* Stephens. *Ann. appl Biol.* **9**: 187–96.

ALTSON, A. M. 1923a. On the Method of Oviposition and the Egg of *Lyctus brunneus* Stephens. *Linn. Soc. Lond. Jour., Zool.* **35** (234): 217–27.

ALTSON, A. M. 1923b. On the Genital System of *Lyctus brunneus* Stephens. With a Note on *Lyctus linearis* Goeze (COLEOPTERA). *Linn. Soc. Lond. Jour., Zool.* **35** (238): 581–97.

ANON. 1938. Lyctus Powder-post Beetles. *Leaflet* No. 3. *Forest Products Research Laboratory*. D.S.I.R. H.M.S.O.

ANON. 1940. Beetles Injurious to Timber and Furniture. *Forest Products Research Bulletin* No. 19. D.S.I.R. H.M.S.O.

ANON. 1941. Pinhole or Powder-post Damage? *Leaflet* No. 17. *Forest Products Research.* D.S.I.R. H.M.S.O.

ANON. 1948. D.D.T. Spray Treatment for prevention of *Lyctus* attack. *Leaflet* No. 43. *Forest Products Research Laboratory*. D.S.I.R. H.M.S.O.

ANON. 1951. Furniture Beetles. British Museum (Natural History) Economic series No. 11.

ANON. 1957. The Kiln Sterilization of *Lyctus*-infested timber. *Leaflet* No. 13. *Forest Products Research Laboratory*. D.S.I.R. H.M.S.O.

BLETCHLY, J. D. 1960. Studying the Eggs of *Lyctus brunneus*. *Timber Technology* **68**: 29–31.

FISHER, R. C. 1928. *Lyctus* powder post beetles. *Bull. Gt. Br. For. Prod. Res.* No. 2: 3.

FISHER, R. C. 1932. Some Aspects of the *Lyctus* Powder-Post Beetle problem in Great Britain. *5th International Congress Entomology*, pp. 757–71.

GAY, F. J. 1953. Observations on the biology of *Lyctus brunneus* Stephens. *Aust. J. Zool.* **1** (1): 102–10.

GERBERG, E. J. 1957. A Revision of the New World species of Powder-post beetle. belonging to the Family LYCTIDAE. *Technical Bulletin* No. 1157. United States Department of Agriculture, Washington, D.C. pp. 1–55.

HATFIELD, I. 1949. *How to Recognize and Control Powder-post Beetles in Wood.* Forest Products Research Society Third Annual National Meeting. Michigan, U.S.A.

HOPKINS, A. D. and SNYDER, T. E. 1917. Powder-post Damage by *Lyctus* beetles to seasoned Hardwood. *Farmers' Bulletin.* 788. United States Department of Agriculture, Washington, D.C.

JOY, N. H. 1932. *Practical Handbook of British Beetles.* Witherby. London.

KLOET, G. S. and HINCKS, W. D. 1945. *A Check list of British Insects.* Stockport.

MUNRO, J. W. 1915. The Larvae of Furniture Beetles—families ANOBIIDAE and LYCTIDAE. *Proc. Roy. Physical Soc. Edinburgh.* **19**: 220–36.

MUNRO, J. W. 1930. Beetles injurious to Timber. *Forestry Commission Bulletin* No. 9. H.M.S.O.

PARKIN, E. A. 1936. A study of the food relations of the *Lyctus* powder-post beetles. *Ann. appl. Biol.* **23**: 369–400.

PARKIN, E. A. 1940. The digestive enzymes of some wood-boring beetle larvae. *Journal of Experimental Biology* **17**: 364–77.

ROSEL, A. 1969. Oviposition, egg development and other features of the biology of five species of Lyctidae (Coleoptera). *J. Aust. ent. Soc.* **8**: 145–52.

TAYLOR, J. M. 1968. Prevention of insect attack in plywood. Part 1: *Lyctus* attack in obeche plywood. *Wood* **33** (2): 35–6.

TOOKE, F. G. C. and SCOTT, M. H. 1944. Wood-boring Beetles in South Africa. *Bulletin* No. 247 (Entomology Series No. 14). Department of Agriculture and Forestry, Union of South Africa, Pretoria.

TOOKE, F. G. C. 1949. Beetles Injurious to Timber in South Africa. Department of Agriculture (Entomology Series No. 28). *Science Bulletin* No. 293. Union of South Africa, Pretoria.

WILSON, S. E. 1933. Changes in the Cell Contents of Wood (xylem parenchyma) and their Relationships to the Respiration of Wood and its Resistance to *Lyctus* Attack and to Fungal Invasion. *Ann. appl. Biol.* **20**: 661–90.

BOSTRYCHIDAE

ANDERSON, R. F. 1960. *Forest and Shade Tree Entomology*. New York.

BEESON, C. F. C. 1933. The Liability of Solid Bamboo Lance Staves to Attack by Borers, *Ind. For.* **59**: 709–12.

BEESON, C. F. C. 1935. Boxwood borers (*Heterobostrychus*). *Ind. For.* **61**: 250–5.

BEESON, C. F. C. 1941. *The Ecology and Control of the Forest Insects of India and the Neighbouring Countries*, Dehra Dun, India.

BEESON, C. F. C. and BHATIA, B. M. 1926. On the biology of the BOSTRYCHIDAE. *Ind. For. Rec. Ent.* **11** (12): 223–320.

BOOCOCK, D. 1960. Powder Post Beetles in Tropical Countries. *Timber Technology* **68**: 382–83.

BRANDSEN, J. BOSTRYCHIDS. *Technical Notes*. Division of Wood Technology. Forestry Commission of N.S.W., Sydney. **12**: 19–23.

BURKE, H. E. and HARTMAN, R. D. 1922. The Leadcable Borer or Short Circuit Beetle in California. *U.S. Dept. Agric. Bull.* No. 1107.

ESSIG, E. O. 1926. *Insects of Western North America*. Macmillan.

FISHER, R. C., CANN, F. R. and PARKIN, E. A. 1932. A Survey of the Damage caused by Insects to Hardwood Timbers in Great Britain. *For. Prod. Res. Bull.* No. 16.

GARDNER, J. C. M. 1933. Immature stages of Indian COLEOPTERA (13), BOSTRYCHIDAE. *Ind. For. Rec. Ent.* **18** (11): 1–19.

HICKIN, N. E. 1953. *Bostrychus capucinus* L. in Furniture. *Ent. mon. Mag.* **89**: 264.

LESNE, P. 1924. *Les Coleopteres Bostrychides ed l'Afrique Tropicale Francaise*. pp. 1–109. Paris.

MALLIS, A. 1969. *Handbook of Pest Control*. 5th Edn. Pittsburgh.

MANSOUR, K. 1934. On the Intracellular Micro-organisms of some Bostrychid Beetles. *Quart. J. Micr. Sci.* **77**: 243–53.

MATHUR, R. N. 1950. The Larva of *Dinoderus bifoveolatus* Woll. *Indian J. Ent.* **10**: 1–4.

MATHUR, R. N. 1955. Identification of Indian LYCTIDAE and *Dinoderus* species. *Ind. For. Bull.* No. 16.

PLANK, H. K. 1948. Biology of the Bamboo Powder-post Beetle in Puerto Rico. *Bull.* No. 44 Federal Experiments Station, Puerto Rico.

CHAPTER SIX

The Smaller Families

The BUPRESTIDAE, LYMEXILIDAE, EUCNEMIDAE, MELANDRYIDAE, OEDEMERIDAE, CURCULIONIDAE and LUCANIDAE

BUPRESTIDAE

This family belongs, together with the ELATERIDAE (Click beetles, Wireworms), to the suborder STERNOXIA. The beetles are elongate and depressed in shape and the head is retracted into the pronotum as far as the eyes. In addition the pronotum possesses distinct lateral margins and fits closely along the posterior margin to the anterior end of the elytra. They are hard in texture and boat-shaped, the elytra gradually converging posteriorly. Some species are only about 2·5 mm in length whilst others are as much as 75 mm. The antennae are serrate. The most noteworthy feature however of buprestid beetles in general is the metallic lustre of a great number of the species. Many are of exceptional brilliance and are often collected by native peoples as items of adornment. They are sometimes called Jewel Beetles in England and are called Splendid Beetles in Germany. The family is widely distributed throughout the world and about 12,000 species have so far been described.

Buprestids are more usually forest insects where they attack weak, sickly or dying trees but some species attack newly felled timber. Again some species oviposit on trees damaged by fire and some of these beetles are strongly attracted to forest fires and will even attempt to crawl on glowing twigs. They often cause a nuisance to fire-fighters by crawling on them and 'biting' them.

In the British list there are five genera containing twelve species but to this must be added the North American *Buprestis aurulenta* L. which emerges from structural and joinery timbers of Douglas fir with such regularity that the student of wood-boring insects must be familiar with it. A number of other species are also imported into Great Britain in timber from various parts of the world. No member of this family can, however, be considered as a serious pest of timber indoors; mostly they are trimmed out in the timber yards, but the emergence of an occasional jewel-like beetle of fair size sometimes causes concern to the householder.

Larvae of BUPRESTIDAE

The prothorax is large and flat—distinctly larger than the other segments of the body (Fig. 9). This characteristic gives rise to the common name of Flat-headed Borer used in North America. On both the pronotum and the prosternum there are distinct plate-like areas of similar size. On the pronotal plate there is a median line or a V- or Y-shaped marking or groove. Buprestid larvae are always legless and the spiracles crescent-shaped. The larval galleries are 'flattened oval' in cross-section and tightly packed with fine frass. The latter is packed in a distinct manner showing an arc-like pattern described in the German literature as 'cloud-like'.

The length of the life cycle in most cases is generally thought to be two years and the majority pass the larval stage in the inner bark and sapwood. Prior to pupation the flat meandering galleries turn into the wood. The larvae and the pupation chambers of the two subfamilies BUPRESTINAE and AGRILINAE can be distinguished. Larvae of the AGRILINAE do not possess the very broad prothorax to the same degree as that shown by the larvae of BUPRESTINAE but the end of the abdomen is armed with a pair of horny spines, absent in the latter subfamily. In the AGRILINAE the larva constructs a very narrow entrance to the pupal chamber and a larger exit gallery is prepared which, when the beetle eclodes, is widened to admit its passage through the bark. The emerging beetle finally makes a semicircular flight hole. On the other hand the larvae of the BUPRESTINAE do not construct an escape gallery, but do orientate themselves before pupation.

Key to the species of BUPRESTIDAE *found in Great Britain*

In this key the species enumerated by Joy have been included but in addition the North American species *Buprestis aurulenta* has been incorporated for reasons given in the text.

1.	Edges of elytra copper-coloured. Length 12–22 mm	*Buprestis aurulenta* L.
–	Edges of elytra not copper-coloured. Usually less than 12 mm in length	2
2.	Pronotum broadest at base, antennae and legs black	*Trachys*, 6
–	Pronotum not broadest at base	3
3.	Antennae clubbed, tarsi short. Length 2·3–3·5 mm. Black with green sheen	*Aphanisticus*, 8
–	Antennae not clubbed, tarsi longer. Length 4–12 mm	4
4.	Elytra more than twice as long as broad . . .	*Agrilus*, 9
–	Elytra less or about twice as long as broad	5
5.	Brilliant metallic green; broader . .	*Anthaxia nitidula* (L.)
–	Entirely black; narrower .	*Melanophila acuminata* (Degeer)
6.	Elytra with sides parallel for anterior two-thirds, with deep irregular impressions. Black with purple or blue sheen and generally with waved transverse bands of white pubescence .	*T. minuta* (L.)

– Elytra with sides nearly parallel for anterior half, without distinct impressions, head and pronotum coppery 7

7. Elytra coppery, with white pubescence often rubbed off . . . . *T. pumila* Ill.

– Elytra bright green, glabrous . . . *T. troglodytes* Sch.

8. Pronotum more transverse, less contracted to base and elytra broader but more dilated behind . *A. pusillus* (Ol.)

– Pronotum less transverse, more contracted to base and elytra narrower and less dilated . *A. emarginatus* (Ol.)

9. Patches of white or yellow pubescence one-third from apex of elytra, near suture as well as on upper side of abdomen (visible obliquely from the side). Bluish or greenish sheen . *A. pannonicus* (Pill. & Mitt.)

– Without patches of light pubescence 10

10. Apex of elytra extending some distance beyond sutural border. Length 6–10 mm 11

– Apex of elytra extending only a short distance beyond the sutural border. Length 4–6 mm 12

11. Rugosity on pronotum and elytra much finer, distinctly finer than on head, entirely coppery or purplish sheen. Length 7–10 mm . . . *A. sinuatus* (Ol.)

– Rugosity on pronotum and elytra much stronger than on head. Entirely blue-green or with coppery pronotum. Length 6–9 mm . . *A. viridis* (L.)

12. ♂ with segments of antennae much dilated, ♀ much less; raised line by hind angle of pronotum beginning a short distance from angle and slightly bent . . . . *A. lacticornis* (Ill.)

– Antennae the same as ♀ *laticornis* in both sexes; raised line on pronotum beginning at angle and straight . . . . *A. angustulus* (Ill.)

Notes on the species

All these Buprestids are rare in Britain and only two are of any economic significance.

Agrilus pannonicus (Pill. and Mitt.) (Fig. 79).

This is the species *biguttatus* (F.) in Joy. In length it is from 8–12 mm, very rare, and found in England as far north as Nottinghamshire. Joy says 'in and around old oak'. In addition, however, it is occasionally imported from Germany and elsewhere in Europe.

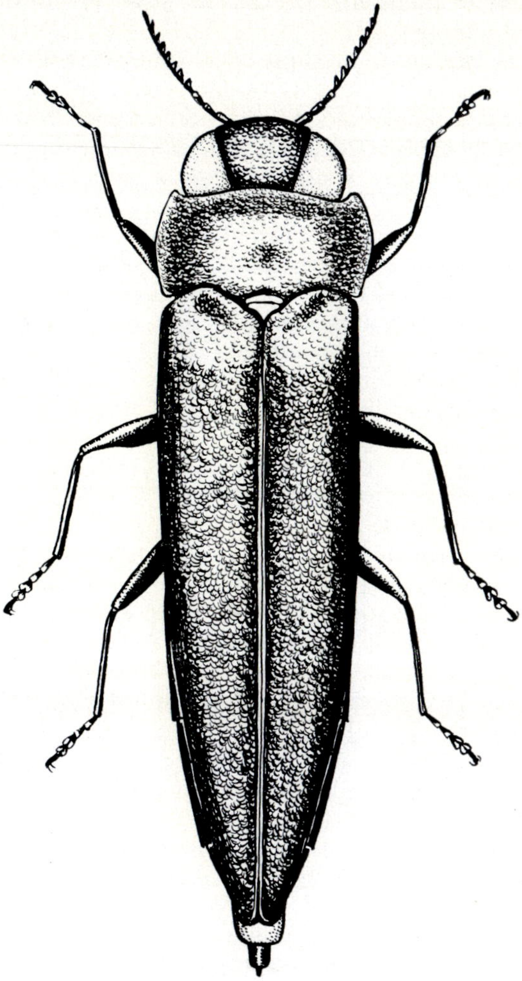

Fig. 79. *Agrilus pannonicus* (Piller and Mitterpacher). Length: 12 mm.

Buprestis aurulenta L. The Golden Buprestid (Fig. 8)

This handsome beetle measures up to about 22 mm in length. It possesses a brilliant yellowish-green or blue-green metallic lustre which is iridescent. The pronotum may be green, blue or copper coloured and there are four pronounced ridges running the length of each elytron. The edges of the elytra are copper coloured. This insect is native to western North America where it is the most common and the most injurious of the *Buprestis* species. It attacks dying trees, stumps and felled logs and Anderson (1960) says that sometimes it attacks partially or fully seasoned wood.

The host tree species are usually Douglas fir, pines, firs, western red cedar, spruce and other conifers.

It is known in Britain on account of its emergence from structural or joinery timbers usually of Douglas fir and often after the lapse of many years. This is the phenomenon of delayed emergence. Shaw (1961) summarized the records of this species in Britain and mentioned the cases of delayed emergence given by Linsley in California and Spencer in British Columbia. The maximum length of larval and pupal period recorded is 25–26 years—a very remarkable phenomenon. Shaw considered that this beetle was more common during the post-war years than hitherto. Actually the recorded instances of its occurrence in Britain represent only a fraction of unrecorded cases, several specimens having come to the notice of the present author each year but have gone unrecorded. Occasional larvae of this species are also found in Britain, usually thought to be 'longhorns' by the inexperienced.

Some other species of BUPRESTIDAE are sometimes introduced in timber into Britain. Linscott has recorded the accidental introduction by the present author of *Chalcophora mariana* (L.) (Fig. 9) and *Dicerca berolinensis* Herbst which emerged in the laboratory in Surrey from maritime pine which contained colonies of the termite *Reticulitermes lucifugus* brought from France. *Chrysobothris chrysostigma* L. (Fig. 80) is occasionally imported from Europe in Oak.

LYMEXYLIDAE

This family, sometimes called the LYMEXYLONIDAE, is placed in the MALACODERMATA where it occupies a place adjacent to the CLERIDAE, the family containing many predators of wood-boring insects.

Two species only are given in the British list, *Hylecoetus dermestoides* (L.) and *Lymexylon navale* (L.). In this family the maxillary palpi in the males are large and much divided and the tarsi are very long and narrow. Another characteristic is the close, short pubescence and the elytra are very finely punctured. The antennae are yellow with the base black and the legs are yellowish. The length of both species varies between 6 and 16 mm. *Hylecoetus dermestoides* may be distinguished from *Lymexylon navale* by the prothorax which is transverse and the antennae serrate, whilst in the latter species the prothorax is quadrate and the antennae filiform.

Notes on the species

Hylecoetus dermestoides (Fig. 81).

In the male the head and thorax are black, the elytra being yellow but black at the apex. The female however is entirely yellow. It is a very local insect but has been recorded from England, Scotland and Ireland but in northern and central Europe it is very common and its distribution range extends as far north as Siberia. It occurs in the timbers and the root stumps

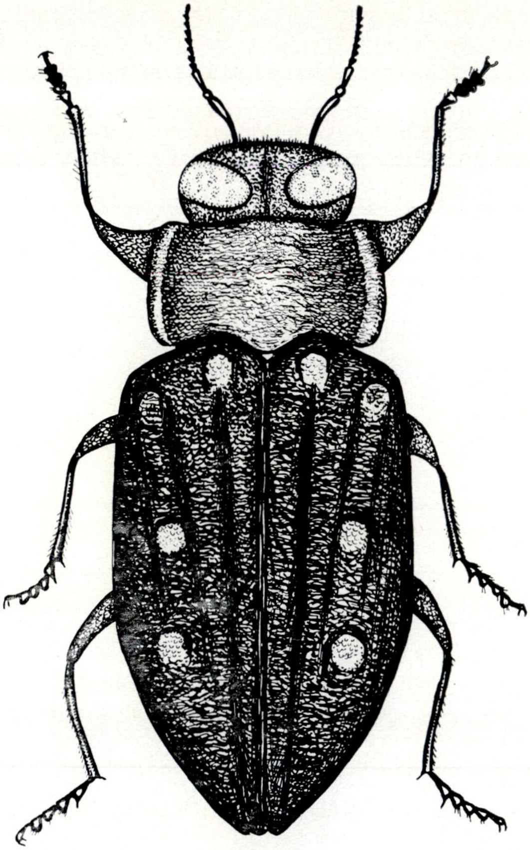

Fig. 80. *Chrysobothris chrysostigma* (L.). Length: 14 mm.

of a variety of hardwood and softwood trees, particular mention is made by Vité of beech and oak but birch, maple, alder, fir, pine, larch, spruce, and Douglas fir may also be infested. In Finland, Saalas reports that it is especially common in birch. Kurir gives a list of over 20 timber species known to be attacked by *H. dermestoides.*

The duration of the adult stage is only a few days and this may take place at any time between the beginning of April and the beginning of July. The female lays her eggs in batches in wood crevices or in the rough bark or occasionally in bore holes. Kurir observed that the egg clutch size varied from 4 to 91 and that the highest number of eggs laid by a female was 146. Seven to fourteen days after the eggs are laid they hatch and the young larvae eat part of the egg-shell to which is attached a symbiotic fungus. For a period the larvae remain together but then they disperse to find suitable places to make ingress holes. Their galleries are unbranched and curved and are found in the sapwood and the heartwood of oak, pine and larch, but the heartwood of beech is said not to be penetrated. Vité says that the larval passages seldom penetrate deeper than 16–20 cm into the wood and with the exception of the ingress hole the larva widens the passage to accommodate its growth. The bore-dust is regularly cleaned out of the passage by the larva, which it does by pushing

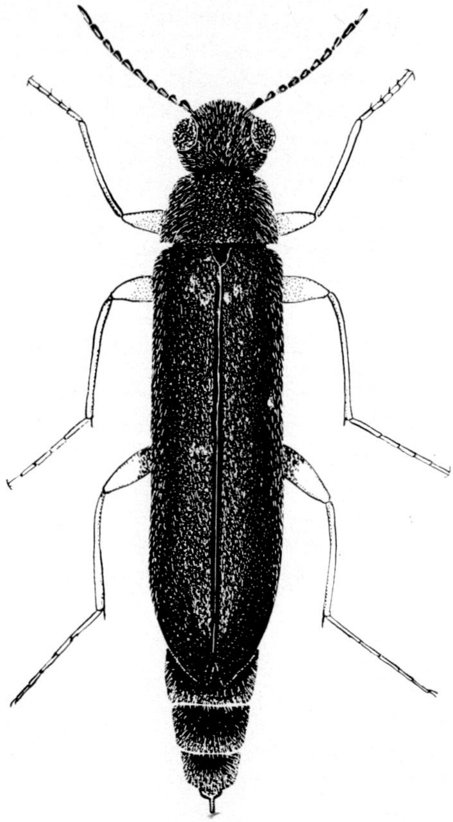

Fig. 81. *Hylecoetus dermestoides* (L.). Length: 15 mm.

it out backwards with its long serrated tail appendage (Fig. 82). The young larva and the larva when ready to pupate possess only small tail appendages. The inflatable 'hood' or collar-shaped prothorax plays an important part in the larva's locomotion. In addition the thoracic legs are well developed. The larva can turn round in its passage in order to help eject bore-dust and the piles of dust, which are sometimes considerable, are characteristic evidence of the presence of attack by this insect. The foodstuff of the larva does not consist of any part of the wood substance but is made up entirely of Ambrosia fungus (*Endomyces hylecoeti* Neg.) growing on the walls of the larval passage and which was introduced by the young larva from its egg-shell. During the winter and until March the entrance to the larval passage is closed by a plug of bore-dust which the larva removes when passage-widening recommences in the spring. Just before pupation the larva constructs a wider chamber near the ingress hole and plugs the internal end with bore-dust. The pupal stage lasts only a week and a few days after eclosion the adult bites its way out through a flight-hole in an already prepared position. Vité says that the adult beetles take no food and that the complete life cycle takes either one or two years but Thompsen (1949) says that in Denmark this takes two if not three years. Richter (1933) often found small mites of the *Tyroglyphopsis* group parasitizing larvae of

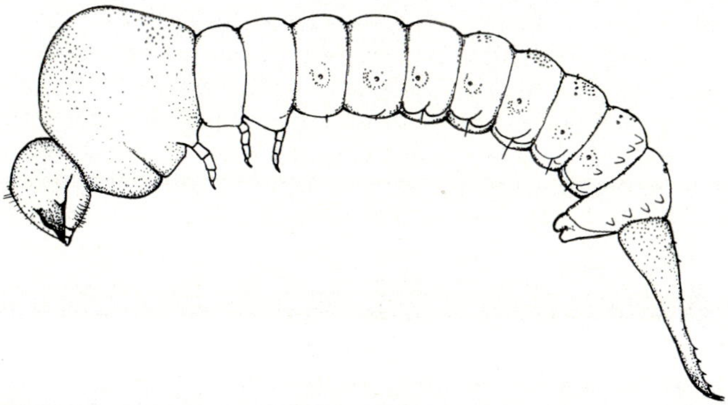

Fig. 82. Larva of *Hylecoetus dermestoides* (L.). Length: 10 mm.

this species—sometimes as many as 20 on the head capsule, but he did not consider that these had much retarding effect on the growth of the larvae.

Hylecoetus dermestoides is so local and uncommon in Britain that it cannot be considered an important pest of timber in this country. In Europe however there have been many instances of extensive damage to valuable timber of oak and beech to the extent that it can only be used as firewood. It is thought also that this insect assists in the spread of wood destroying fungi.

Hylecoetus flabellicornis Schn.

This is closely related to the previous species. It is not found in Britain but occurs in northern Europe where it attacks the fresh stumps of spruce.

Lymexylon navale

This is found in northern, eastern and central Europe but it is usually less common than *Hylecoetus dermestoides*. It differs also from the latter species in being confined to oak. Attack may occur in standing trees, felled trunks or in cut stumps but always where bark has been damaged or removed, thus drying out, to some extent, the underlying sapwood. In such places the eggs are laid from the end of May to July in clutches of five to fifteen. The young larvae bore very narrow galleries straight into the centre of the trunk and indeed the galleries are so narrow that in the German language the larvae are known as 'hair-worms'. As the larvae increase in size they enlarge the diameter of the gallery and also construct auxiliary galleries or feeding passages which are perpendicular to the main horizontal gallery. They may be traced upwards, downwards or to the sides and are from only a few millimetres to several centimetres in length. The horizontal galleries are very difficult to detect in the cut wood as they are mostly filled with bore-dust which however may be slightly different in colour from the sound wood—either very slightly whiter or very slightly yellower.

In contrast to *Hylecoetus dermestoides*, *Lymexylon navale* obtains its nourishment from cellulose, or substances of a like nature and does not cultivate nor feed on fungi. The galleries usually have clean walls but nevertheless may be stained due to secondary fungal invasion. Reissig says that the full-grown larva returns to make its pupal chamber immediately underneath the outer wood surface close to its original ingress point but makes a new gallery for this purpose. The entire gallery of one larva may be as much as 2 m which is an extraordinary length for one animal to bore in wood.

From the pupal chamber a short gallery extends towards the outside of the wood and the entrance to the inner gallery is plugged with bore-dust. Reissig mentions variations in the manner in which the galleries are bored. In fresh wood the larva will sometimes follow horizontally one annual ring instead of cutting across them. The life cycle is at least of one year's duration.

Because of its relatively rare occurrence *Lymexylon navale* cannot be said to be of great economic importance, but it can be of considerable nuisance value where, following an accidental blow to a good sound tree a fungal infection follows an attack by *L. navale*.

Melittomma albitarsis Blair.

The larva of this species, which bores the wood of *Bombax malabaricum* in India, is figured by Beeson. It possesses an extraordinary anal extremity, this being truncate, concave and with a raised margin with brush-like crenulations which are obviously used for brushing and cleaning refuse from the galleries and for plugging.

In North America, where they are known as timber worms, two species of some little importance are *Melittomma sericeum* (Hann) and *Hylecoetus lugubris* Say. The former species infests most blight-killed chestnut trees.

Melittomma insulare Fairm.

Although technically not a timber beetle, this species is a serious pest of coconuts in some parts of the world. The beetle is slender and dark brown varying in length from 8 to 18 mm. The adults live for only a few days and females lay their eggs in cracks in the trunk of coconut palms, most commonly at the base. The larvae bore into the soft central tissues of the palm bole but each day the larva backs towards the tunnel entrance to remove accumulated debris, the caudal end being specially adapted for the purpose, being sclerotized with a serrated margin. Around the gallery and in advance of it is a zone of wood tissue infected with micro-organisms. The end result of such a state of affairs is that the coconut palm, weakened at the base by a large cavity, falls in a high wind.

EUCNEMIDAE

This family contains many wood-borers in warmer climates, but in Britain it is represented by three species, of which *Melasis buprestoides* (L.) is fairly common in the southern half of the country. It lives in rotten wood of beech, alder and other trees. The larvae (Fig. 83) are unusual in being modified to live an inactive life; legs are vestigial or completely absent, and the head is very much reduced in size. The larvae resemble those of some Buprestids.

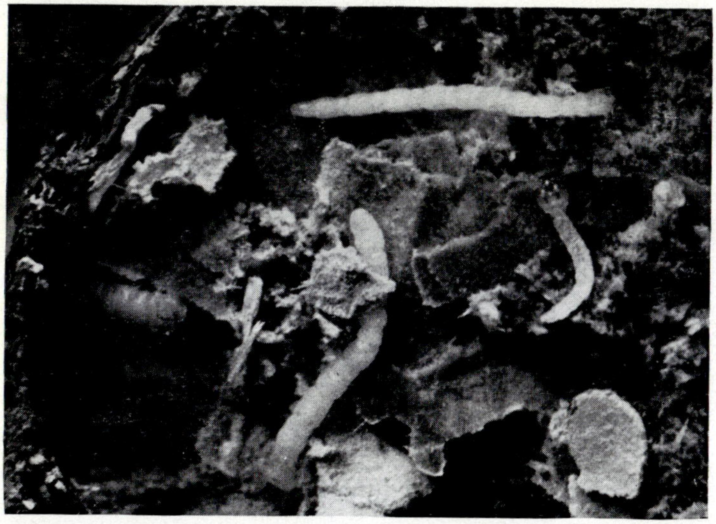

Fig. 83. Branch of alder split open showing larvae and larval chambers of *Melasis buprestoides*.

MELANDRYIDAE

The family MELANDRYIDAE is, in general, north temperate in distribution although some are found elsewhere. Beeson lists five species found in India. In the British list there are nine genera containing twenty-one species but all are rare and of no direct economic importance.

Serropalpus barbatus Schall. (Fig. 84).

This beetle is not an indigenous British insect. It is, however, distributed over the greater part of Europe and extends to the east as far as Amur in Siberia and to the south as far as northern Italy and Greece. It is also found in North America but doubts have been expressed as to whether this insect is identical with the European species.

Previous to 1949 there were only two known occurrences of *Serropalpus barbatus* in Britain but Bletchly (1955) has described how from that date

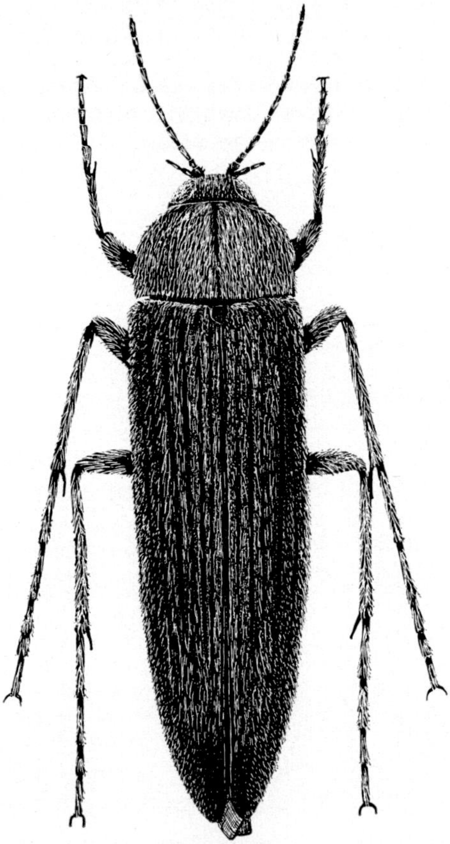

Fig. 84. *Serropalpus barbatus* Schall. Length: 13 mm.

to 1953 there were seventeen records of larvae or adults being received for identification at the Forest Products Research Laboratory at Princes Risborough. The following account is very largely taken from the work of Bletchly and to a lesser extent from that of Vité.

The blackish-brown adult beetle is elongate and narrow with the antennae and legs reddish-brown. It possesses large maxillary palps and the terminal joint is club-shaped. The antennae of the males reach up to half the length of the body. In length the beetles vary from 8 to 18 mm. They fly late in the evening in summer, the actual months seem to vary greatly according to the geographical area, in Poland July to September, in Finland June to August and in the Leningrad area the end of June to mid July.

It appears from Bletchly's summary that silver fir (*Abies alba*) is the preferred host timber of *Serropalpus barbatus* in Europe and of the seventeen records of the species in Britain given, four are definitely from this timber species and another four rather more indefinitely from *Abies* sp. Two species of spruce (*Picea abies* and *P. orientalis*) are also attacked, more particularly in East Prussia and in the Leningrad area and *Juniperus communis* is also recorded as a host timber. Balsam, hemlock, larch and spruce are said to be susceptible to attack in America.

Eggs are laid at night usually on the lower parts of the trunks, on felled trees or more rarely on the stumps.

The larva (Fig. 85) varies in length from 12 to 25 mm and in shape is

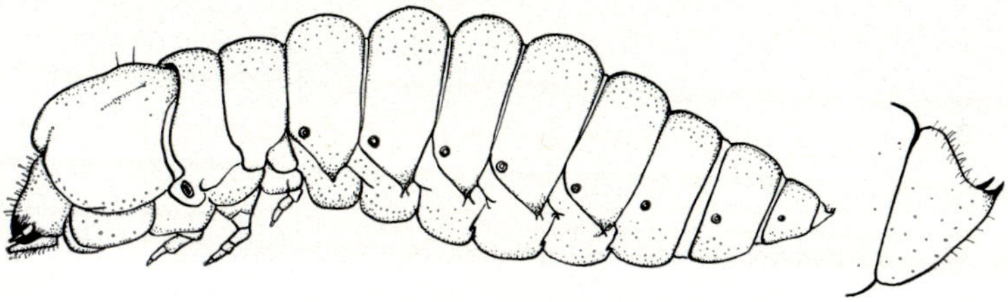

Fig. 85. Larva of *Serropalpus barbatus* Schall. Length: 21 mm, and terminal segment (enlarged).

characteristic as it tapers towards each end, more particularly towards the hinder end, and is widest at the middle. The head is globular and the antennae are three-segmented. The prothoracic segment is distinct and greater in width than the meso- and meta-thorax combined. The latter resemble the abdominal segments in appearance. The larva possesses well-developed legs. On the terminal segment of the abdomen there are two brown curved urogomphi.

The larvae bore gradually into the wood (Fig. 86), the round section gallery growing correspondingly in size. The gallery is U-shaped, curving round so that when the larva is fully grown it arrives back at the surface of the wood.

OEDEMERIDAE

Another family in the HETEROMERA series with a fairly important wood-boring beetle in Britain is the OEDEMERIDAE. The species of this widely distributed family are elongate, rather soft-bodied and usually characterized by fairly bright or metallic coloration. The British list consists of four genera containing seven species in all, but one only is of significance in decay of timber indoors—*Nacerdes melanura* (Linnaeus). This species has been given the common name of 'Wharf borer' on account of the

Fig. 86. Damage by larvae of *Serropalpus barbatus*.

habit of the larvae of boring into wharf timbers, piling, harbour and dock works, and other wood work at the high tide mark. In much of the literature this insect is included in the genus *Anoncodes* of Redtenbacher (1845) but more recently it is included in Faldermann's genus *Nacerdes* (1836) although the form *Nacerda* is sometimes employed.

Nacerdes melanura appears to have a fairly wide distribution in the coastal and estuarine areas in the North Temperate Zone but in addition it has been recorded by Walker (1936) in Gibraltar, Greece, Sardinia, Tangier, New Zealand, Syria, Shanghai, Korea, Japan, South Africa, Brazil, the Argentine, Costa Rica, California, North America and Europe where the rather special conditions for larval development are present. In North America, Balch (1937) says that this insect is recorded from much of the United States and Canada particularly along the coastal areas. They found the larvae in wharf timbers between flooding and the high-water level.

Vité states that this oedermerid is distributed by ship traffic over all coastal areas of the temperate zone. In Britain, although found in England, Scotland, and Ireland *Nacerdes melanura* is commonest in the Midlands and in the estuaries in the south and south-east, the Thames estuary and London being notorious for attacks by this species.

The adult beetles (Fig. 87) vary in length from 7 to 14 mm, and are rather soft in texture. In colour they are yellowish brown but the apices of the elytra are black. The antenna of the male has twelve segments, whilst in the female there are eleven only. The eyes, sides of the thorax, legs and ventral parts generally are blackish and the whole body is covered with dense yellow pubescence.

A very common beetle of the countryside—*Rhagonycha fulva* the 'soldier beetle'—very closely resembles the wharf borer and is mistaken for it on many occasions each season so that it is important to note the essential differences between these insects. Taking the prothorax first, in

Fig. 87. *Nacerdes melanura* (L.). Length: 12 mm.

Nacerdes melanura this is wider nearer the head, rather more convex and does not have the lateral flanges as shown in *Rhagonycha fulva*. Next, taking the elytra, these show three longitudinal lines (the two nearer the central elytral suture are more prominent) which are absent in *R. fulva*. Finally in *Nacerdes melanura* there are only four tarsal segments in each hind leg, whereas in *Rhagonycha fulva* there are five. Another insect sometimes confused with *N. melanura* is the cerambycid *Phymatodes testaceus* (L.), see p. 251.

The larva (Fig. 88) is slender, from 12 to 30 mm in length and of a greyish-white colour. There are distinct intersegmental grooves which give the larva a superficial resemblance to a string of beads. The head is large and yellowish with the mouthparts pitchy-black in colour. The pronotum is large and together with the next two thoracic and the following two abdominal segments carries dome-like protrusions covered with short spinules. The legs are moderately well developed and in addition there are two pairs of pseudopods on the third and fourth abdominal segments.

Infestations by this insect take place mainly in softwood but occasionally in oak which is damp or wet and in the process of fungal decay. Other timber species are however sometimes attacked and Walker (1936) records *Eucalyptus*. The coastal and estuarine habitats for this insect presuppose that the beetle prefers woodwork wetted with salt water. On the other hand the insect is sometimes very common in inland localities such as in the banks of the Birmingham canals and the River Avon at

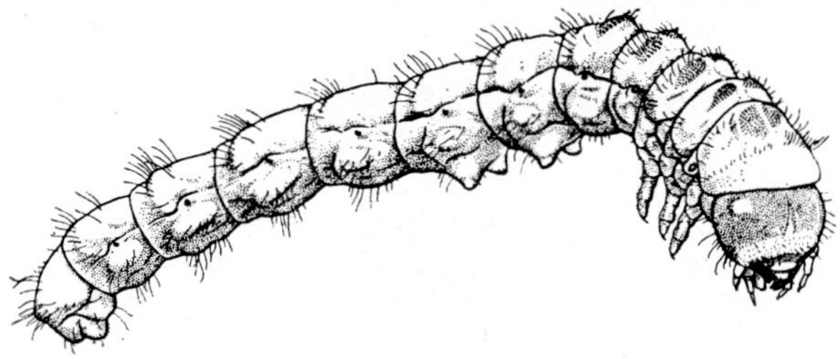

Fig. 88. Larva of *Nacerdes melanura*. Reproduced by permission of the Trustees of the British Museum.

Stratford on Avon and moreover it was very common indeed on the bombsites of London in situations where salt water was absent. It is still common in damp basements and cellars, particularly in London.

Fisher (1935) repeats the observation of Schwarz that 'Larvae have been found in the wood at the bases of telegraph poles where dogs had urinated, also in the wood of fences similarly saturated. Wood wetted with salt water or urine seems more liable to attack'. It has been found by the present author around the base of lavatory pans in properties in London.

Fisher thought that the insect was most prevalent in timber at a height 2–4 feet above the line of permanent saturation by the water. The present author once dropped a few larvae into a jar of water and found that they rapidly drowned so that timber in which the larvae are actually developing must not exceed a moisture content much above the fibre saturation point.

The larvae do not bore definite sized or shaped galleries but tend to work in rather indeterminate spaces which are plugged here and there with long torn fibres of the wood. These boring spaces are often adjacent to sound wood which confirms Fisher's statement that 'the general conclusion to be drawn . . . is that the larva of *Anoncodes melanura* is not a primary pest of sound timber but bores into decayed wood which is in a damp condition. It appreciably accelerates the destruction of timber in which fungal decay has already started.'

CURCULIONIDAE

The family series RHYNCHOPHORA contains four families, the ANTHRIBIDAE (= PLATYSTOMIDAE), BRENTHIDAE, CURCULIONIDAE and the SCOLYTIDAE. In the first three the head of the adult is usually produced into a rostrum and in almost all cases the tarsi are apparently four-jointed but rarely they are five-jointed and three-jointed. The larvae of all families feed upon plant tissues of all sorts including wood and almost all are legless, the exception being the BRENTHIDAE and some of the ANTHRIBIDAE. The BRENTHIDAE consist of narrow elongate beetles almost all of which are tropical. The larvae bore in wood but little is known of their biology except that a few species are predaceous in the adult stage on the larvae of other wood-boring insects. The ANTHRIBIDAE are also mostly tropical and usually found in fungal-decayed wood or in fungi, but they are represented in Britain by ten species. Although species of the families ANTHRIBIDAE and BRENTHIDAE are sometimes accidentally introduced with timber they are of no economic importance in Britain as pests of timber in buildings. The SCOLYTIDAE, however, are of great importance to the forester not only in Europe but elsewhere and some species are dealt with in a later chapter.

The family CURCULIONIDAE, is remarkable in that it is certainly the largest in number of species in the whole of the animal kingdom. Sharp estimated that the minimum number of species was 200,000 of which only something just over a tenth have been described. They are known as 'weevils' and many are serious pests of crops and stored food products. A number are important forestry pests but in Britain four species only occur in woodwork indoors in numbers that constitute an economic problem. These are: *Pentharthrum huttoni*, *Euophryum confine*, *E. rufum* and *Caulotrupis aeneopiceus*.

Key to the small reddish-brown weevils found indoors in Great Britain

1. Elytra only slightly longer than pronotum (1·2 to 1·5:1). Tips of elytra not completely covering abdomen. Length of pronotum plus elytra 3·0–3·3 mm . . . *Sitophilus* spp. (Grain weevils)

– Elytra distinctly longer than pronotum (1·7 to 2·0:1). Tips of elytra cover abdomen. Length of pronotum plus elytra, 2·2–3·6 mm . . Wood-boring weevils, 2

2. Elytra visibly wider than pronotum. No constriction in anterior quarter of pronotum. Eyes not protruding. Dark brown to black . *Caulotrupis aeneopiceus* (Boh.)

– Elytra not wider than pronotum. Distinct constriction in anterior quarter of pronotum. Eyes bulging. Reddish-brown 3

3. Metasternum finely and densely punctured, and with a narrow groove in midline (Fig. 94). Tips

of elytra without explanate sides (Fig. 90). Length of pronotum plus elytra, 3·0–3·6 mm . *Pentarthrum huttoni* (Woll.)

– Metasternum sparsely covered with coarse punctures, and without a median groove (Fig. 94). Tips of elytra with distinct explanate sides (Fig. 89) 4

4\. Pronotum with a median, longitudinal, unpunctured line, often in centre only and quite short; punctures relatively coarse (Fig. 96) . *Euophryum confine* (Broun)

– Pronotum completely covered with relatively fine punctures; no trace of an unpunctured median line (Fig. 96) . . . *Euophryum rufum* (Broun)

Notes on the species

Pentarthrum huttoni Wollaston 1854.

This insect, of wide distribution in England and Ireland, is a weevil of typical appearance with the head prolonged into a snout-like projection (Fig. 90). It is brownish or pitchy-black in colour and cylindrical. In length

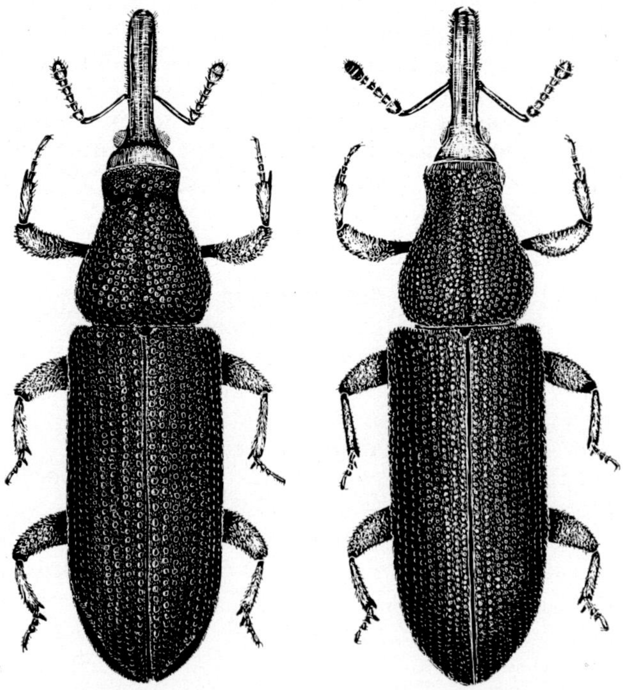

Fig. 89 (left). *Euophrum confine* (Broun). Length: 3·8 mm.
Fig. 90. (right). *Pentarthrum huttoni* Woll. Length: 4·0 mm.

it varies from about 3 to 5 mm. Adults and larvae, both of which carry out boring, are found within softwoods and hardwoods including plywoods of birch, alder and oak, in ill-ventilated situations where fungal decay has commenced. Often the species of fungus concerned is *Coniophora cerebella*, the cellar fungus. Infestations by this insect can be differentiated from those of *Anobium punctatum* by the frass (Fig. 91) which is finer in

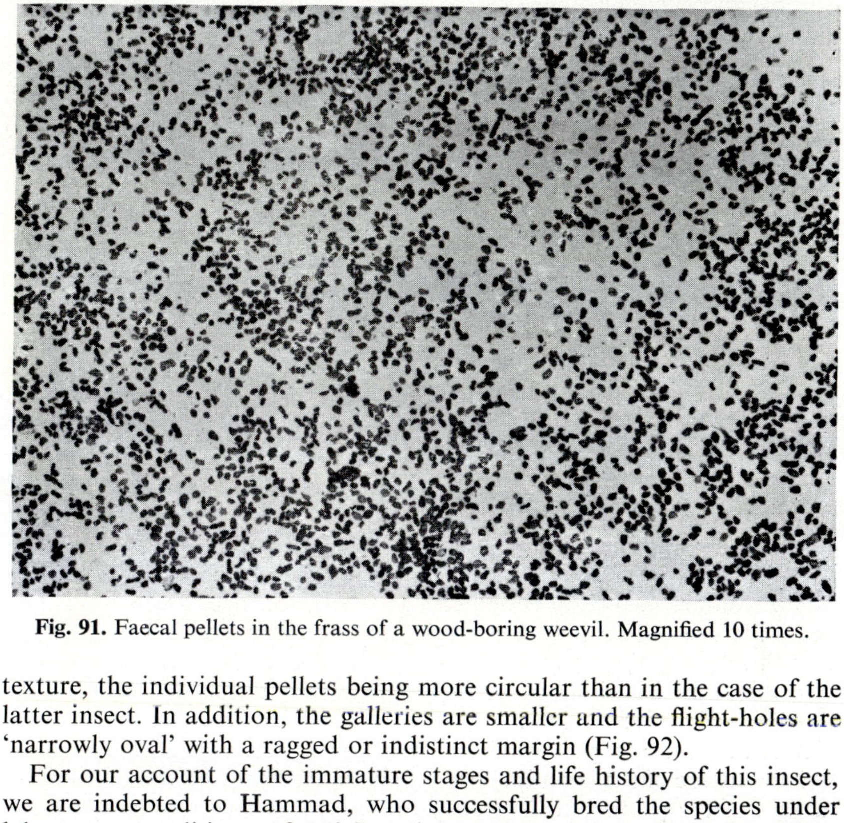

Fig. 91. Faecal pellets in the frass of a wood-boring weevil. Magnified 10 times.

texture, the individual pellets being more circular than in the case of the latter insect. In addition, the galleries are smaller and the flight-holes are 'narrowly oval' with a ragged or indistinct margin (Fig. 92).

For our account of the immature stages and life history of this insect, we are indebted to Hammad, who successfully bred the species under laboratory conditions of 25° C and 95–100 per cent relative humidity. Egg-laying commences about four days after mating. The female bites out holes in the wood with her mandibles and lays single eggs in such sites or in cracks and crevices and then seals them with a semi-transparent whitish layer of excretion from the ovipositor. Hammad found that two females laid fifty eggs in 80 days. The eggs, which are shining white, somewhat flexible and flattened at one end, hatch after 16 days. The young larvae first consume the whole of the shell, which takes the first twenty-four hours. The length of the first instar larva is about 0·5 mm with a diameter of about 0·24 mm. The head is circular in outline and very pale in colour. The young larva makes a hole in the exterior of the wood and then bores

Fig. 92. Damage to timber by a wood-boring weevil, showing irregular flight-holes and surface damage.

inwards at an angle of about 45° and shortly afterwards constructs its tunnel parallel to the surface of the wood. Five instars take place and the larva pupates in from six to eight months after hatching. The larva constructs a pupal cell which becomes lined with fungal hyphae. The body of the mature larva is about 3·3 mm in length and about 1·0 mm wide (not 0·1 mm as printed in Hammad's account). It is nearly white in colour and widest at the middle of the third thoracic segment (Fig. 93). When viewed laterally the body is strongly convex dorsally, much less convex ventrally and the whole cuticle is covered with extremely fine but conspicuous sclerotized spines. A pronotal sclerite is present which covers almost the entire tergite and there is also a prosternal sclerite. A characteristic feature of the larva is a cervical sclerotized ring around the anterior margin of the prothorax. Normally the head is deeply retracted into the thorax but about two days before pupation the head becomes evaginated. The pupal stage lasts about 16 days and the pupa is about 3·5 mm long and 1·0 mm wide (not 0·1 mm as printed in Hammad's account). It is white and the ninth abdominal segment is furnished with a pair of strong spines which emerge from fleshy processes.

In the adult stage the females may be separated from the males by the rostrum which is thinner and longer in the former sex. The weevils measure 3·0–5·0 mm in length and construct irregular and ragged flight-holes at an angle of approximately 45° to the surface. The adult weevils live for about 16 months after emergence. The habit of feigning death has been well developed and Hammad has remarked that although the males

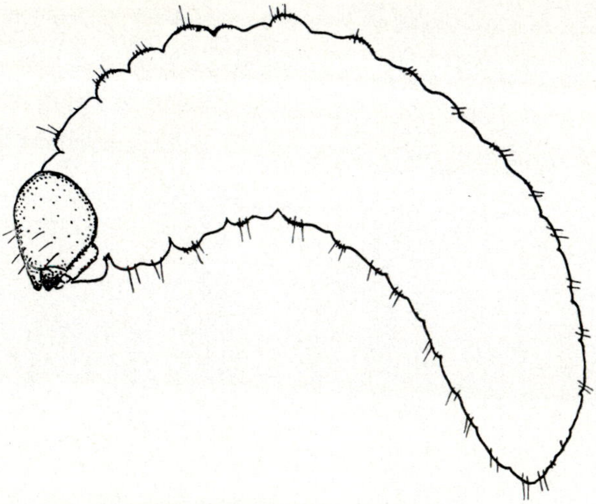

Fig. 93. Larva of a wood-boring weevil.

may be easily removed from the wood surface, the females resist removal by clasping the surface with the tibial spines. The extent to which the adults can copulate and maintain a series of generations within the wood—common to a number of rotting wood COSSONINAE—is not known.

Euophryum confine (Broun 1881).

The history of this insect in Britain is nothing short of remarkable. In the first place it is indigenous to New Zealand where it was first described by Broun in 1881 under the name of *Pentarthrum confinis*. Then in 1908 he placed it in the genus *Euophryum* as *confinum* later corrected by Grensted to *confine* in 1944. It does not appear to be a pest in its native country.

It was first found in Britain in 1937 and in the ten following years it was recorded a number of times in London and the home counties. Allen found it in a dead willow tree in 1940 and it was found in oak and willow in Epping Forest by Buck. In 1946 it was found by Hinton and McKenny Hughes in Glasgow. The latter authors considered that by 1946 the species was well established in many London houses. Hickin recorded it in his own home in Surrey in 1949 and from Ireland in 1952. It is now widespread throughout England and is extremely common in London and its importance in this regard is discussed in a later chapter.

The beetle (Fig. 89) is typically weevil-like with a rather broad snout. There is a deep constriction immediately behind the eyes and the tips of the elytra are dilated, which distinguish it from *P. huttoni*. In addition, Linscott (1967) has recorded that the sculpturing of the metasternum is finer in *Pentarthrum* than in *Euophryum*, approximately in the ratio of 16:11 (Fig. 94). The life history of *E. confine* does not appear to have been studied.

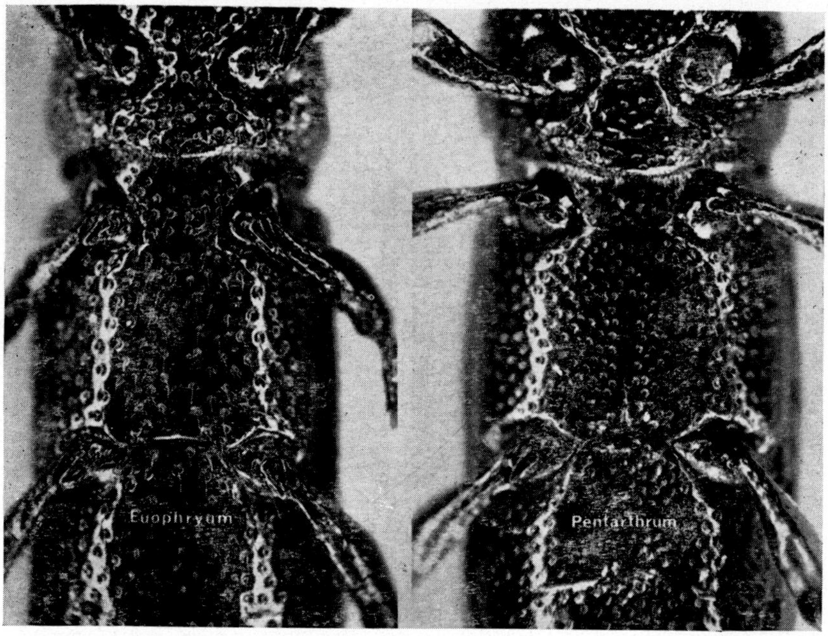

Fig. 94. Metasternites of *Euophryum confine* and *Pentarthrum huttoni*.

Euophryum rufum (Broun).

This is another New Zealand insect (Fig. 95), which was first recorded in Britain in 1934. It is very closely related to *E. confine*, and has no doubt been confused with that species on many occasions. The difference is seen in the pattern of punctures on the pronotum (Fig. 96). That of *E. rufum* is confusedly punctured over the entire area, whereas in *E. confine*, there is a narrow central strip which is without punctures. This smooth area may run the entire length of the pronotum (as in Fig. 96), or it may be confined to a small area in the centre of the pronotum, where it is often difficult to see.

Linscott (1967) suggested that *E. rufum* is slightly smaller than *E. confine*. To check this, 138 weevils from various sources were measured, using the length of the pronotum plus elytra. (The length of the head varies with the sexes and was not included.) Seventy-four specimens of *E. rufum* were measured; the average length was 2·7 mm, and the range was 2·2–3·3 mm. The average length of 64 *E. confine* was 2·9 mm, the range being 2·5–3·4 mm (Edwards, unpublished). Thus, there is so much overlap in size that this feature cannot be used in identification. It appears that further work is required to establish beyond doubt whether these two insects are, in fact, different species.

The biology of *E. rufum* has not been studied, but it is assumed to be

Fig. 95. *Euophryum rufum* (Broun).

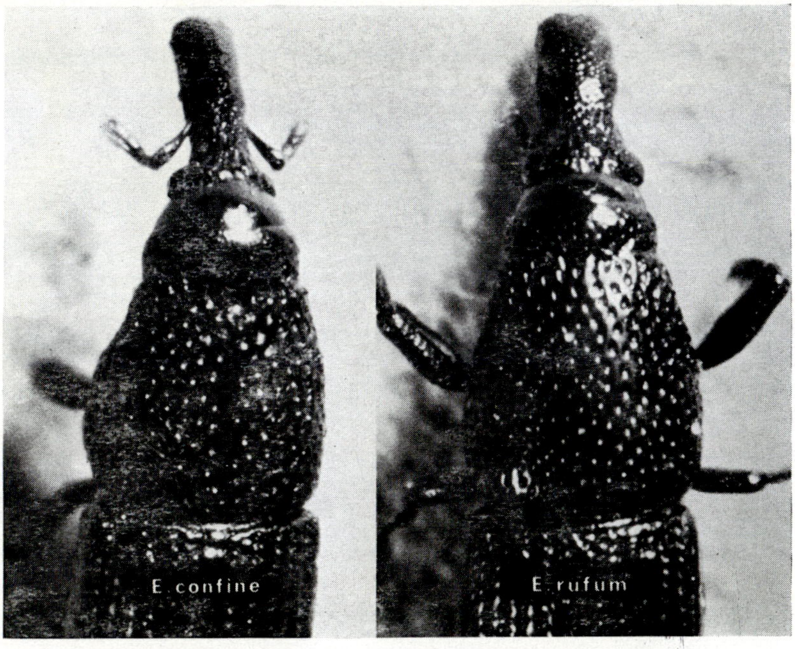

Fig. 96. Pronotum of *Euophryum confine* and *E. rufum*.

similar to that of *E. confine* and *P. huttoni*. The weevil is widespread in Britain (Buck 1948).

Caulotrupis aeneopiceus (Boheman 1844).

This weevil is less frequently encountered in buildings than the previous three species. It occurs in the same type of damp, rotten wood as *Pentarthrum* and *Euophryum*. One instance has been noted of all four wood-boring weevils coexisting in hardboard wall panelling in a damp London basement (Edwards, unpublished). Becker has recorded this species attacking wine casks and joists in cellars.

Those weevils that are forestry pests are usually twig, leaf or blossom destroyers but a few (*Pissodes*, *Magdalis*, etc.) are cambium feeders, usually only superficially grooving the sapwood. These are not described in this present work. Another group of weevils however are associated with dead or fungally decayed parts of the branches or trunk of a standing tree or fallen or buried wood and very occasionally these species have some slight economic significance in Britain.

Cossonus parallelepipedus (Herbst).

Found in decaying elm, poplar, oak, fir, pine and sometimes in buried timber. C. *linearis* (F.) is found in dead poplar and willow.

Codiosoma spadix Herbst.

Occasionally found in pine wood in buildings in Europe but, although a British insect, is seldom found indoors in Britain.

Eremotes

Three species of *Eremotes* are found in Europe damaging damp wood, but in Britain only *E. ater* (L.) is known. It is very rare in buildings.

Rhyncolus culinaris Germ.

This species is known as the pit-prop beetle in central Europe where it often damages wet wood in mines and also in houses. It is not found in Britain.

Rhyncolus truncorum Germ.

Occasionally found damaging wet wood indoors (Linscott 1962). There are two other British species, *R. gracilis* Rosenhauer and *R. lignarius* (Marsham), about which little appears to be known of their occurrence in buildings, but the latter has been recorded in elm and dead ivy.

Cryptorhynchus (= *Cryptorhynchidius*) *lapathi* (L.).

This British species is widely distributed in Europe, Asia, and North America. It is found attacking the trunks of alder, poplar, willow and birch

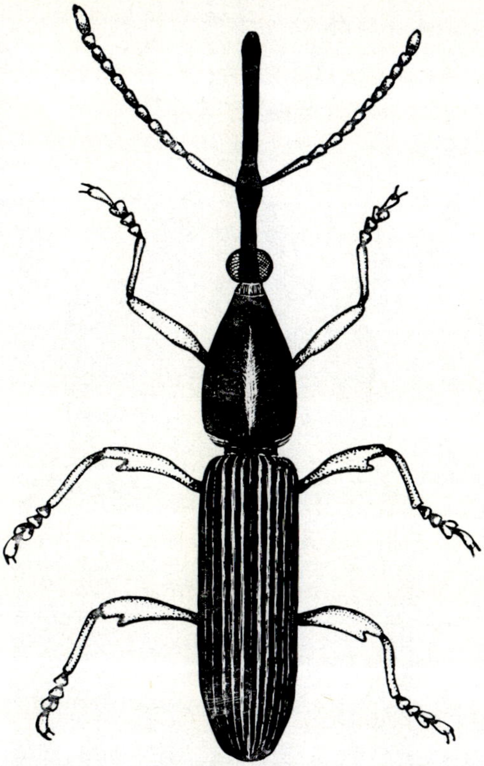

Fig. 97. *Pseudoceocephalus picipes* Ol. Length: 14·2 mm.

sometimes even when the trees are apparently healthy. It is seldom found indoors.

Pseudoceocephalus picipes (Olivier) (Fig. 97).

This long, thin weevil has been introduced to Britain from Mozambique in a batch of chamfuta timber.

Sipalinus squalidus Kolbe (Fig. 98).

Imported with the same batch of chamfuta from Mozambique.

LUCANIDAE

Although the enumeration and description of insects found in wood in an advanced state of fungal decay, which are not likely to be found indoors, is beyond the scope of the present work, it is of interest to note that there is one record of an infestation of a wooden beam and rafter in a old house by the beetle *Dorcus parallelipipedus* (L.).

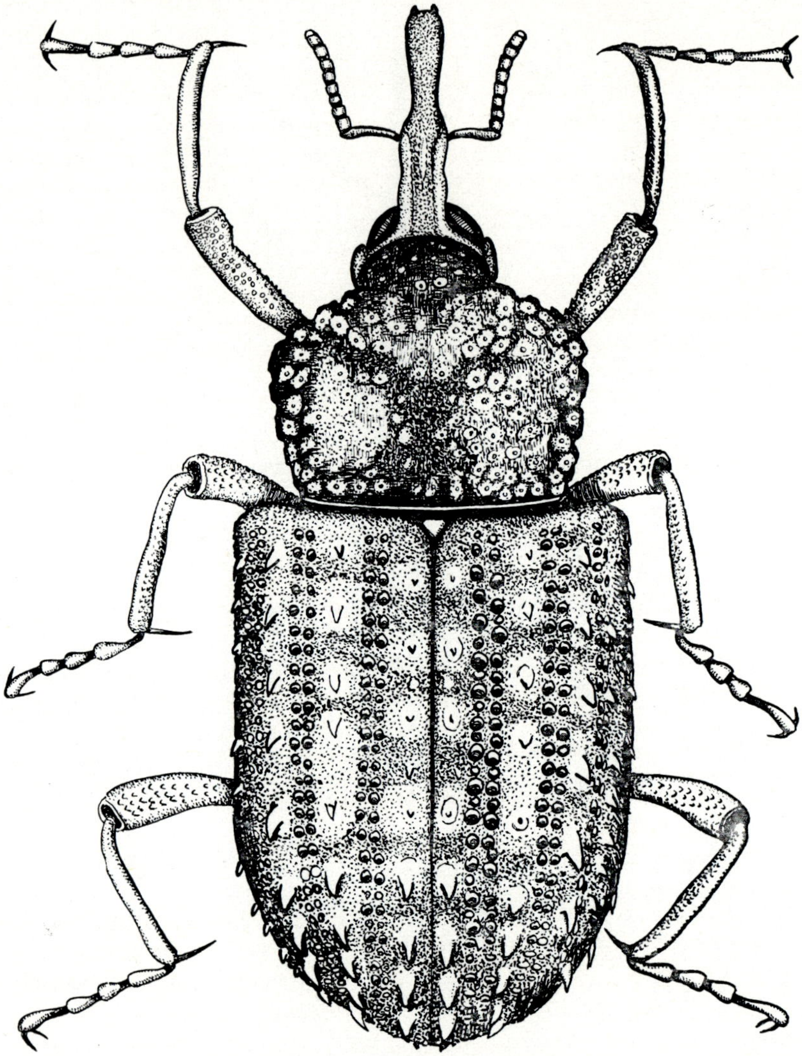

Fig. 98. *Sipalinus squalidus* Kolbe. Length: 25 mm.

REFERENCES

BUPRESTIDAE

ANDERSON, R. F. 1960. *Forest and Shade Tree Entomology*. Wiley. New York.

BEAL, J. A. 1932. The Turpentine Borer, *Buprestis apricans* Herbst. *U.S. Dept. Agric. Cir*. **226.**

BEESON, C. F. C. 1941. *The Ecology and Control of the Forest Insects of India*, pp. 110–19. Dehra Dun.

BURKE, H. E. 1917. *J. econ. Ent.* **10**: 406–7.

DIBB, J. R. 1948. *Field Book of Beetles*. Brown. Hull.

GRAHAM, S. A. 1962. *Forest Entomology*. McGraw-Hill. New York.

SHAW, M. W. 1961. The Golden Buprestid *Buprestis aurulenta* L. (COL. BUPRESTIDAE) in Britain. *Ent. mon. Mag.* **97**: 97–8.

VITÉ, J. P. 1952. *Die Holzzerstörenden Insekten Mitteleuropas*, pp. 53–4. Musterschmidt. Göttingen.

LYMEXYLIDAE

KURIR, A. 1972. Zweiter Beitrag zur Bionomie des Sägehörnigen Bohrkäfers. *Holzforschung u Holzverwertung* **24**: 127–35.

RICHTER, G. 1933. Beitrag zur Biologie von *Hylecoetus dermestoides* L. *Mitt. Forstwirtsch. Forstwiss.* **4**: 545–60.

SAALAS, U. 1923. Die Fichtenkäfer Finnlands, II. *Ann. Akad. Sci. Fennicae* **22**: 169–72.

THOMPSEN, M. 1949. Contributions to the Biology of *Xyloterus domesticus* L. and *Hylecoetus dermestoides* L., two Wood-boring COLEOPTERA. *Proc. VIII Intern. Kongr. Entomol.* 1948. Stockholm.

VESEY-FITZGERALD. D. 1941. *Melittomma insulare* Fairm. (COL. LYMEXYLIDAE.) A serious pest of coconut in the Seychelles. *Bull. ent. Res.* **31**: 383–402.

VITÉ, J. P. 1953. *Die Holzzerstörenden Insekten Mitteleuropas*. Musterschmidt. Göttingen.

MELANDRYIDAE

BLETCHLY, J. D. 1955. A little-known Borer in Imported Softwood: *Serropalpus barbatus* Schall. (COLEOPTERA, MELANDRYIDAE). Review of Literature and Recent Occurrences in Great Britain. *Forestry* **28**: 67–72. (This paper contains 19 references.)

OEDEMERIDAE

BALCH, R. E. 1937. Notes on the Wharf Borer *Nacerda melanura* Linn. *Canadian Ent.* **69** (1): 1–5.

BLAIR, K. G. 1936. A further note on *Anoncodes melanura* L. *Ent. mon. Mag.* **72**: 17–18.

BRITTON, E. B. 1961. *Domestic Wood-boring beetles*. Brit. Mus. (N.H.) Economic series No. 11A, p. 29.

FISHER, R. C. 1936. The habitat of *Anoncodes melanura*. L. *Ent. mon. Mag.* **72**: 41–2.

HICKIN, N. E. 1954. *Woodworm*. Simpkin Marshal. London.

LAING, F. 1936. *Anoncodes melanura* L. as a Destructive Insect. *Ent. mon. Mag.* **72**: 41–2.

WALKER, J. J. 1936. Supplementary notes on *Anoncodes melanura* L. *Ent. mon. Mag.* **72**: 43.

WOOD-BORING WEEVILS

ALLEN, A. A. 1942. *Pentarthrum huttoni* Woll. (COL. CURCULIONIDAE) in South-East Herts. *Ent. mon. Mag.* **78**: 117.

ALLEN, A. A. 1944. A new British species of COSSONINI (COL. CURCULIONIDAE) recorded in error as *Pentarthrum huttoni* Woll. *Ent. mon. Mag.* **80**: 120.

BUCK, F. D. 1947. *Euophryum confine* Broun. (COL. CURCULIONIDAE) in Epping Forest. *Ent. mon. Mag.* **83**: 131.

BUCK, F. D. 1948. *Pentarthrum huttoni* Woll. and some imported Cossoninae. *Ent. mon. Mag.* **84**: 152–4.

GRENSTED, L. W. 1944. Editorial note on G. V. Hudson's note. *Ent. mon. Mag.* **80**: 292.

HAMMAD, S. M. 1955. The Immature Stages of *Pentarthrum huttoni* Woll. (COLEOPTERA: CURCULIONIDAE) *Proc. R. ent. Soc. Lond.* (A) **30**: 33–9.

HICKIN, N. E. 1949. A New Wood-boring Weevil *Euophryum confine* Broun in Surrey. *Ent. mon. Mag.* **85**: 186.

HICKIN, N. E. 1952. *Euophryum confine* Broun the Wood-boring Weevil in Ireland. *Ent. mon. Mag.* **88**: 157.

HINTON, H. E. and MCKENNY-HUGHES, A. W. 1947. A New Pest in Houses. *Mon. Bull. Minist. Hlth.* 173, Oct. 1947.

HUDSON, G. V. 1944. *Euophryum confine* Broun not Brown. (COL. CURCULIONIDAE). *Ent. mon. Mag.* **80**: 292.

KEMP, S. 1944. *Pentarthrum huttoni* Woll. (COL. CURCULIONIDAE) in South Devon. *Ent. mon. Mag.* **80**: 177.

LINSCOTT, D. 1967. Woodboring Weevils in Great Britain. *Proc. 2nd. Brit. Pest Control Conf.* Notts. April 1967.

CHAPTER SEVEN

The Longhorn Beetles

The family CERAMBYCIDAE, including *Hylotrupes bajulus*

This account of the longhorn beetles found in the British Isles is written from the point of view of the student of wood decay. Because of this, many species introduced in timber from overseas are included, and a large number of illustrations is given to aid in identification. Every species described in the text is illustrated by a picture of either the whole insect or characteristic parts of it, so that the adult stage can be identified by a student who is not necessarily a trained entomologist. Descriptions of the larvae however are not included because identification would cause difficulties to the non-entomologist and because of the excellent detailed monographs by Duffy.

General form

On the whole, the CERAMBYCIDAE are rather more than medium size, but some species such as *Macrotoma heros* and *Titanus giganteus*, are amongst the largest known insects, with a body size considerably larger than that of a small mammal. On the other hand, some species are very small and a few are even minute. The form of the beetle is fairly well defined, although there is wide variation. The body is usually elongate with the elytra wider than the pronotum (Figs. 99 and 100). The most characteristic feature of the CERAMBYCIDAE, however, shared by all but a few species, is the extreme length of antennae, which are usually as long as, or longer than, the body. In *Batocera kibleri* Schwartz they reach a length of 23 cm. The eyes are large and frequently bow-shaped. The invagination is produced by the large tubercle from which the antenna arises (Fig. 101). The antennae are usually filiform, but sometimes, as in the British species *Prionus coriarius* Mull., they are serrate. They are occasionally also pectinate and sometimes they are ornamented with tufts of hair. The legs of CERAMBYCIDAE are moderately large and in a few species the front legs are produced to an exaggerated degree, both the femur and the tibia being as long as, or longer than, the body. In every case there are five tarsal segments, although usually the fourth segment is very much reduced and only four segments are visible until the insect is carefully examined. The third segment has a very definite bilobed appearance.

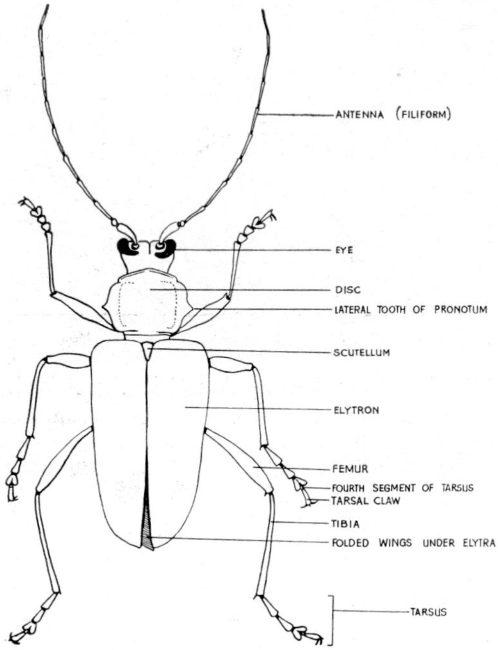

Fig. 99. Typical cerambycid beetle, showing the main parts. From above.

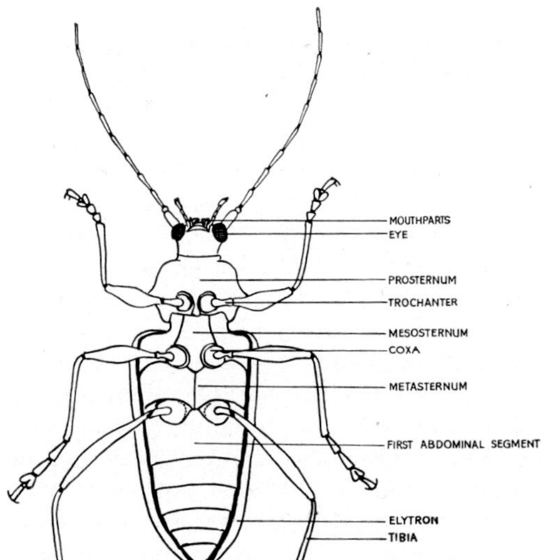

Fig. 100. Typical cerambycid beetle, showing the main parts. From below.

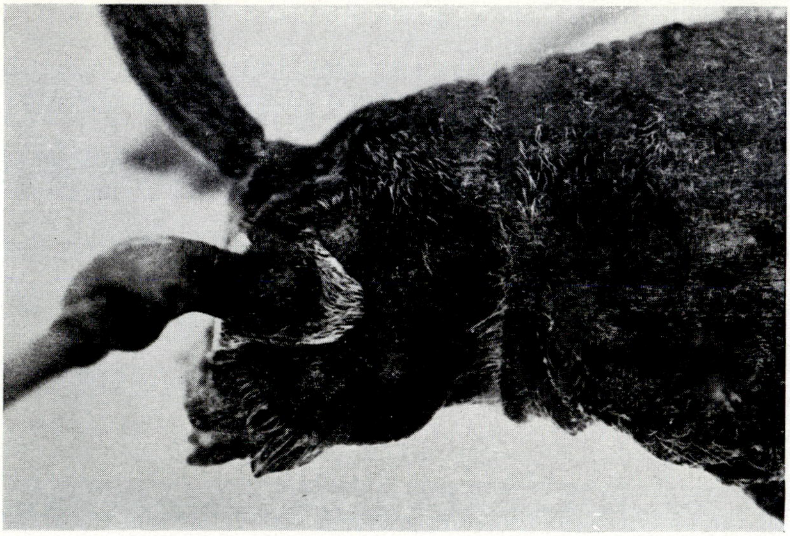

Fig. 101. Eye and antenna of *Eburia quadrigeminata* Say.

World distribution

The CERAMBYCIDAE is a family of beetles of about 20,000 named species. They are found throughout the world, wherever trees or bushes grow and indeed wherever timber is transported or used.

The number of British indigenous species is usually given as rather more than 60. Duffy lists 64, but gives a further 46 species found as larvae in imported timber. In tropical areas the number of species is very much greater than in temperate regions. Beeson states that over 1,200 are known in the Indian region and of these 200 were described as new from 1914 to 1941. No doubt this list is now much longer and will grow for many years. Duffy describes the larvae of, or makes reference to, 603 species from the African region, but remarks that many thousands of African species are known. In America also, there is a multiplicity of species, and, more especially in South America, some are of gigantic proportions and often they are of remarkable form and colour.

Food plants

The great majority of the larvae of CERAMBYCIDAE infest some part of the woody tissue of woody plants. There are exceptions, however, as a few species are found consuming the roots, and others the stems, of herbaceous plants. The larvae of a few cerambycids are found only in the pith of young shoots. Beeson has tabulated the situations inhabited by the larvae, and in addition to those given above, he lists bark, under bark and grooving the sapwood (obviously cambium feeders), in sapwood, in heartwood (he differentiates between living and dead heartwood, but actually all heartwood can be considered as dead), in cones, in soil, and partly in liane and partly in tree.

In addition, the extent to which fungal decay occurs in the woody tissue largely determines which species of CERAMBYCIDAE will attack it. *Hylotrupes bajulus* will attack the dead sapwood of *Pinus sylvestris* when fungal decay is absent, but the dead stumps, heavily fungal infested, are attacked by *Rhagium* and some other genera. Many CERAMBYCIDAE attack only one species of plant, but some are polyphagous, attacking a large number of plants. Beeson gives the interesting information that the tree *Shorea robusta* (a member of the genus from which the timbers meranti and seraya are converted), is attacked by no fewer than 38 cerambycids, whereas the beetle *Stromatium barbatum* is known to attack 311 tree species.

Economic importance

In the broad sense, of course, all insects are of economic importance. All play some part in establishing a situation which can be construed as the level of productivity when viewed from the economic standpoint of man. Amongst the British cerambycid beetles, however, one stands out as being fairly important as a pest of softwood building timber in a localized area. This is *Hylotrupes bajulus*, found in N.W. Surrey and probably not established elsewhere, although it turns up from time to time in other districts including Ireland. On the other hand, this same insect is a major wood-destroying pest in many countries overseas, including the whole of Europe and South Africa. In recent years it appears to have increased its range considerably so that although of minor significance in the United Kingdom, it must be assessed as a pest from a much broader point of view.

A number of other species damage sound wood, including *Clytus arietis*, which, although not normally infesting furniture, will sometimes complete its development and emerge from wood which has been converted and manufactured. The same is true of *Eburia quadrigeminata*; in this case the timber is infested whilst growing in North America, but is cut and imported into the United Kingdom. This species is also an excellent example of 'delayed development'. Although in its natural habitat the life cycle is short—probably about two years—when the tree is felled and the timber dried, the completion of the life cycle can be accomplished, but this may take as long as 30 years.

On the other hand, many cerambycids can be considered as beneficial to forestry, as Duffy has pointed out, because the larvae of these species tunnel into fungal-decayed wood and thus break up the stumps and rotting trunks and bring this unproductive wood substance back into the earth, where it can provide the nutrients and the water-holding properties for a new generation of forest trees, or for some other biotic purpose.

Classification

The CERAMBYCIDAE or LONGICORNIA are included together with the

LEGEND OF COLOUR PLATE

A. *Plagionotus arcuatus* (L.)

B. *Hylotrupes bajulus* (L.)

C. *Saperda carcharias* (L.)

D. *Anaglyptus mysticus* (L.)

E. *Aromia moschata* (L.)

F. *Purpuricenus hochleri* (L.)

G. *Saperda scalaris* (L.)

H. *Rosalia alpina* (L.)

I. *Gracilia minuta* (Fabricius)

Length of line = 12 mm in each case except 'I' (*Gracilia minuta*) in which it is 6 mm.

NOTE: *Purpuricenus hochleri* and *Rosalia alpina* are rare immigrants.

D

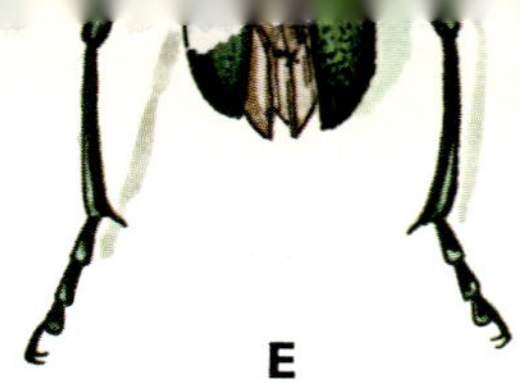
E

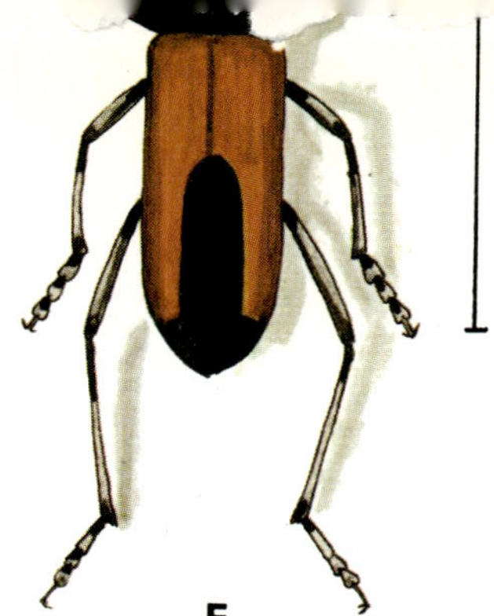
F

G

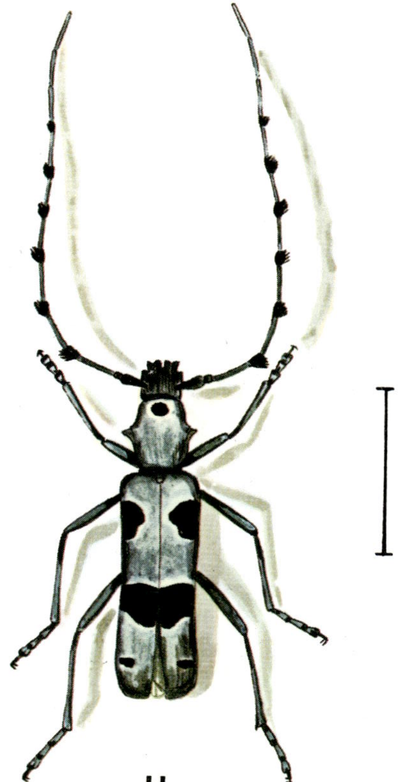
H

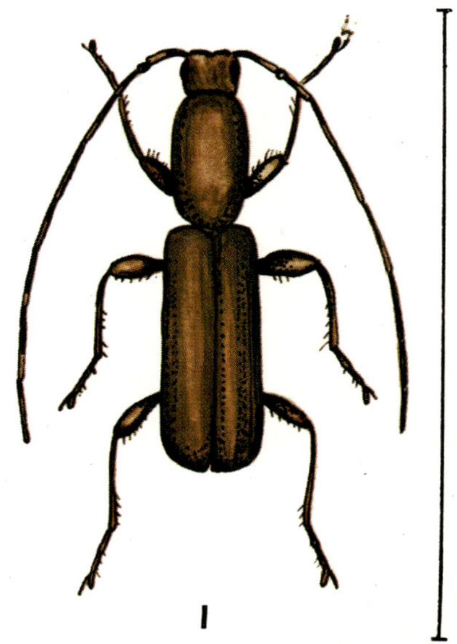
I

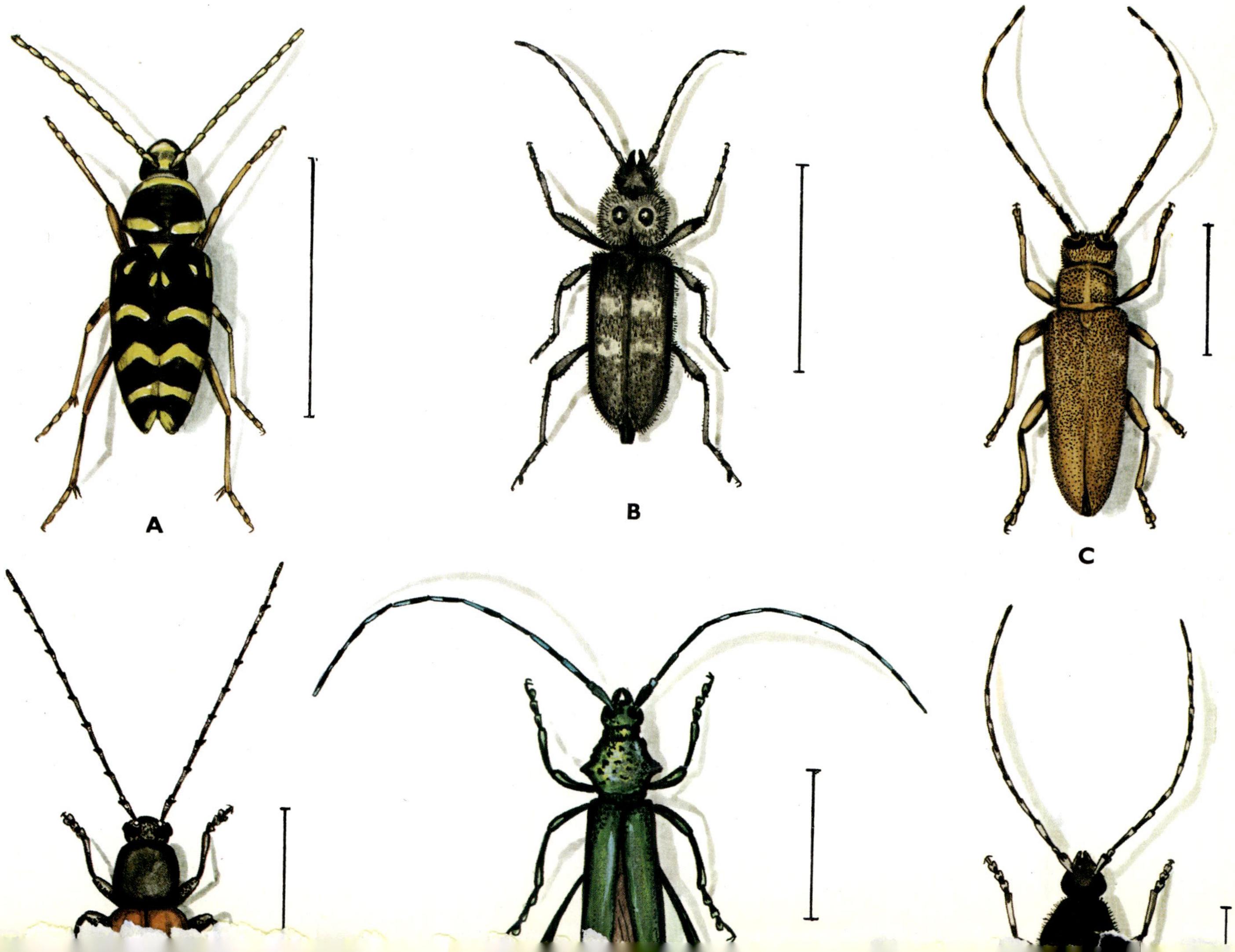

BRUCHIDAE and CHRYSOMELIDAE in the superfamily CHRYSOMELOIDEA. All these insects are xylophagous or phytophagous and possess six cryptonephric tubules. The larvae are characterized by their leg development, the mandibles lacking a mola and the antennae having three segments. The superfamily is grouped within the suborder POLYPHAGA and a shortened version of Crowson's key to superfamilies is given by Imms.

CERAMBYCIDAE which may occur in Britain are divided into six subfamilies and a key for their separation is given below. This key mostly follows that of Duffy which in turn was based on those of Fowler (1889) and Reitter (1912), but in order to make the key intelligible to students of wood decay who are not necessarily trained entomologists, some of it has been rewritten. Many entomological terms are often confusing so that in the descriptions in the keys which follow, only entomological descriptive terms are used which are given by Torre-Bueno (1950), or which have the meaning given in the *Shorter Oxford English Dictionary*.

1. Fourth tarsal segment distinct, not being covered wholly or in part by the third tarsal segment which is of normal size (Fig. 102). Chestnut colour, bright and shining. Size 18 mm — PARANDRINAE (p. 194)
– Fourth tarsal segment indistinct, covered wholly or in part by the third tarsal segment, which is usually enlarged or strongly lobed 2
2. Sides of the prothorax have a distinct margin and are armed with three or more teeth. Front coxae transverse. Prosternum produced as a definite blunt process behind the anterior coxae. Length at least 23 mm . . . . PRIONINAE (p. 195)
– Sides of the prothorax without a distinct margin, lateral teeth usually absent, but sometimes one pair only present. Front coxae round, oval or conical. Prosternum not produced as a definite blunt process behind the anterior coxae 3
3. Front of head vertical to longitudinal axis of body, or bent strongly inwards. Anterior tibiae obliquely grooved on their inner side. Apical segments of labial palpi tapering or pointed . LAMIINAE (p. 262)
– Front of head obliquely or nearly vertically inclined to the longitudinal axis of body. Anterior tibiae not grooved on their inner side. Apical segments of labial palpi enlarged 4
4. Anterior coxae distinctly conical. Sides of head

strongly constricted behind the temples which are definitely protuberant. Elytra usually taper posteriorly . . . . . . LEPTURINAE (p. 198)

– Anterior coxae round or oval, but never conical. Sides of head never strongly constricted behind the temples which are hardly protuberant. Elytra parallel or tapering 5

5. Femora distinctly club-shaped or stalked apically, or prothorax with lateral tubercles, or elytra terminating in a sharp point. Hind femora usually extending beyond the apices of elytra CERAMBYCINAE (p. 224)

– Femora not or scarcely club-shaped apically. Prothorax without lateral tubercles. Elytra never terminating in a sharp point. Hind femora not extending beyond the apices of elytra. Body entirely brown or black. From CONIFERAE only ASEMINAE (p. 219)

PARANDRINAE

The PARANDRINAE is the subfamily usually placed first in the CERAMBYCIDAE by taxonomists. Its sole claim, however, to be included in this book arises from the rare importation of the species *Parandra brunnea*

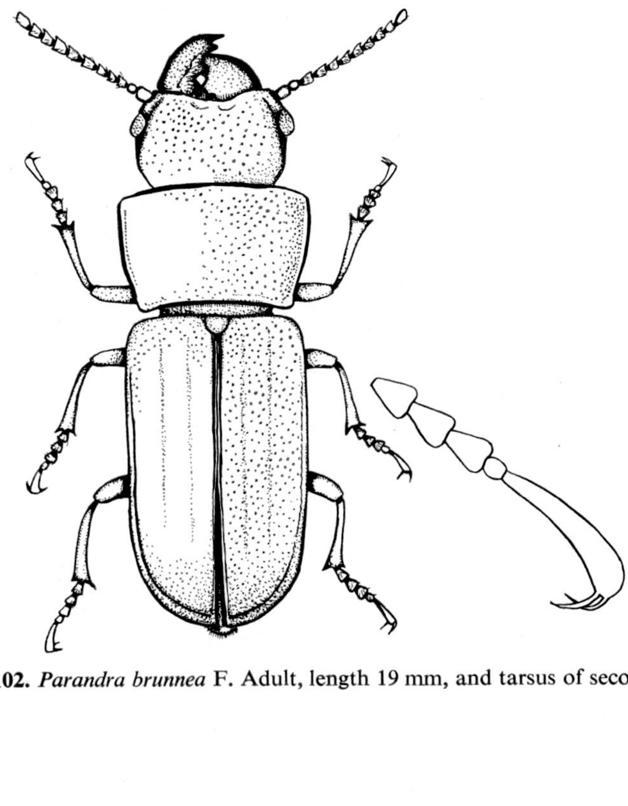

Fig. 102. *Parandra brunnea* F. Adult, length 19 mm, and tarsus of second leg.

Fabricius from North America and the even rarer importation of *P. polita* Say from the same region. These species are most un-cerambycid-like in appearance (Fig. 102). They are shining and glossy, chestnut coloured, and with antennae that do not reach as far as the elytra. The lateral margins of the prothorax are devoid of spines and in the male the mandibles are large and sickle-shaped.

Parandra brunnea F. damages nearly all timber species in the eastern states of North America. Fruit trees appear to be specially vulnerable to attack, which usually takes place towards the base of the tree. The larval galleries are characteristically entirely packed with the fine, powdery frass. It has been concluded that this species probably causes more severe damage (in North America, of course) to a greater variety of trees than any other North American cerambycid.

PRIONINAE

The three genera and four species in the PRIONINAE found in Britain can be differentiated as follows:

1. Lateral margins of prothorax approximately parallel with a few blunt teeth and with the anterior angles acute. Large jaws. Length about 45 mm. *African* . . . . *Mallodon downesi* Hope
- Lateral margins of prothorax tapering, or if approximately parallel, then with three pairs of lateral spines 2
2. Prothorax with three pairs of lateral spines, the middle pair the longest. Third segment of antenna less than twice as long as first. Apices of elytra slightly dentate at sutural angle. Length 23–40 mm . . . . *Prionus coriarius* (Linnaeus)
- Prothorax with numerous spines distributed along whole length of lateral margin. Third segment of antenna more than twice as long as first. Apices of elytra terminate in a point at sutural angle 3
3. Lateral spines on prothorax shortest in middle. Length 28–60 mm. *North American* . *Ergates spiculatus* Leconte
- Lateral spines on prothorax longest at middle with one pair of much larger spines just behind middle. Length 25–45 mm. *European* . *Ergates faber* (Linnaeus)

Notes on the species

Mallodon downesi Hope

This is a large species occasionally found in timber from Africa where it infests *Ceiba, Eriodendron, Hevea* (Rubber Tree), *Coffea* and *Theobroma.* Lepesme states that it attacks only decaying wood, but Duffy quotes Rae as saying that only sound wood is attacked. Stumps of iron-wood (*Lophita alata*) are also said to be attacked.

An illustration of this species, made from a specimen from the West India Dock, London, which accompanied a cargo of makore (*Mimusops heckelii* Hutch. and Dalz.) from West Africa appears in *Entomologists' Gazette* **10**:17.

Prionus coriarius (L.) (Fig. 103)

This is the only truly indigenous British species of PRIONINAE. It is found from July to October in the southern half of England and is pitch-black, sometimes tinged with reddish-brown. In size it varies from 25–40 mm and the larvae are usually found in the roots (Fig. 104) of a wide variety of broad-leaved and coniferous trees (which is unusual for a cerambycid). Old parkland trees seem to be preferred. The larva is up to 76 mm in length and lives for a minimum of three years before completing its development. Duffy has given a full account of the immature stages and biology. It has been recorded as damaging telegraph poles and the only

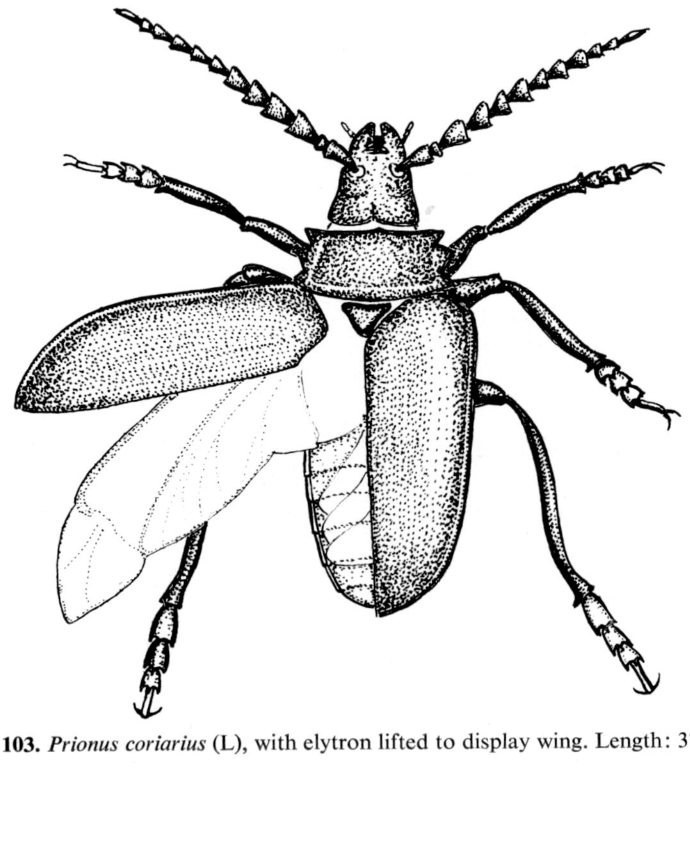

Fig. 103. *Prionus coriarius* (L), with elytron lifted to display wing. Length: 37 mm.

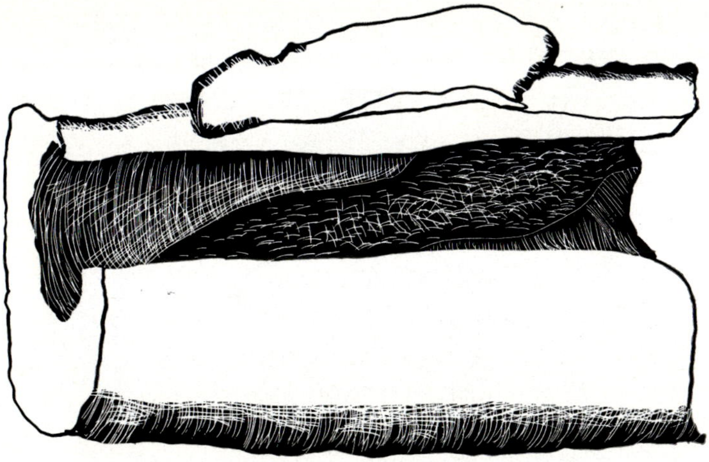

Fig. 104. Damage to tree-root by larva of *Prionus coriarius*.

specimen of this beetle collected by the writer was taken from such a situation.

Ergates

Ergates faber (L.) (Fig. 105) is a European and North African species imported from time to time into the United Kingdom. The sight of such a large insect often excites interest in the timber yard.

The specimen figured, however, was recorded by Fraser and it seems certain that it must have been a British insect, inasmuch as it had developed from eggs laid in a baulk of Columbian pine at Heysham pier, Lancs. The larvae grow to as much as 90 mm in length and are found only in coniferous wood. The presence of fungal decay accelerates development, which usually takes from four to five years, but may, in some circumstances, take very much longer. Part of the pupal chamber is shown in Fig. 106. Adults emerge July to September and have been recorded as boring through lead to effect their release from wood.

The male has the head, pronotum and legs black, with the elytra reddish-brown, whilst the female is uniformly black. In the specimen figured, however, although female, the elytra were reddish-brown and in addition the single lateral projection on the prothorax was absent, but there were a pair of short spines on the anterior half of the prothorax and a group of three on each side near the posterior margin. The larvae require the wood to be very damp, and Becker has recorded their occurrence in buildings where such conditions prevail. They have also been recorded as attacking telegraph poles.

Ergates spiculatus Leconte. This large species is sometimes imported from North America, where it is found infesting the coniferous trees

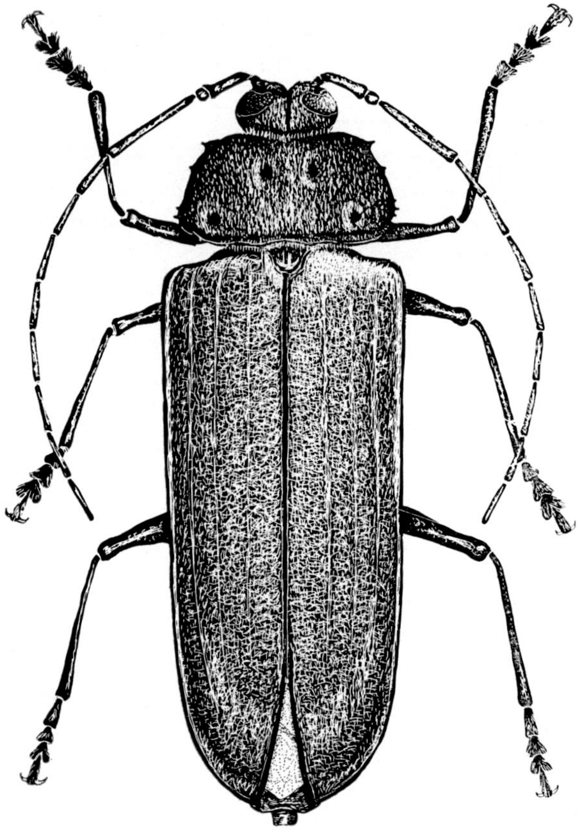

Fig. 105. *Ergates faber* (L.), drawn from specimen which completed the whole of its life cycle in England. Length: 44 mm.

Pinus ponderosa and *Pseudotsuga mucronata.* The damage caused by one single larva, after burrowing deep into the heartwood, is said to be considerable. This species is separated from *Ergates faber* by the more numerous spines on the lateral margins of the prothorax, which are shortest in the centre, becoming long both anteriorly and posteriorly, although the figure which Duffy gives is not convincing in this respect.

LEPTURINAE

The subfamily LEPTURINAE consists mostly of medium-sized to small insects—from about 20 mm in length to about 6 mm—and the larvae are usually found only in wood in an advanced stage of fungal decay. The adult beetles often visit the flowers of UMBELLIFERAE in the hot sun. *Rhagium mordax* and presumably *R. inquisitor* are attracted when flying to white objects, such as clothing.

The following key will differentiate the genera and the isolated species:

1. Thorax with a conspicuous pair of lateral tubercles 2
– Thorax without lateral tubercles 3
2. Antennae short, not extending as far as middle of elytra; third segment shorter than first. Lateral tubercle of prothorax sharp, not pubescent at tip. Prosternal process reaching posterior part of anterior coxae. Posterior tarsi with first segment depressed, strongly broadened apically, not longer than second and third together . . *Rhagium*. 8
– Antennae longer, extending beyond middle of elytra; third segment much longer than first. Lateral tubercle of prothorax blunt, covered with pubescence mainly arising from the base. Prosternal process not reaching posterior part of anterior coxae. Posterior tarsi with first segment subcylindrical, scarcely broadening apically, much longer than second and third together. Length 15–24 mm. Elytra black or brown *Stenocorus meridianus* (Linnaeus

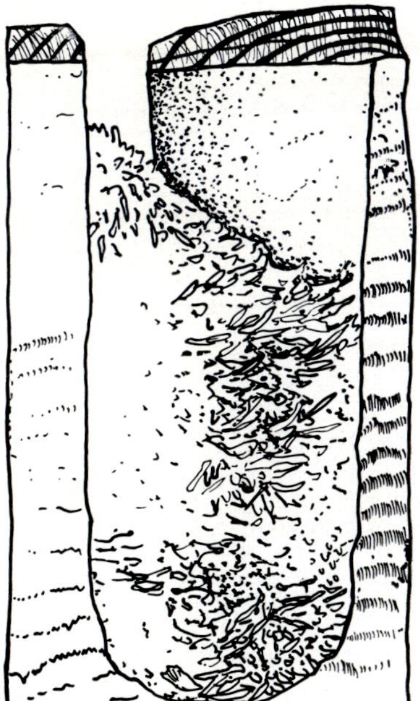

Fig. 106. End of pupal chamber of *Ergates faber* in Columbian pine.

3. Posterior angles of thorax more or less right-angled or evenly rounded 4
– Posterior angles of thorax acutely produced into a distinct spine-like process 5
4. Elytra parallel-sided and broadly rounded apically. Thorax red. Elytra bluish-black. Length 6–9 mm . . . . . . *Acmaeops collaris* (Linneaus)
– Elytra slightly to strongly tapering posteriorly. Thorax usually black, but if red then elytra also red . . . . *Leptura*. 10
5. Elytra broadly rounded apically 6
– Elytra obliquely truncate and marginate or mucronate apically. Elytra always strongly tapering posteriorly . . . . . *Strangalia*. 15
6. Elytra distinctly tapering posteriorly; yellow with black markings . . . . . *Judolia*. 20
– Elytra parallel-sided; yellow or black, never both . . . 7
7. Antenna with first segment shorter than third segment. Eyes almost touching front margin of head. Elytra black . . . . *Grammoptera*. 21
– Antenna with first segment as long as third. Eyes placed well behind front margin of head. Elytra yellowish-brown. Length 6–8 mm . *Alosterna tabacicola* (Degeer)
8. Antennae long, extending well beyond shoulders of elytra; very finely pubescent. Elytral pattern extremely variable but usually consisting mainly of a pair of oblique yellow bands which may coalesce in extreme cases; shoulders, and some of area between yellow bands, reddish-brown. Length 14–18 mm . . . *R. bifasciatum* Fabricius
– Antennae shorter, not or only just reaching shoulders of elytra; with dense greyish pubescence. Elytral pattern consisting of one or two black marks bordered by yellow bands. Shoulders black 9
9. Temples parallel-sided behind eyes. Head with a deep longitudinal median impression between eyes. Head and thorax covered with golden pubescence. Elytral pattern with one black mark. Length 14–19 mm. . . . *R. mordax* (Degeer)
– Temples distinctly converging behind eyes. Head without an impression between eyes. Head and

thorax covered with greyish-white pubescence. Elytral pattern with two black marks. Length 12–15 mm . . . . . *R. inquisitor* (Linneaus)

10. Elytra shining, entirely yellowish brown; apices broadly rounded (Fig. 112b). Length 6–9 mm. *L. livida* Fabricius

– Elytra dull, entirely black, red or yellow, or black and yellow; apices obliquely truncate or emarginate 11

11. Elytra black, each with three orange or yellow marks which are sometimes confluent (Fig. 113). Length 8–11 mm. Southern species . . *L. sexguttata* Fabricius

– Elytra never black with orange or yellow marks . . . 12

12. Elytra entirely black. Scutellum covered with very dense golden pubescence. Length 14–18 mm. Chiefly from *Fagus* . . . *L. scutellata* Fabricius

– Elytra mainly or entirely red or yellow. Scutellum sparsely covered with fine greyish pubescence 13

13. Elytra yellow with apices black. Length 10–14 mm. (Fig. 114) . . . . . . *L. fulva* Degeer

– Elytra entirely yellow or red 14

14. Pubescence on elytra black. Legs entirely black; first segment of hind tarsus more than four times as long as second (Fig. 112a). Length 9–12 mm. . . . . . *L. sanguinolenta* Linnaeus

– Pubescence on elytra pale yellow. Legs with tibiae and tarsi yellow; first segment of hind tarsus less than four times as long as second. Length 12–20 mm . . . . . *L. rubra* (Linnaeus)

15. Elytra black, with four transverse yellow or orange bands 16

– Elytra entirely black, or almost entirely black, red or yellow; never banded 17

16. Legs and basal segments of antennae red. Front and hind margins of thorax densely fringed with yellow pubescence (Fig. 115a). Length 13–18mm *S. aurulenta* (Fabricius)

– Legs and basal segments of antennae black. Front and hind margins of thorax not densely fringed with yellow pubescence. Length 13–18 mm . . . . . . *S. quadrifasciata* (Linnaeus)

17. Antennal segments with basal third yellow.

Elytra yellow with black markings (Fig. 116). Length 15–17 mm . . . . . *S. maculata* (Poda)

– Antennal segments entirely black or red. Elytra never yellow with black markings except at apex 18

18. Head, thorax and legs red. Thorax with a pair of small mediolateral protuberances (Fig. 117). Length 9–15 mm . . . . . *S. revestita* (Linnaeus)

– Head, thorax and legs black. Thorax without a pair of lateral protuberances 19

19. Elytra entirely black (Fig. 115b). Length 7–9 mm *S. nigra* (Linnaeus)

– Elytra yellow (male) or red (female) with apices and inner margins black (Fig. 118). Length 7–9 mm . . . . . . *S. melanura* (Linnaeus)

20. Black, with three transverse yellow bands. Setae on head and prothorax grey and erect. Length 8–11 mm. *Northern species* . . *J. sexmaculata* (Linnaeus)

– Yellow with black markings. Setae on head and prothorax yellow and decumbent. Length 7–11 mm. *Southern species* . . . *J. cerambyciformis* (Schrank)

21. Femora and tibiae entirely reddish testaceous. Elytra densely covered with golden pubescence except at apex which is black and shining (Fig. 121a). Length: 6–7 mm. . . . . *G. ustulata* (Schaller)

Femora and tibiae partly or entirely black. Elytra completely but sparsely covered with greyish-yellow pubescence 22

22. Front tibiae red. Segments of at least apical half of antennae red basally, black apically (Fig. 121b). Length 4·5–6 mm . *G. ruficornis* (Fabricius)

– Front tibiae black. Segments of at least apical half of antennae unicolorous 23

23. Segments of at least apical half of antennae entirely black (Fig. 121c). Last three abdominal sternites of female black. Femora usually completely black. Length 4·5–6 mm *G. ruficornis* var. *holomelina* Poole

– Segments of at least apical half of antenna entirely reddish-brown (Fig. 121d). Last three abdominal sternites of female red. Femora usually red basally, but occasionally completely black. Length 6–9 mm . . . *G. variegata* (Germar)

Notes on the species

Rhagium

Rhagium bifasciatum Fabricius (Fig. 107). Said by Kaufmann to be the most widely distributed longicorn beetle in the British Isles, occurring in almost all the vice-counties of England, Scotland, and Wales. It has not been collected from the central part of Ireland, but as it occurs where most of the collecting has already taken place it is certain to turn up there

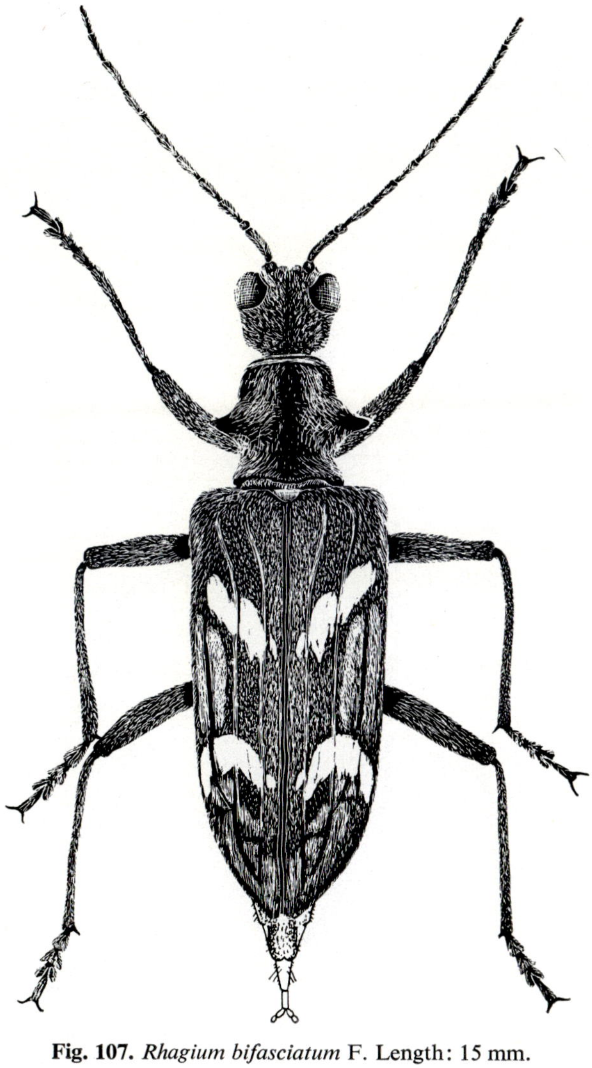

Fig. 107. *Rhagium bifasciatum* F. Length: 15 mm.

also. It seems to prefer the decaying logs and stumps of Scots pine, but it has also been found in fir, spruce, ash, oak, beech, birch, hornbeam, alder, mountain ash, and willow. It has been found also in old posts and palings.

The elytral pattern of the adult is extremely variable and the reader is referred to the papers of Blair and Kaufmann for detailed treatment of this subject. About seventeen varieties of elytral colour pattern have been named. Kaufmann also gives an extensive bibliography concerning this species.

Rhagium mordax (Degeer) (Fig. 108). The glabrous black mark on each elytron with an oblique yellow band on each side is distinctly eye-like and readily distinguishes this species from other CERAMBYCIDAE. The larvae feed on a variety of timbers, preferring the cambium and outer

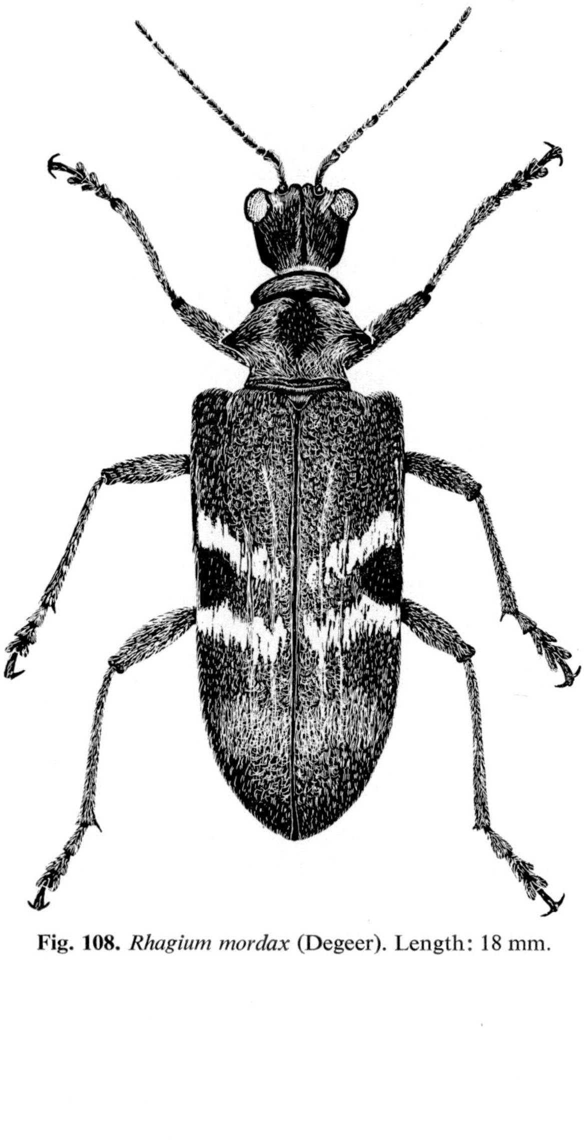

Fig. 108. *Rhagium mordax* (Degeer). Length: 18 mm.

sapwood of rotting boles or stumps. *Quercus*, *Castanea*, *Pinus*, *Abies*, *Picea*, *Larix*, *Alnus*, *Betula*, *Fagus*, *Juglans*, *Sorbus*, *Fraxinus*, *Crataegus*, *Ilex*, *Carpinus*, *Tilia*, *Pyrus aucuparia* and *P. malus*, have been recorded as host plants, but *Quercus* is the most common. Although the adult beetle metamorphoses in August, it remains in the pupal cell until the following Spring.

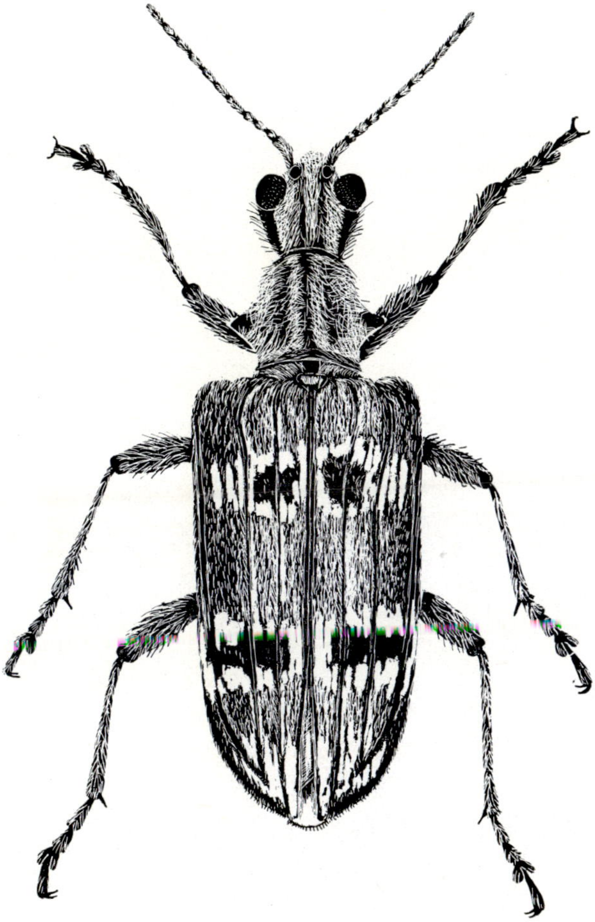

Fig. 109. *Rhagium inquisitor* (L.). Length: 14 mm.

Rhagium inquisitor (Linnaeus) (Fig. 109). In this species there are two black marks on each elytron surrounded by yellowish bands. Duffy's key and also the key given by Joy is misleading in this respect. This species has a northern distribution and the larvae prefer to feed on the wood of conifers. *Pinus*, *Picea*, *Larix* and *Cedrus atlantica* have been listed as hosts, as also has *Betula* and *Quercus*. Duffy states that the larvae prefer less decayed wood than do those of *R. mordax*, which probably accounts for their not infrequent importation in softwoods from North America.

Stenocorus meridianus (Linnaeus) (Fig. 110).

The host plants of this variable beetle (both in size and colour) are given

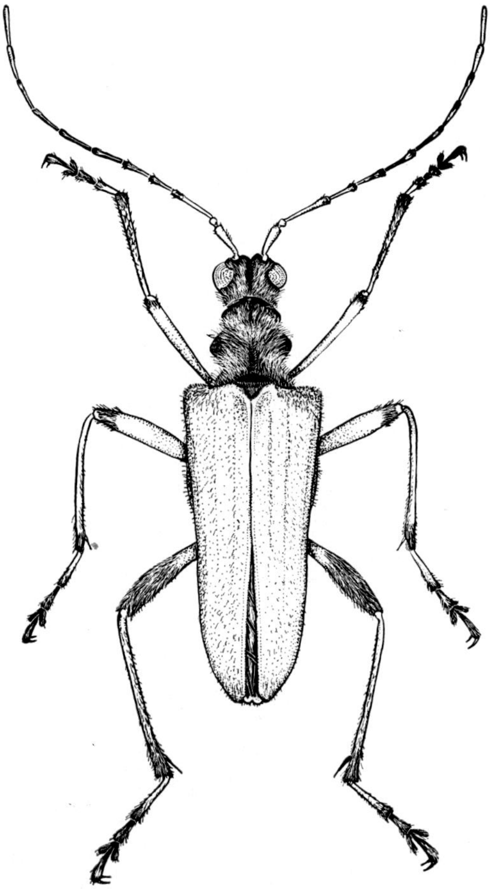

Fig. 110. *Stenocorus meridianus* (L.). Length: 17 mm.

as *Prunus cerasus*, *Fraxinus* and *Salix*, and Duffy states that in captivity eggs were laid on stumps of *Quercus, Pinus, Fagus, Populus, Alnus, Pyrus malus* and *Betula*. The larvae, which spend two years in this stage, feed in stumps and roots and they emerge as adults in May and June.

Joy (who gives the incorrect spelling *Stenochorus*) describes the variety of coloration of this local species as 'head and thorax black, elytra entirely black, or reddish-yellow, or with apex and suture black, or base only yellow, antennae black or red'.

Acmaeops collaris (Linnaeus) (Fig. 111).

This small cerambycid has the prothorax chestnut red, with the basal third parallel-sided. The elytra are shining black or blue-black, pitted and covered with long outstanding pubescence as are the legs and prothorax. The head is black also.

The adult beetles are found on flowers in woods, but little was known of their biology until larvae were found by Duffy and Fraser. These are not typical longhorn larvae, but are active fast-moving larvae able to travel up smooth vertical walls, the legs being extremely long and sclerotized. The above authors give two years as the length of life cycle and they found the larvae under the loose dry bark of an upturned stump of *Quercus*, but Perris gives *Castanea* as the host plant.

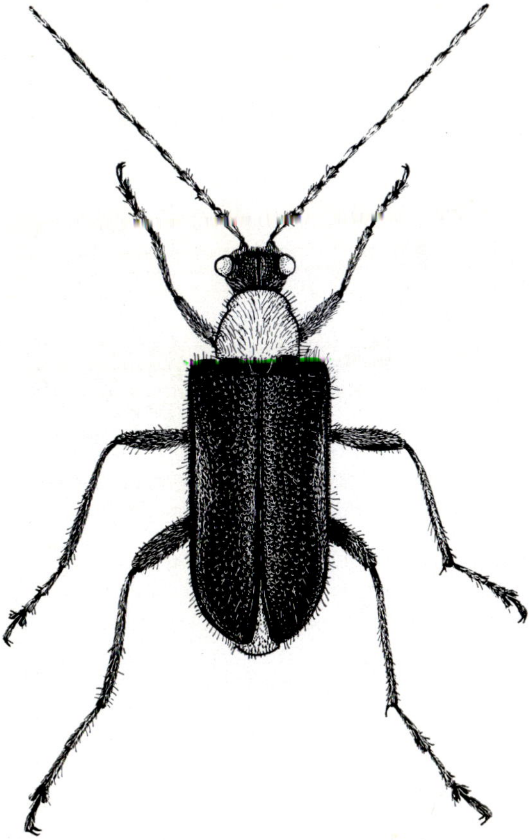

Fig. 111. *Acmaeops collaris* (L.). Length: 6 mm.

The larvae are thought to feed on the frass in the old galleries of other cerambycid beetles, probably those of *Clytus arietis*, or perhaps on fungal growths in these situations, and Duffy considers that this is the only species of cerambycid larva which normally exists without the protection and support of a gallery. The adults emerge in May and June.

Joy gives the distribution of this species as being local in southern England and found as far north as Lancashire.

Leptura (Fig. 112).

There are six British species in this genus. One of the characters used by Duffy in his key concerns the degree of tapering of the elytra. In the

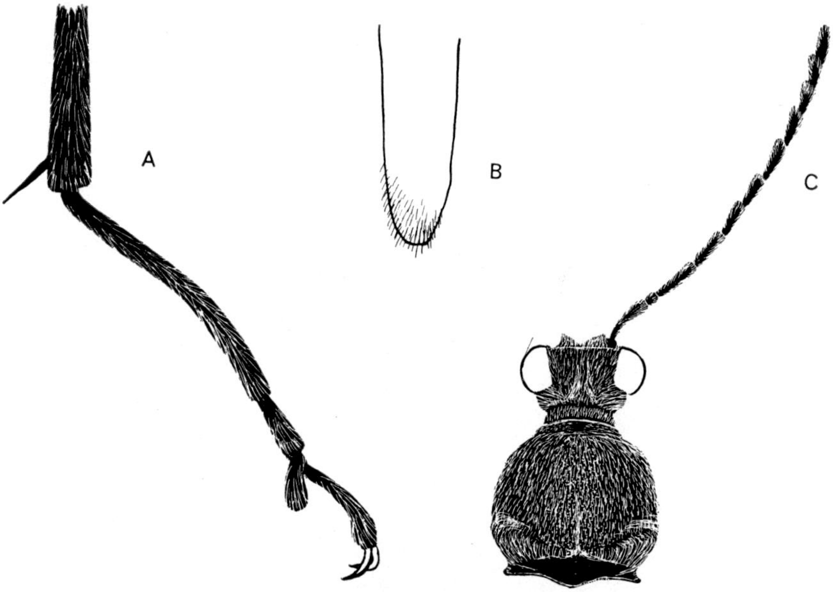

Fig. 112. *Leptura*. A, hind tarsus of *L. sanguinolenta*; B, apex of left elytron of *L. livida*; C, pronotum and head of *L. rubra*.

accompanying illustrations one species (*Leptura sexguttata*) is shown with only slightly tapering elytra (Fig. 113) and one (*L. fulva*) with distinctly tapering elytra (Fig. 114). However, such a character, based on degree, used in a key is almost always unsatisfactory, especially if an isolated specimen is to be identified without the aid of a collection.

Leptura livida Fabricus is one of the smaller species of the genus. It is locally distributed in southern England as far north as Staffordshire. Little is known however of its biology and host plants.

Leptura sexguttata Fabricius (Fig. 113) is a rare species about which little is known. It is found only in the south of England. The posterior pair of yellow marks on each elytron are sometimes confluent. A specimen exhibiting this character is shown in the figure. The elytra are occasionally entirely black. The thorax has long pubescence.

Leptura scutellata Fabricius is another rare southern species. The host plants are given as *Carpinus*, *Fagus* (chiefly), *Betula* and *Quercus*. Duffy states that the biology is similar to that of *Strangalia quadrifasciata*

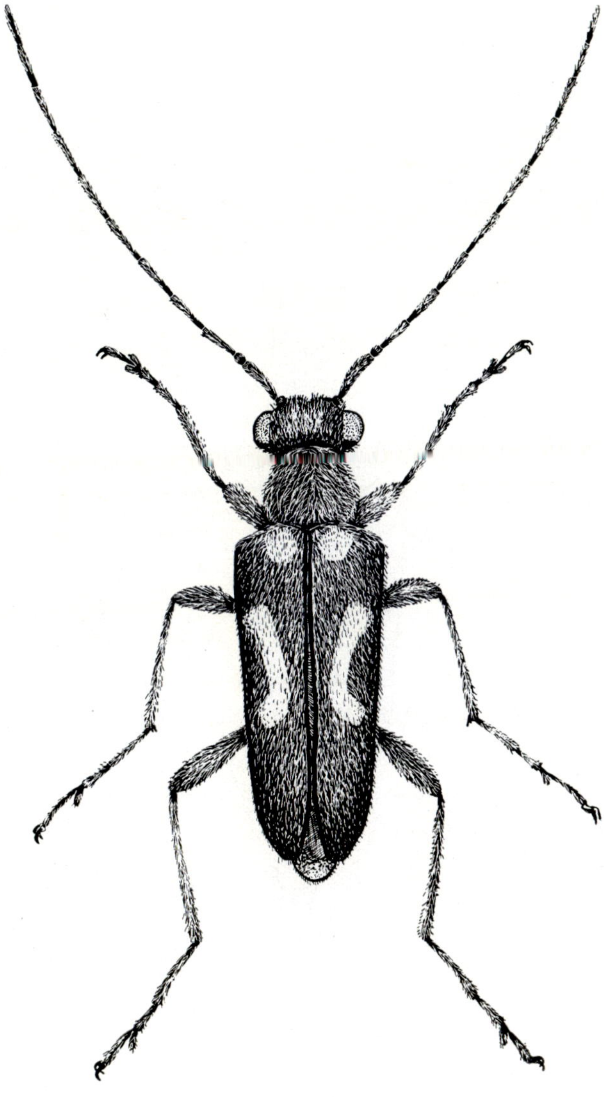

Fig. 113. *Leptura sexguttata* F. Length: 6·3 mm.

(Linnaeus)—see below. This is the only species in the genus with the elytra entirely black.

Leptura fulva Degeer (Fig. 114) is easily distinguished from other species in the genus by the black margin to the apices of the elytra. The pubescence on the elytra is yellow, the thorax is broadest before the base. It is another

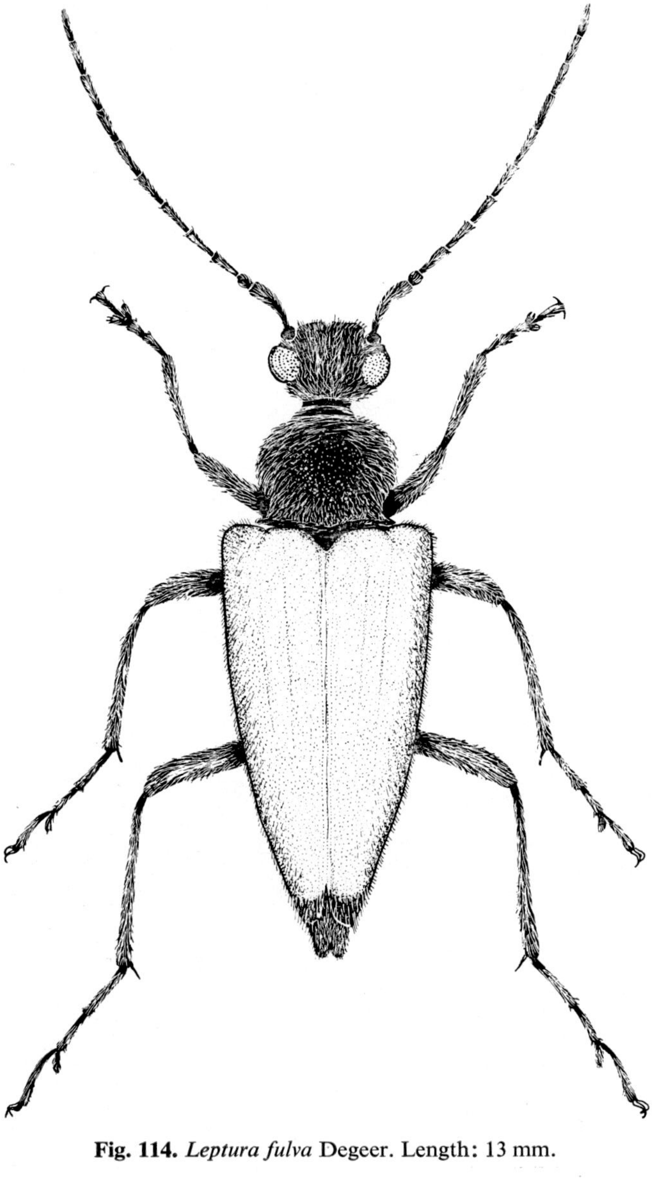

Fig. 114. *Leptura fulva* Degeer. Length: 13 mm.

rare species found in the south of England and South Wales and practically nothing is known of its biology.

Leptura sanguinolenta Linnaeus. Pubescence on the elytra is black and the thorax is broad at the base. In the male, however, the apices of the elytra are black, the rest being yellowish whilst those of the female are entirely reddish. This is a rare species, but is found in Scotland as well as England. *Pinus* and *Abies* are given as hosts and Duffy states that eggs are laid usually below ground level in stumps and boles. Dead pines with the bark charred by fire are especially susceptible to infestation by this species.

Leptura rubra (Linnaeus). The characters given in the key are sufficient to distinguish this species. In addition, at the widest part of the pronotum there are definite shoulders associated with a depression on the inside, but the long pale yellow pubescence directed forwards normally covers them (Fig. 112c). The apical segments of the antennae are serrate.

In the male the prothorax is black and the elytra are reddish yellow whilst in the female the prothorax and elytra are red. In England it is very rare and local, being found only in fir stumps in Norfolk, but on the Continent *Larix*, *Picea*, *Pinus*, and *Abies numidica* are given as host plants. Its biology is said to be similar to that of *Strangalia quadrifasciata* and it has been reported as attacking telegraph poles.

Strangalia

Strangalia aurulenta (Fabricius) (Fig. 115a). The larvae are found in dead and decaying stumps of *Quercus*, *Alnus*, *Fagus*, *Salix*, *Betula*, *Populus*

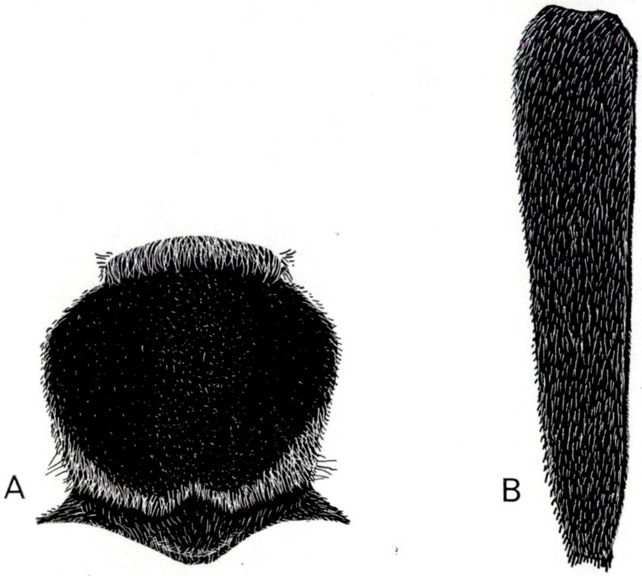

Fig. 115. *Strangalia.* A, pronotum of *S. aurulenta*; B, left elytron of *S. nigra*.

tremula, *Prunus cerasus*, *Ulmus*, *Aesculus*, *Castanea* and *Juglans*. Duffy states that the biology is similar to that of *S. quadrifasciata*. This is a very rare insect found in southern England and South Wales. It emerges from May to July.

Strangalia quadrifasciata (Linnaeus) is another rare species. The larvae are found in dead and decaying stumps of *Quercus*, *Alnus*, *Populus canadensis*, *Betula*, *Fagus*, *Salix* and *Picea*. Duffy says that the larvae make meandering galleries deep in the sapwood and pack the galleries with fibrous shavings. The adult beetles are found from May to July.

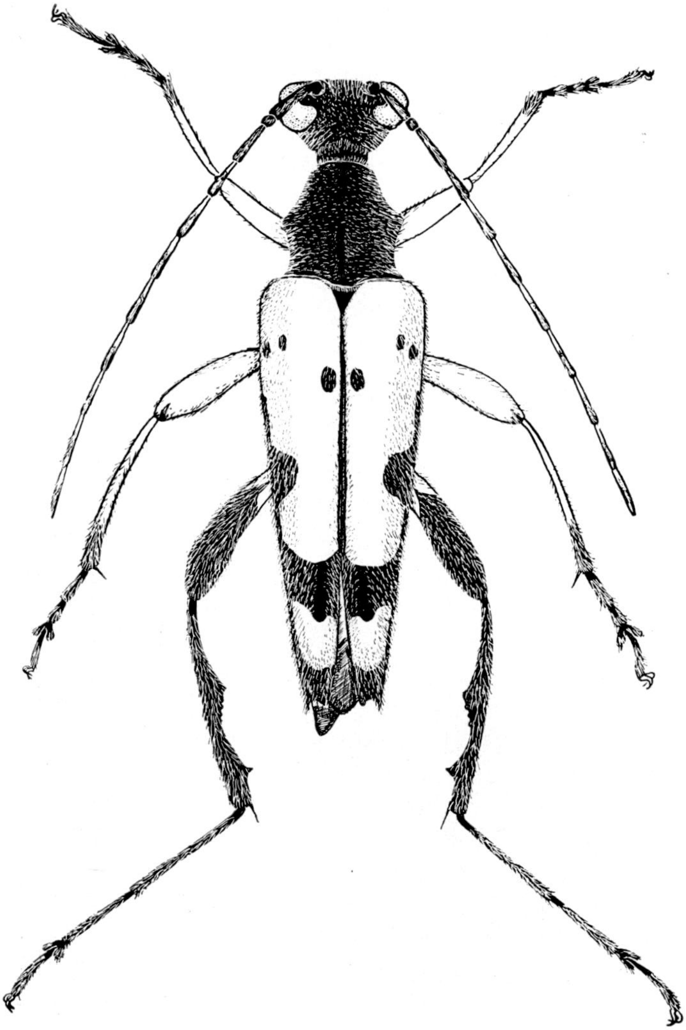

Fig. 116. *Strangalia maculata* (Poda). Length: 13 mm.

Strangalia maculata (Poda) (Fig. 116). The markings on the elytra of this locally distributed beetle are variable and in extreme cases the elytra may be entirely black, or the anterior markings may be sparse or even absent. On the inner edge of the hind tibiae of the males are two large and a few smaller tooth-like projections. The larvae are found in stumps of *Quercus*, *Betula*, *Salix*, *Fagus*, *Fraxinus*, *Populus tremula*, *Carpinus*,

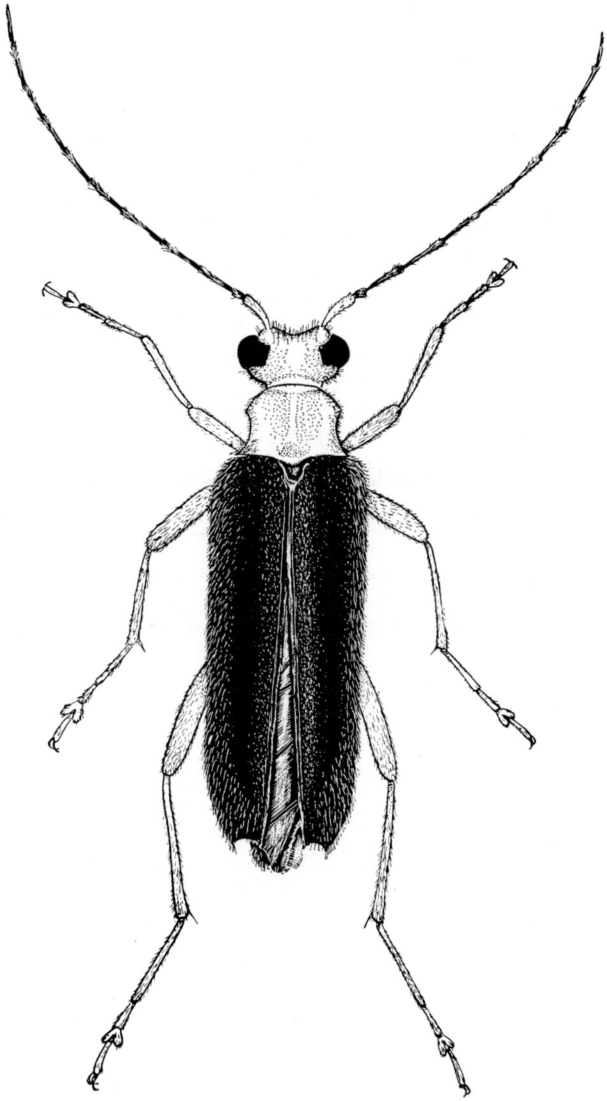

Fig. 117. *Strangalia revestita* (L.). Length: 10 mm.

Corylus, *Pinus* and *Castanea*. A larva has also been found feeding in a pine cone from which it was successfully reared.

Strangalia revestita (Linnaeus) (Fig. 117). Little is known of this rare insect. The host plant is given by Mendizabal as *Prunus avium*. Mr. Duffy kindly lent a specimen for figuring.

Strangalia nigra (Linnaeus) (Fig. 115b). Little is known of the biology of this insect. Joy states that it is found in southern England extending as far north as Worcester.

Strangalia melanura (Linnaeus) (Fig. 118). The black inner margins of the elytra are often extensive, occupying about a quarter of the width of the elytra. The pubescence, especially along the outer margins of the

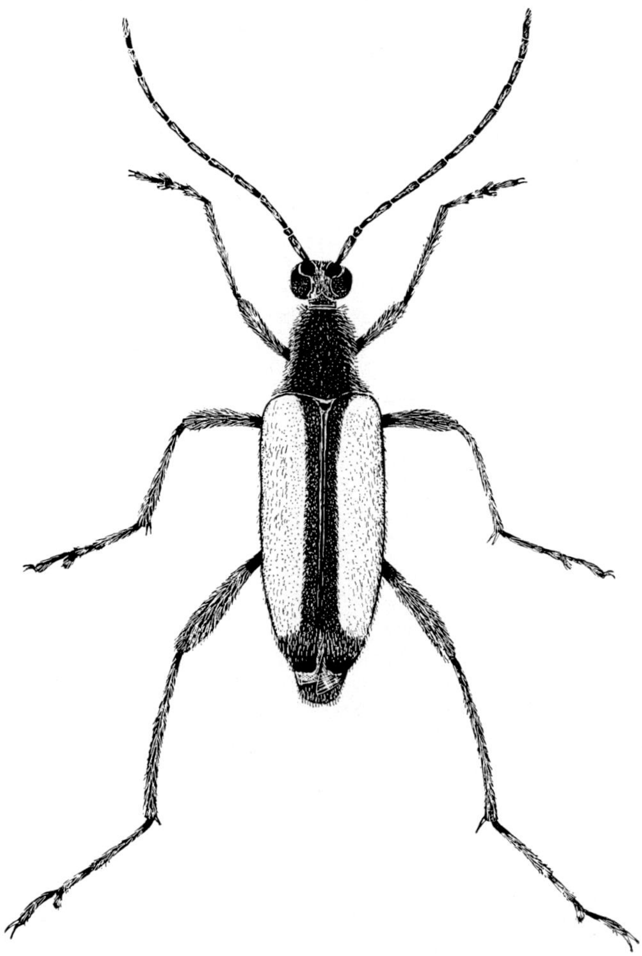

Fig. 118. *Strangalia melanura* (L.). Length: 9·2 mm.

elytra and at the apices, is stiff and almost spine-like. Host plants are given as roots of *Cytisus* and the slender, very rotten branches of *Quercus* and *Acer pseudoplatanus*.

Strangalia attenuata (Linnaeus). This insect is stated by Duffy to be found very rarely in imported timber. The larvae are found in *Aesculus*. No examples are in the author's collection.

An interesting point concerning the coloration of the British species of *Leptura* and *Strangalia* is the example of evolutionary divergence which it shows. Thus in each genus, species have evolved with wasp patterns—transverse yellow and black bars. This is well shown in several strangalids

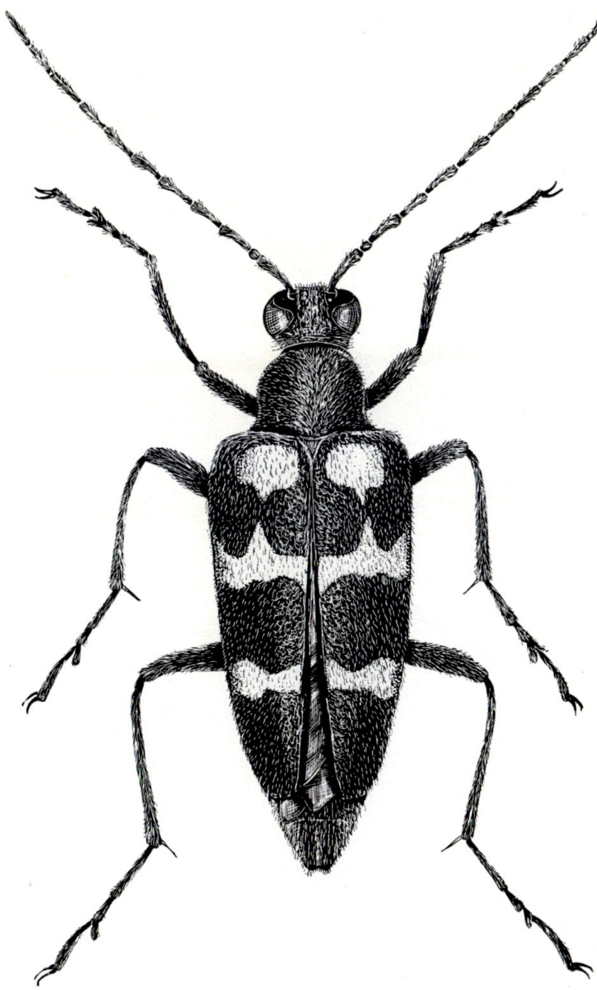

Fig. 119. *Judolia sexmaculata* (L.). Length: 10 mm.

and to a lesser degree by *Leptura sexguttata*. There is an entirely black species in each genus (*L. scutellata* and *S. nigra*), a red species (*L. rubra* and *S. melanura* female) and a species with black apices to the elytra (*L. fulva* and *S. melanura*).

Judolia

The two species of *Judolia* have been confused in the past, but they are very different in appearance, as can be seen from the illustrations.

Judolia sexmaculata (Linnaeus) (Fig. 119). The larvae probably infest

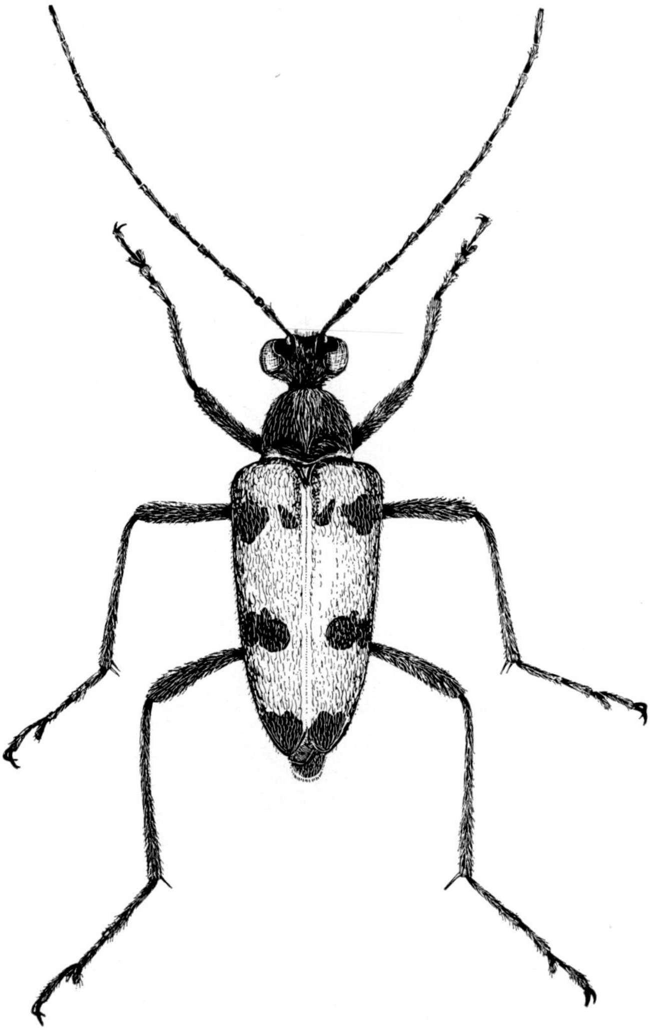

Fig. 120. *Judolia cerambyciformis* (Schrank). Length: 8 mm.

old exposed roots of *Pinus* and *Abies* and the adults emerge from June to August. Mainly found in Scotland.

Judolia cerambyciformis (Schrank) (Fig. 120). Duffy has studied the biology of this species and found it to infest the exposed roots of recently dead or blown trees of *Quercus*, *Betula* and *Castanea*. The larvae are able to move through the earth from one root to another and they make a cocoon several inches deep in the earth. The adults emerge in May and June, tunnelling through the earth to do so for as much as 30 cm. This very local species is more often found in the southern counties. It appears in Joy in the genus *Pachytodes*.

Grammoptera (Fig. 121)

Grammoptera ustulata (Schaller) larvae are found in thin dead twigs and smaller branches of *Quercus* and *Castanea* where they feed on the outer

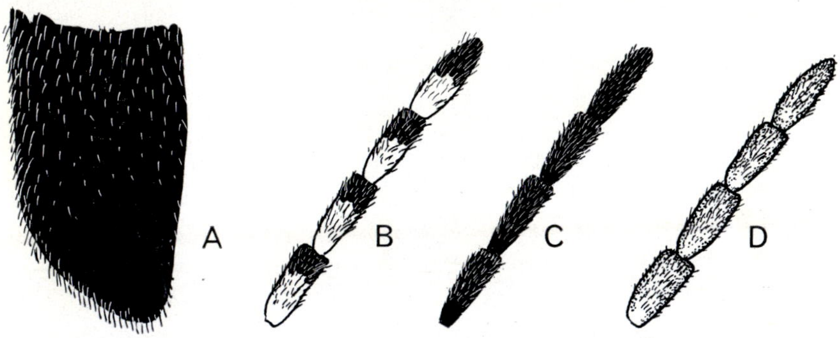

Fig. 121. *Grammoptera.* A, apex of elytron of *G. ustulata*; B, antenna of *G. ruficornis*; C, antenna of *G. ruficornis* var. *holomelina*; D, antenna of *G. variegata*.

sapwood. Adults emerge during May and June, completing the life cycle in one year. It is a very local species, occurring in Buckinghamshire and Hampshire.

Grammoptera ruficornis (Fabricius) (Fig. 122). The biology is similar to the previous species, but has a very much wider range of host plants. Duffy records the following, *Crataegus*, *Hedera*, *Hibiscus syriacus*, *Populus*, *Ilex*, *Pyrus malus*, *Aesculus*, *Prunus cerasus*, *Fraxinus*, *Ulmus*, *Quercus*, *Pyrus aucuparia*, *Salix*, *Acer pseudoplatanus*, *Prunus spinosa*, *Corylus*, *Sambucus nigra* and *Rhamnus*. It is not recorded from CONIFERAE and is more widespread than the previous species.

Grammoptera ruficornis var. *holomelina* Poole has appeared in previous works as a distinct species *Grammoptera holomelina* Poole, but Duffy, who has examined a long series, considers that it is merely a variety of *G. ruficornis*. Its biology therefore is likely to be similar to the above.

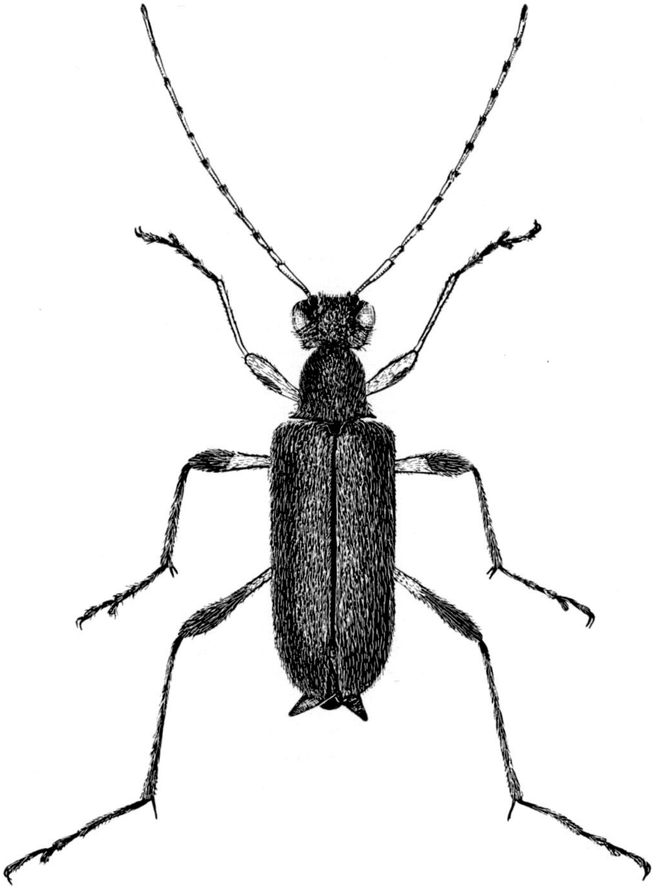

Fig. 122. *Grammoptera ruficornis* (F.). Length: 6 mm.

Grammoptera variegata (Germar) larvae are found feeding on the outer sapwood of thin twigs and branches of *Quercus* and *Castanea*, but they bore rather deeper into the sapwood to pupate. The adults emerge during April, May and June. It is rather a rare insect found in southern England as far north as Yorkshire.

Alosterna tabacicolor (Degeer) (Fig. 123).

The larvae infest old, damp, rotten stumps of *Pinus, Quercus, Corylus* and *Acer campestris*. They often tunnel deeply into the sapwood. The adults emerge from April to June. In Kloet & Hincks and Joy, the spelling is given as *tabacicolor* and the latter author places it in the genus *Leptura*. Duffy (1953), however, gives the spelling as *tabacicola* in the text and *tabacicolor* in the legends to the text figures. It is a rare species.

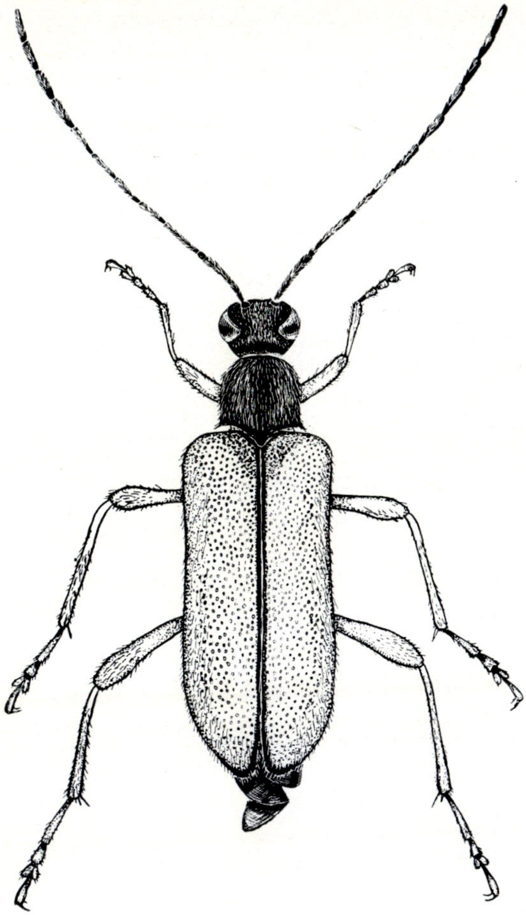

Fig. 123. *Alosterna tabacicolor* (Degeer). Length: 6·3 mm.

Ulochaetes leoninus Leconte.

This species, a member of the NECYDALINI is occasionally found in timber of *Abies grandis, Pseudotsuga, Pinus* and *Picea* imported from America. Craighead states that the larvae extend their galleries through the sapwood deep into the heartwood. The galleries are tightly packed with fine shredded frass. No specimens are in the author's collection.

ASEMINAE

In the subfamily ASEMINAE there are five indigenous species. A number of others are however occasionally introduced in timber from the European continent or from North America. In Kloet and Hincks it constitutes the first part of the subfamily CERAMBYCINAE and in Joy it is represented as the family TETROPIIDAE.

The three indigenous genera and five species can be identified by means of the following key:

1. Eyes with inner margin deeply emarginate so that they are almost divided into two parts (Fig. 124a). Elytra smooth, with striae indistinct or absent. Length 10–15 mm. Chiefly from *Larix* . *Tetropium*. 3
– Eyes not deeply emarginate; kidney-shaped. Elytra rugose, striae distinct 2
2. Elytra with only two pairs of distinct striae. Head, thorax and elytra brown. Eyes very coarsely faceted (each facet distinctly visible with a ×15 lens). Length 12–26 mm . . . *Arhopalus*. 4
– Elytra with at least four pairs of distinct striae. Head and thorax black. Elytra normally black, but light brown in var. *agreste* Fabricius. Eyes finely faceted (each facet scarcely visible with a ×15 lens). Length 10–18 mm . *Asemum striatum* (Linnaeus)
3. Disc of thorax smooth, shining and finely punctured (Fig. 124b). Elytra black. Length 10–15 mm . . . . *T. gabrieli* Weise
– Disc of thorax rugose, dull and coarsely punctured (Fig. 124c). Elytra reddish brown. Length 10–14 mm . . . . *T. fuscum* Fabricius
4. Eyes with numerous short setae (Fig. 126b). Third tarsal segment divided almost to base (Fig. 126d). Length 12–16 mm. Northern species . . . . *A. rusticus* (Linnaeus)
– Eyes without setae (Fig. 126c). Third tarsal segment much less strongly divided (Fig. 126e). Length 13–26 mm. Southern species . . *A. ferus* (Mulsant)

Notes on the species

Tetropium

Tetropium gabrieli Weise (Fig. 124b). This is the only species of the genus given by Kloet and Hincks as indigenous. It is found in southern England as far north as Leicester. In the normal form the tibiae and tarsi are reddish-brown, the femora being mainly black, but in the aberration *crawshayi* Sharp the tibiae and tarsi are much darker in colour. The adult beetles are usually found under the bark of felled *Larix*—the usual host of the larvae, but *Pinus*, *Picea* and *Abies* are also given as host plants. The principal work on the biology of this beetle has been done by Crawshay and Schimitschek, but Duffy and Kaufmann have also contributed.

Batches of about six eggs are laid between the scales of the bark of

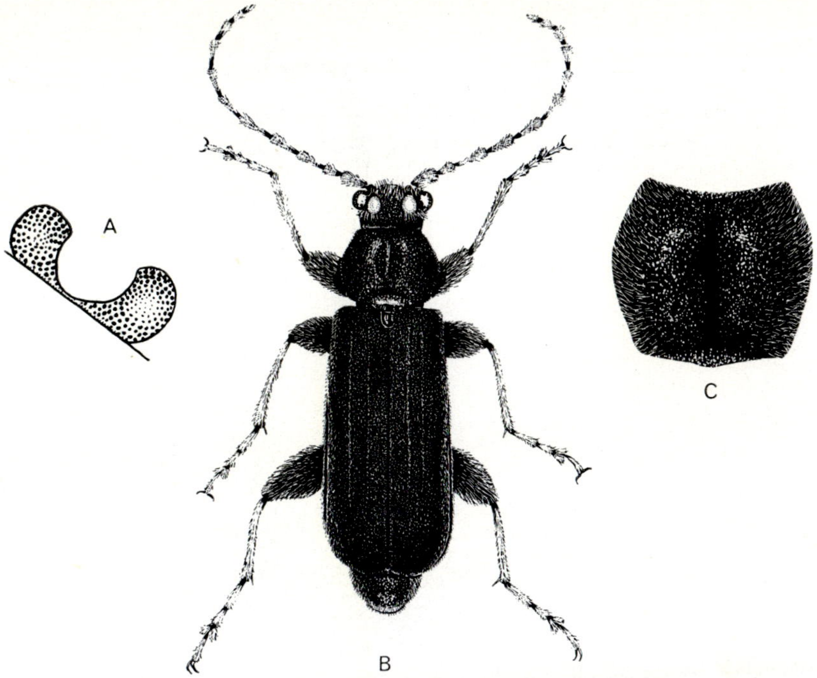

Fig. 124. *Tetropium.* A, eye of *T. gabrieli* showing almost complete separation; B, adult *T. gabrieli*, length 11 mm; C, pronotum of *T. fuscum*.

injured, unhealthy, or recently felled trees. When the young larva is about two or three weeks old it has penetrated as far as the inner bark. The soft bast, the cambium layer and sometimes the outer sapwood are eaten. When pupation takes place in the sapwood, the larva bores a vertical elliptical hole in a horizontal direction for a distance of 25–65 mm and then bores downwards for approximately the same distance. The pupal chamber is made and a plug of fibres is used to seal it off (Fig. 125).

Usually an annual life cycle takes place, but in exceptionally hot summers there may be two emergences during the year, normally during May and June and exceptionally in late August.

It is an important forestry insect as unhealthy trees may be rapidly destroyed by it, but in felled trees the damage is only superficial. It has been recorded as damaging lead by Laing.

Tetropium fuscum Fabricius. This insect, the larvae of which attack *Picea* and *Pinus*, is sometimes imported from Europe with the timber. The economic importance and biology are similar to those of *Tetropium gabrieli.*

Tetropium castaneum Linnaeus. This beetle is also sometimes introduced

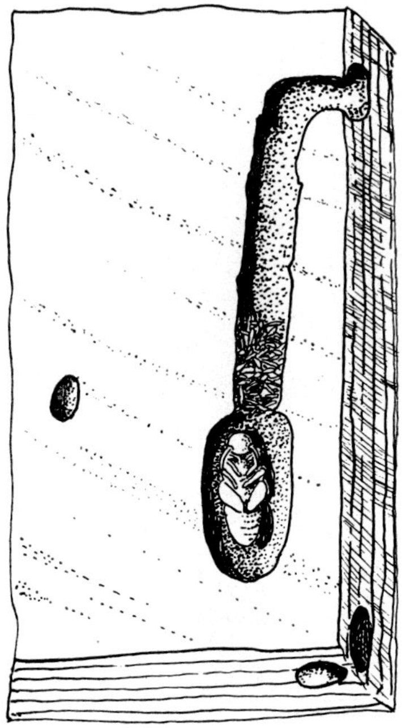

Fig. 125. Pupa of *Tetropium gabrieli* Weise with plug of torn wood fragments in tunnel in freshly cut larch. The pupal chamber is 20 mm long.

into British timber yards with timber from Europe. The specimens in the author's collection were found by Mr. M. S. Fraser at Bootle, Lancashire, in 1948, and Mr. A. Price from Finnish *Picea* in Berkshire in 1961.

Tetropium cinnamopterum Kirby. This is another species very like *T. gabrieli* in biology, but imported with timber (*Abies*, *Pinus* or *Picea*) from North America.

Tetropium velutinum Leconte. Mr. A. Price has sent the author some specimens introduced with *Picea* into a timber yard in Berkshire from Canada.

These last three species are similar to *Tetropium fuscum* in having the elytra reddish-brown.

Arhopalus

Arhopalus fetus (Mulsant) = *Criocephalus polonicus* Mots. (Fig. 126a). This is a rare insect found in southern England, extending as far north as Nottinghamshire, although in some districts in Surrey it is a common insect, sometimes being misidentified as *Hylotrupes bajulus*. Duffy gives *Pinus* and *Picea* as host plants, but Xambeu also records *Prunus cerasus*. Although occasionally this species attacks apparently healthy trees, more usually the infestation occurs in decaying or dead stumps, or dead

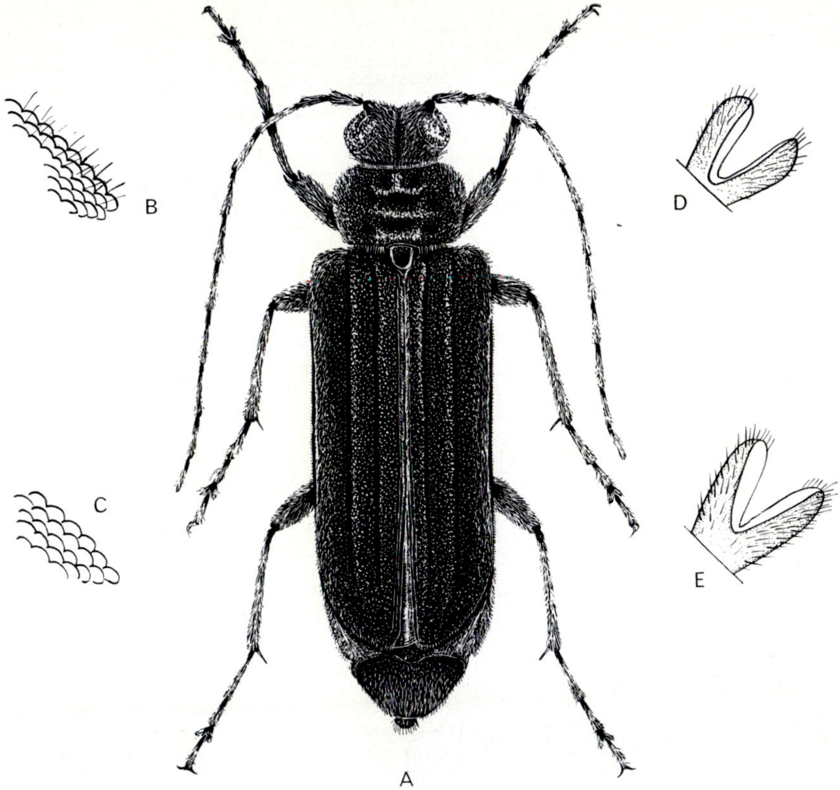

Fig. 126. *Arhopalus.* A, adult *A. ferus* (Mulsant), length 19 mm; B, part of eye of *A rusticus*; C, part of eye of *A. ferus*; D, third tarsal segment of *A. rusticus*; E, third tarsal segment of *A. ferus*.

standing trees, especially those scorched by fire. The larvae make galleries similar to those of *Asemum striatum*, and indeed they often occur together in the same timber. The pupae are active (like *A. striatum*), moving up and down the cell (Fig. 127). Adults emerge from July to August.

Arhopalus rusticus (Linnaeus) = *Criocephalus rusticus* Linnaeus. In Great Britain this beetle is found only in eastern Scotland, in pine stumps, and its biology follows closely that of the foregoing species. In France it is sometimes found damaging structural woodwork in buildings. The emerging adults will bore through zinc and solder.

Asemum striatum (Linnaeus) (Fig. 128).

The pronotum is usually wider nearer the base whereas in *Arhopalus* it is wider nearer the front. There are two distinct foveae situated on the disc in *Asemum striatum*, but in *Arhopalus* the foveae are divided and more complex. Larvae of this species are usually found in stumps of *Pinus*, but

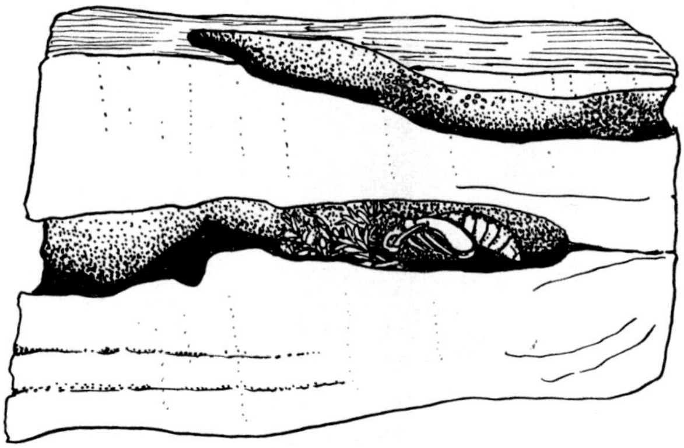

Fig. 127. Pupa of *Arhopalus ferus* (Mulsant) in piece of conifer stump. The plug of torn wood fragments is seen in the tunnel, which appears to be following a natural fissure. The pupal chamber is 25 mm long.

additionally *Picea*, *Abies* and *Larix* are recorded as host plants and exceptionally Kaufmann has recorded *Quercus*. According to Duffy the eggs are inserted between the bark scales of stumps, boles and roots of slightly decaying or recently dead trees. He mentions that scorching of the roots by fire renders the timber especially susceptible to attack.

Standing trees are sometimes attacked, but only when damage has been sustained. Duffy records finding larvae in a number of telegraph poles which they had severely damaged.

The young larva first feeds on the bark, then tunnels into the outer sapwood, making a winding gallery vertically downwards, tightly packed with frass and more coarsely torn fibrous particles. When the larva is about half-grown one or more small perforations are made through the bark to the exterior and frass is forced through them. The tunnel is then made in an inwards direction and when it is about 20–40 cm below the surface of the stump, the end is enlarged. The larva then tunnels vertically upwards making a straight tunnel, kept free from frass, and just underneath the surface the pupal cell is constructed. The pupa is active and can travel up and down the cell and the adults emerge during June or July.

CERAMBYCINAE

The fourth subfamily, the CERAMBYCINAE, with twenty-five genera is the largest of the five subfamilies and a key, based mostly on that of Duffy, is given below:

1. Elytra much shorter than abdomen, the latter at least partly covered by the extended wings 2
– Elytra completely covering abdomen except sometimes the pygidium 3

Fig. 128. *Asemum striatum* (L.). Length: 15 mm.

2. Eyes oval. Pronotum quadrate. Front and middle femora clavate and bearing a few short, silky setae. Length 3–5 mm. Usually from wickerwork . . . *Leptideella brevipennis* (Mulsant) 5

– Eyes comma-shaped, their inner margins strongly emarginate. Pronotum elongate. Front and middle femora petiolate apically and bearing numerous long bristly setae . . *Molorchus*. 25

3. Pronotum with a conspicuous pair of lateral or dorso-lateral spines or tubercles. Length at least 20 mm 4
– Pronotum without lateral spines or tubercles. Length usually much less 9
4. Elytra metallic green, blue or copper 5
– Elytra dull black, brown or blue 7
5. Antennae serrate or pectinate. Elytra each with an apical tooth or spine. Legs and antennae red or black. African species 6
– Antennae simple and filiform. Elytra without apical teeth or spines. Legs and antennae greenish-blue. Length 30–32 mm. Chiefly from *Salix* (Colour Plate, E) . . *Aromia moschata* (Linnaeus)
6. Femora strongly petiolate (as in *Molorchus*). Antennae strongly pectinate. Elytra tapering posteriorly and each bearing an apical spine. Chiefly from *Khaya* . . . . *Cordylomera*. 26
– Femora not petiolate. Antennae serrate. Elytra parallel-sided, truncate and dentate apically. Chiefly from *Aucoumea* and *Khaya* . . *Plocaederus*. 27
7. Elytra pale blue, with black transverse bands and spots. Antennae blue, with apices of segments bearing conspicuous tufts of black hairs. Length 20–32 mm. Chiefly from *Fagus* (Colour Plate, H) . . . . . . *Rosalia alpina* (Linnaeus)
– Elytra brown or black. Antennae brown or black without tufts of black hairs 8
8. Elytra with apices each produced into a single short sutural spine, or evenly rounded; dark brown or black, shining and without markings. Pronotum coarsely transversely rugose; disc without a pair of tubercles. European species. From deciduous trees . . . . *Cerambyx*. 28
– Elytra with apices each produced into a pair of long spines (one on sutural angle, the other on outer angle); light brown, with two pairs of paired, shining, white, elongate spots, one pair at base, the other just beyond middle. Pronotum densely punctured and disc with a pair of paramedian dark brown tubercles. Length 18–26 mm. North American species. . *Eburia quadrigeminata* Say
9. Apices of elytra produced into one or two conspicuous spines. Elytra never with yellow design . . . 10

– Apices of elytra without apical spines (except CLYTINI *pars*, but then elytra with a yellow design) 12

10. Elytra with sutural angle produced apically into a single stout spine. Elytra shining and very strongly punctured. Length 16–30 mm. Oriental species. From seasoned timbers of many kinds; often in plywood packing-cases *Stromatium barbatum* (Fabricius)

– Elytra with apices each produced into a pair of stout spines. Elytra dull and indistinctly punctured 11

11. Elytra with outer apical spine shorter than sutural spine. Pronotum deeply transversely rugose and finely punctured. Elytra with shining golden pubescence which is distributed in rather obscure patches arranged more or less in three transverse broad bands. Antennae with each segment serrate on outer margin. Length 24–26 mm. East African species. Chiefly from *Afzelia* *Pachydissus hector* Kolbe

Elytra with outer apical spine longer than sutural spine. Pronotum not transversely rugose; coarsely punctured. Elytra with dull yellow pubescence evenly distributed over entire surface. Antennae with each segment with a short apical spine on inner margin. Length 20–30 mm. North American species. Chiefly from *Quercus* *Romaleum rufulum* Haldeman

12. Head broader than pronotum. First abdominal sternite elongate, as long as remaining sternites combined. Anterior coxal cavities closed behind. Thorax elongate. Elytra brownish-yellow . . *Obrium*. 29

– Head narrower than pronotum. First abdominal sternite less elongate, shorter than remaining sternites combined. Anterior coxal cavities open behind 13

13. Legs unusually short, with hind tibiae not more than two-thirds as long as femora, which are very strongly clavate. Body strongly depressed. Elytra testaceous, strongly punctured. Antennae much shorter than body. Length 7–10 mm. North American species. Chiefly from seasoned *Quercus* and *Carya* . . . *Smodicum cucujiforme* Say

– Legs of normal proportions, with hind tibiae

at least as long as femora. Body never strongly depressed 14

14. Pronotum distinctly elongate. Antenna with third segment not longer than fourth. Elytra light brown. Length 4·5–6 mm. Chiefly from dead twigs, wattle hurdle and wickerwork *Gracilia minuta* (Fabricius)

– Pronotum quadrate to strongly transverse. Antenna with third segment longer than fourth 15

15. Anterior coxal cavities transverse, angulated externally. Distance between inner margins of eyes above antennae slightly less than distance between bases of antennae. Elytra never with conspicuous yellow bands or grey at apex 16

– Anterior coxal cavities rounded, not angulated externally. Distance between inner margins of eyes above antennae slightly more than distance between bases of antennae. Elytra with conspicuous yellow bands or grey at apex (except *Trinophylum*) 20

16. Anterior coxae separated by at least their combined width. Sides of pronotum densely covered with long outstanding greyish-white hairs; disc with a conspicuous pair of round black, shining tubercles. Length 8–20 mm. From dry coniferous timbers, especially in roof voids and attics of houses and in packing-cases *Hylotrupes bajulus* (Linnaeus)

– Anterior coxae subcontiguous. Sides of pronotum sparsely covered with short, usually dark hairs; disc without black tubercles 17

17. Sides of pronotum strongly angled medially. Elytra densely covered with scarlet pubescence. Length 9–11 mm . . *Pyrrhidium sanguineum* (Linnaeus)

– Sides of pronotum rounded. Elytra never covered with scarlet pubescence 18

18. Elytra and pronotum metallic blue or violet. Disc dull and very strongly rugose. Length 11–13 mm. From dead or seasoned coniferous timber . . . . . *Callidium violaceum* (Linnaeus)

– Elytra never metallic, but if bluish, then pronotum red. Disc shining and sparsely or finely punctured 19

19. Elytra uniformly yellow, blue or brown. Length more than 6 mm . . . . . *Phymatodes*. 30
– Elytra with basal third red and the remainder black with two oblique white stripes. Length less than 6 mm . . . . *Poecilium alni* (Linnaeus)
20. Elytra densely pubescent, with a conspicuous pattern of contrasting colour. Eyes finely faceted 21
– Elytra very sparsely setose, brown, shining; densely and coarsely punctured. Eyes coarsely faceted. Length 12–15 mm. Indian species, but now possibly established in Britain. Chiefly from *Quercus* . . . . *Trinophylum cribratum* Bates
21. Pronotum with at least three short, transverse carinae on disc. Femora acutely spined apically. North American species . . . . . *Neoclytus* spp.
– Pronotum without carinae. Femora without apical spines 22
22. Apices of elytra each produced into a pair of spines and deeply emarginate between. Bases of elytra around scutellum yellow. Length 10–16 mm. Oriental species. Chiefly from bamboo *Chlorophorus annularis* Fabricius
– Apices of elytra not produced into spines. Bases of elytra around scutellum black or red . . . 23
23. Pronotum transverse, with sides strongly rounded; with a yellow transverse band across middle (often incomplete). Hind margin of head with a transverse yellow band. Elytra black with yellow bands . . . . . *Plagionotus*. 31
– Pronotum quadrate to slightly elongate, with sides slightly rounded; without a median transverse band. Hind margin of head without a transverse yellow band 24
24. Legs yellow. Elytra with conspicuous yellow markings; apices covered with bright yellow pubescence. Pronotum with front and hind margins yellow. Length 9–13 mm . *Clytus arietis* (Linnaeus)
– Legs black or brown. Elytra without yellow markings; apices covered with greyish pubescence. Pronotum completely black . . . *Anaglyptus*. 32
25. Elytra each with an oblique whitish mark broadest in the posterior half. Male antennae

12-segmented. Length 8–14 mm. From coniferous trees . . . . . . *M. minor* (Linnaeus)

– Elytra unicolorous. Male antennae 11-segmented. Length 5·5–8 mm. From deciduous trees . . . . . *M. umbellatarum* (von Schreber)

26. Femora almost completely red. Length 15–20 mm . . . . . . *C. spinicornis* (Fabricius)

– Femora entirely black. Length 15–20 mm var. *suturalis* Chevrolat

27. Basal third of elytra closely and strongly punctured, the punctures as close and strong as those on median third. Prosternal process truncate posteriorly. Pronotum with two distinct oblique impressions on disc. Length 20–30 mm *P. basalis* Gahan

– Basal third of elytra finely and very sparsely punctured, the punctures indistinct or at least finer and less numerous than those on median third. Prosternal process bearing posteriorly a distinct median tubercle. Pronotum without oblique impressions on disc. Length 20–25 mm *P. viridipennis* (Hope)

28. Elytra strongly tapering towards apex and sutural angles with an apical spine (Fig. 135a); black with at least apical third reddish-brown. Length 25–50 mm . . . . *C. cerdo* Linnaeus

– Elytra not tapering towards apex and sutural angles without an apical spine (Fig. 135b); entirely black. Length 16–28 mm . . *C. scopoli* Füessly

29. Disc of prothorax shining. smooth. Elytra orange testaceous. Length 6–10 mm . *O. cantharinum* (Linnaeus)

– Disc of prothorax dull, finely punctured. Elytra dark brown. Length 4·5–6 mm . *O. brunneum* (Fabricius)

30. Elytra finely and sparsely punctured. Antennae reddish-yellow. Sides of thorax yellow. Length 8–13 mm . . . . *P. testaceus* (Linnaeus)

– Elytra very strongly punctured, almost rugose. Antennae dark brown. Sides of thorax black. Length 6–8 mm . . . . . *P. lividus* (Rossi)

31. Apical third of elytra yellow with two transverse black bands (the posterior one often incomplete) and apex emarginate. Scutellum brown. Length 13–17 mm . . . *P. detritus* (Linnaeus)

– Apical third of elytra black, with two transverse yellow bands and apex rounded. Scutellum bright yellow. Length 9–18 mm . . . *P. arcuatus* (Linnaeus)

32. Antennae with apex of third segment with long spine. Femora brown with a conspicuous thin white band of pubescence along the inner and outer margins. Length 6–9 mm. N. American *A. verrucosus* Olivier
- Antennae without spine. Femora unicolorous. Length 9–22 mm. Elytra with basal third red, but black in var. *hieroglyphicus* Hbst. . *A. mysticus* (Linnaeus)

Notes on the species

Leptideella brevipennis (Mulsant) (Fig. 129).

This very small species was formerly only imported from time to time from the Continent in wickerwork but has now been established in England for

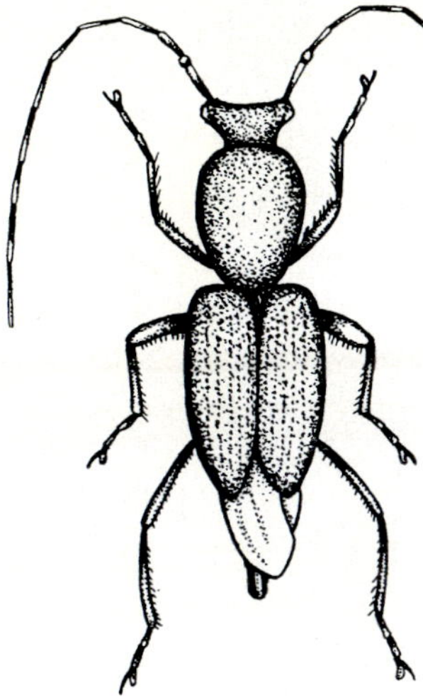

Fig. 129. *Leptideella brevipennis* (Mulsant). Length: 4·5 mm.

some years. Joy (under *Leptidea brevipennis*) gives it as being found in Southern England as far north as Yorkshire. The larvae are found in dead dry twigs of *Salix* and *Rosa canina*, but on the Continent larvae have been found in *Castanea, Quercus, Corylus, Cupressus, Robinia, Ficus, Juglans, Ceratonia* and *Pinus*. When the adult beetles find their way into barns and store-sheds they can do great damage to wickerwork and recently a unique collection of rural crafts was almost entirely destroyed by this insect, the wickerwork being progressivley reinfested until virtually nothing was left except a thin outer veneer filled with powder. The beetles emerge in June and July and sometimes the pronotum is somewhat lighter in colour than the elytra.

Molorchus

Molorchus minor (Linnaeus) (Fig. 130a). The larvae of this species are found in the damaged or recently cut branches of *Pinus*, *Picea*, *Larix*, and *Betula*, but rarely they are also found in exposed roots and boles. Duffy states that until the larvae are almost two-thirds grown they score

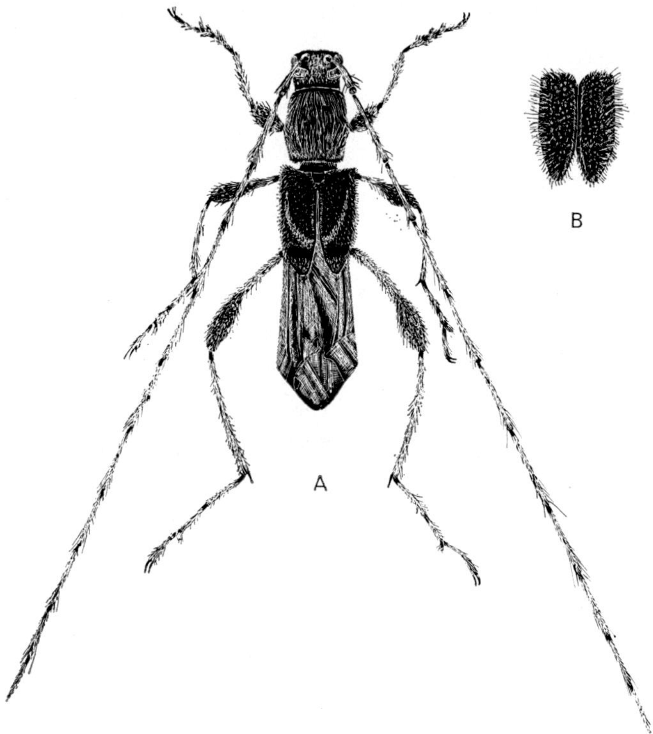

Fig. 130. A, *Molorchus minor* (L.). Length, 14 mm. B, elytra of *Molorchus unbellatarum* (von Schreber).

out broad shallow grooves on the outer face of the sapwood (Fig. 131). They then enter the wood, forming a curving gallery which, before pupation is plugged at two points—at the entrance to the cell gallery and again halfway along. The life cycle is completed in two years, and, although the final metamorphosis takes place during August and September, the adult beetles remain in the pupal cell until the following May or June. It is found locally in southern England as far north as Yorkshire.

Molorchus umbellatarum (von Schreber). The biology of this very rare species is similar to that of *M. minor*, but Duffy gives *Rubus*, *Pyrus malus* and *Rosa canina* as host plants and quotes Reineck's records of *Picea* and *Pinus*. Duffy also states that pupation takes place during April and May,

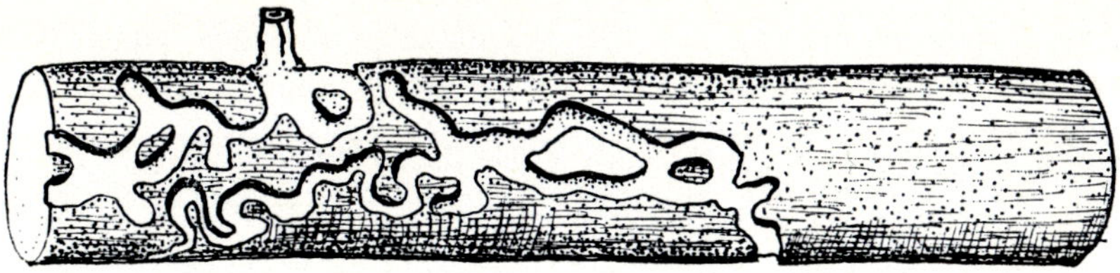

Fig. 131. Larval tunnel of *Molorchus minor* (L.) in larch branch, shown when bark has been removed. The piece of wood is 21 mm in diameter.

the adults emerging during May and June. In the author's specimens the elytra are much more strongly ciliated than in *Molorchus minor*, whilst the so-called oblique whitish mark seems an unsatisfactory character in some specimens (Fig. 130b).

Cordylomera spinicornis (Fabricius) (Fig. 132).

This insect, and the black variety, *suturalis* Chevrolet (previously thought

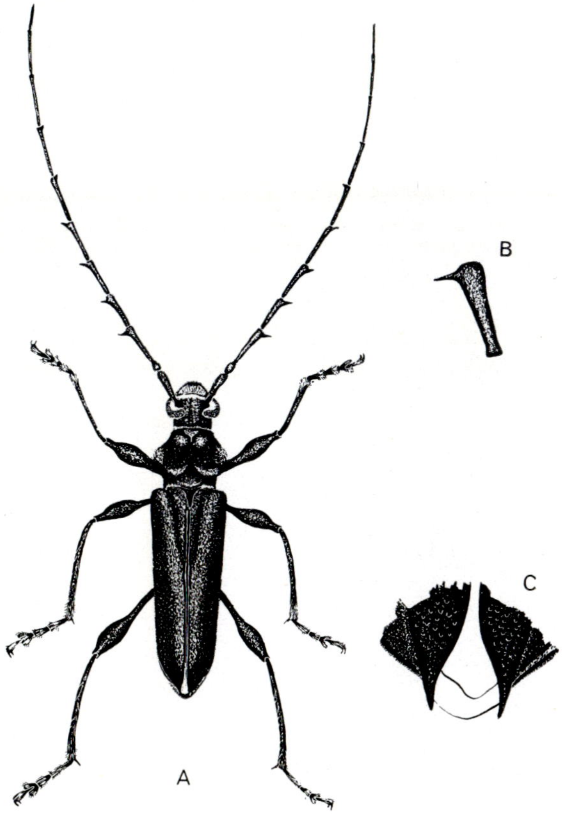

Fig. 132. *Cordylomera spinicornis* var. *suturalis* Chev. A, adult, length 19 mm; B, fourth antennal segment; C, apices of elytra.

to be a separate species), are often accidentally imported from Africa. Both are found in *Khaya* (African mahogany and other timber species included in this trade term). Fraser records var. *suturalis* flying and copulating in the summer sunshine in a Lancashire timber yard.

Plocaederus viridipennis (Hope) (Fig. 133).

Several species of *Plocaederus* are imported into Britain from time to time in timber, *P. viridipennis* (Hope) being the most common. The shining elytra of this insect are rather more bluish and not so tapering as those of *Philematium natalense* Bates. In addition, in the former insect the pronotum is black and dull but is green and shining in the latter. *P. viridipennis* is very much more robust than *Cordylomera sutalis* Chev. and the femora are not dilated which they are in the latter species. The eyes nearly meet dorsally. The larvae are found burrowing in the outer sapwood of mature trees of West African mahogany (*Khaya*), iroko (*Chlorophora excelsa*) and afara (*Terminalia superba*). The inner bark is perforated here and there, the channels being kept free of frass. The pupal chamber is usually in the sapwood but occasionally it occurs in the heartwood. Before pupation, however, the flight-hole is constructed and then plugged with long fibres and finally a complete calcareous cocoon is secreted. Although

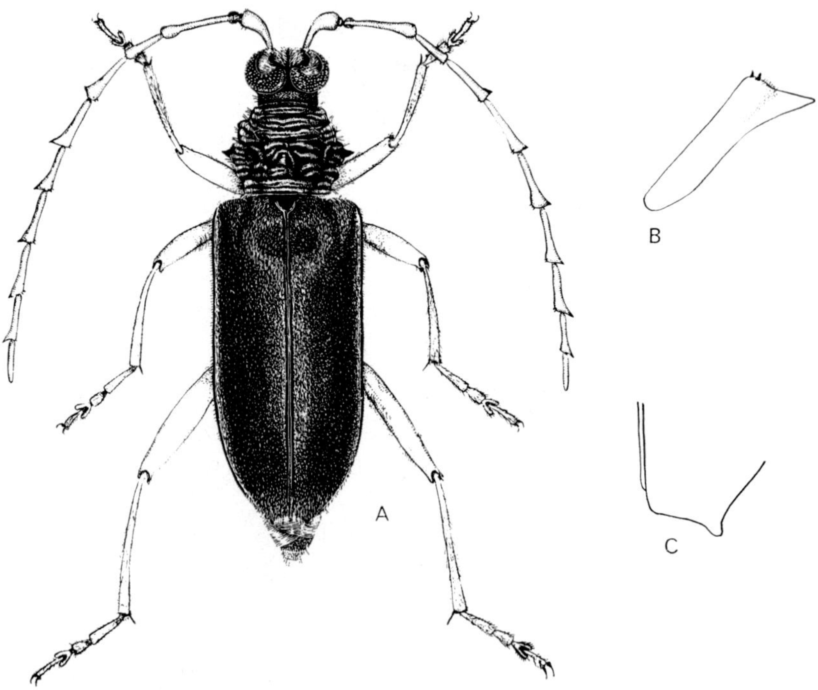

Fig. 133. *Plocaederus viridipennis* (Hope). A, adult, length 25 mm; B, ninth antennal segment; C, apex of right elytron.

restricted to the sapwood, damage by this species can be severe. There is little information available concerning *Plocaederus basalis* Gahan except that its country of origin is West Africa and that it attacks the timber of *Aucoumea klaineana* (gaboon or okoume). *Plocaederus fucatus* Thompson has recently been imported in timber of *Lovoa klaineana*. *Plocaederus denticornis* (F). has been found in timber of *Pycnanthus kombo* from West Africa.

Cerambyx

The genus *Cerambyx* contains two European species which although not indigenous to Britain are occasionally introduced in parcels of timber from Europe and these large handsome Longhorn Beetles are then prizes for the beetle collectors.

Cerambyx cerdo Linnaeus (Fig. 134). This is 'Le grand Capricorne' of

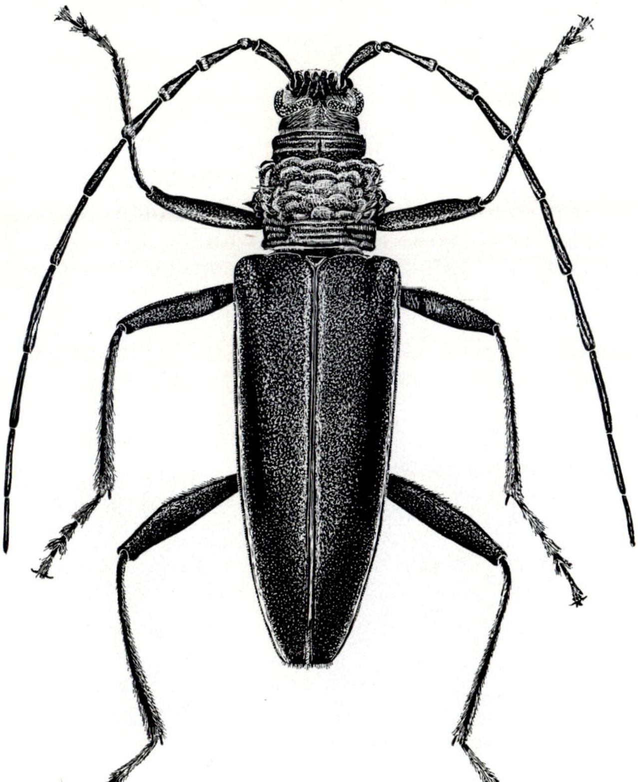

Fig. 134. *Cerambyx cerdo* (L.). Length: 42 mm.

the French and the 'Grösser Eichenbock' of the Germans. It is one of the largest wood-boring insects likely to be found in Britain, measuring from 25–50 mm, although large specimens of *Ergates faber* (L.) measure up to 55 mm. The head and pronotum are shining black but the elytra, although black proximally, gradually become reddish-brown towards the apices (Fig. 135a). The beetle is to be found only at night on the trunks of old oak trees and is common generally throughout Europe. In the environs of Paris it may be found in the Bois de Boulogne, Fôret de Saint-Germain and the Fôret de Fontainbleau. Although *Quercus* is the preferred host timber it is also recorded in *Pyrus malus*, *Ulmus*, *Fagus*, *Robinia*, *Salix* and *Carpinus*. Although plantation trees are sometimes attacked, the usual focus of infestation is an isolated mature or over mature tree. The value of such trees from an economic timber point of view may be entirely ruined by this insect. Before pupation, which takes place in August, the larva secretes a calcareous operculum which seals off the entrance to the pupal chamber from the larval gallery. The pupal cell is 80 mm in length and 25 mm in diameter. The final metamorphosis takes place in the late autumn but the adult does not bore its way out until the following May, June or July. The length of the life cycle is probably two to three years.

The immature stages appear to be particularly susceptible to predation and parasitization and it is of interest that the large hymenopteron *Rhyssa persuasoria* (L.), which attacks the larvae of the Wood Wasp, parasitizes this species. It is recorded also that the larvae of the Goat Moth, *Cossus cossus* (L.) are often found in association with the larvae of *C. cerdo* and the latter usually come off worst in an encounter.

Cerambyx scopolii Füessly. This is 'Le petit Capricorne' of the French and the 'Buchenspiessbock' of the Germans. The latter name is taken for its habit of attacking the beech tree. It varies from 16 to 28 mm in length

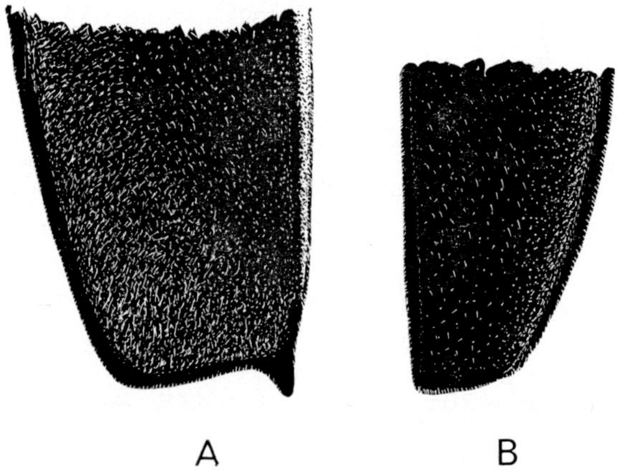

Fig. 135. *Cerambyx*. A, apex of left elytron of *C. cerdo*; B, apex of right elytron of *C. scopolii*.

so that although much smaller than *C. cerdo* on average, there is some slight overlap in size. It is entirely black in colour (Fig. 135b) and is also generally common throughout Europe. The beetle is found during the daytime on flowers of ROSACEAE, UMBELLIFERAE and the elder. The recorded host timbers are *Prunus cerasus*, *P. avium*, *Pyrus malus*, *Quercus*, *Ulmus*, *Castanea*, *Fagus*, *Ribes*, *Carpinus*, *Acer* and *Fraxinus*. The biology is similar to that of *Cerambyx cerdo* but its economic importance is apparently greater in connection with orchards than in mature oaks. At first the larvae are cambium feeders, then they enter the sapwood and finally the heartwood in which they make broad irregular galleries. The length of the larval stage is two to three years and the pupal cell is 30–40 mm in length and 10 mm in width.

Eburia quadrigeminata Say (Fig. 136).

The information given in the key is a sufficient description of this very characteristic beetle. This North American insect occasionally emerges from furniture or cabinet woodwork constructed of timber originating from that continent. The flight holes are large and oval. Craighead has recorded the following timbers as being attacked, *Quercus, Carya, Robinia, Fraxinus, Castanea, Ulmus, Fagus, Acer* and *Cupressus*. It is from the first named timber, however, that most beetles emerge in the United Kingdom, American oak being by far the commonest of the North American hardwood timbers imported.

The young larvae gain access to the standing mature oak tree by way of wounds and scars where the bark has been removed. Thence they bore into the heartwood, packing the large contorted galleries tightly with granular frass. *Eburia quadrigeminata* exhibits to a marked degree the phenomenon of delayed development. When the timber is converted and seasoned the young larvae survive and at a later date emerge from furniture fabricated from it, sometimes after an almost unbelievable interval. Seven records from the literature varying from 5 to 40 years (this latter specimen was said to have emerged from *Betula*, a species not recorded as a host timber by Craighead) are summarized by Duffy. It cannot, however, be called a common insect in Great Britain, scarcely one specimen annually being received for identification by Rentokil.

In America the life cycle takes two or more years and the beetles emerge in the summer.

Stromatium barbatum (Fabricius).

This insect is an important pest in India, Ceylon, Burma, Mauritius and Madagascar where it is found not only in the forests but in all types of woodwork indoors, from packing-cases to window frames and doors. About 350 species of timber are known to be attacked by this beetle. Often development of the larvae is delayed, the beetles emerging some years after the timber has been utilized—the longest recorded being 10 years. The complete life cycle can take place in plywood 4 mm in thickness. Specimens

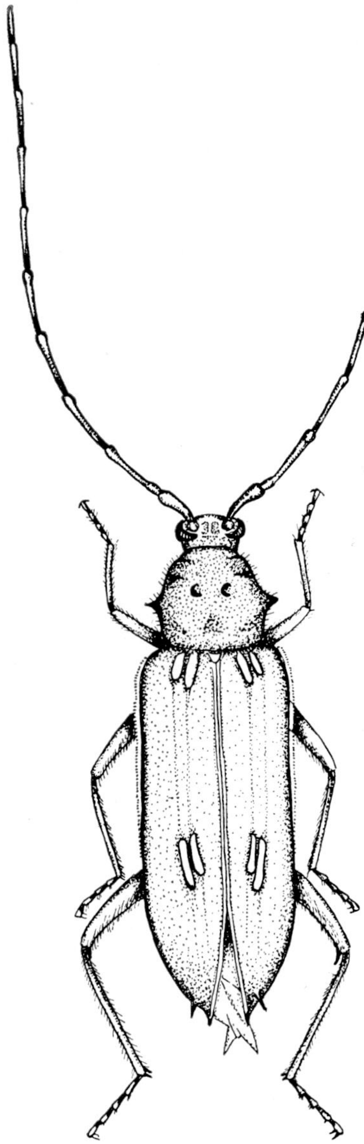

Fig. 136. *Eburia quadrigeminata* Say. Length: 18 mm.

are imported into Britain from time to time usually in tea chests or in packing-cases. The adult varies from 12 to 28 mm in length and from reddish-brown to almost black, covered with tawny pubescence. The pronotum and the elytra are closely punctured. The beetles are active at night, hiding during daylight.

The very irregular larval tunnels which are tightly packed with fine powdery frass, cross and recross until all the wood has been consumed, only a thin veneer of sound surface wood remaining untouched—reminiscent of a heavy attack by *Hylotrupes bajulus*. It is thought probable that the larval gut contains the enzyme cellulase and thus may have the power of digesting cellulose. Two other species are recorded as being occasionally imported, *Stromatium longicorne* Newman from Oriental countries and *S. fulvum* (Villiers) from the Mediterranean and South America.

Pachydissus hector Kolbe (Fig. 137).

This fairly large but somewhat sombre coloured insect is occasionally imported into this country in the timber of *Afzelia africana* from West Africa. It bears a superficial resemblance to a *Cerambyx* but a vague elytral pattern serves to distinguish it. This pattern in produced by patches of fine

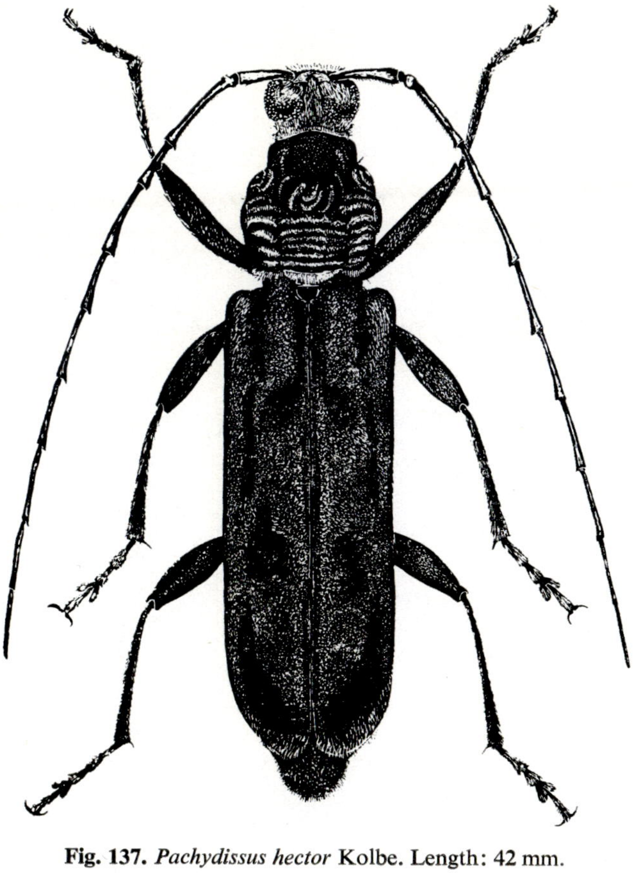

Fig. 137. *Pachydissus hector* Kolbe. Length: 42 mm.

pubescence irregularly placed which is, in addition, somewhat iridescent, so that the pattern changes according to the angle at which the insect is observed. The larvae are cambium feeders but the pupal cell is constructed either deep in the sapwood or in the heartwood. The pupal cell is capped but not lined, with a calcareous secretion. Duffy has seen a number of logs completely destroyed by this insect. *Pachydissus regius* Aurivillius has also been imported from West Africa in *Tectona grandis*.

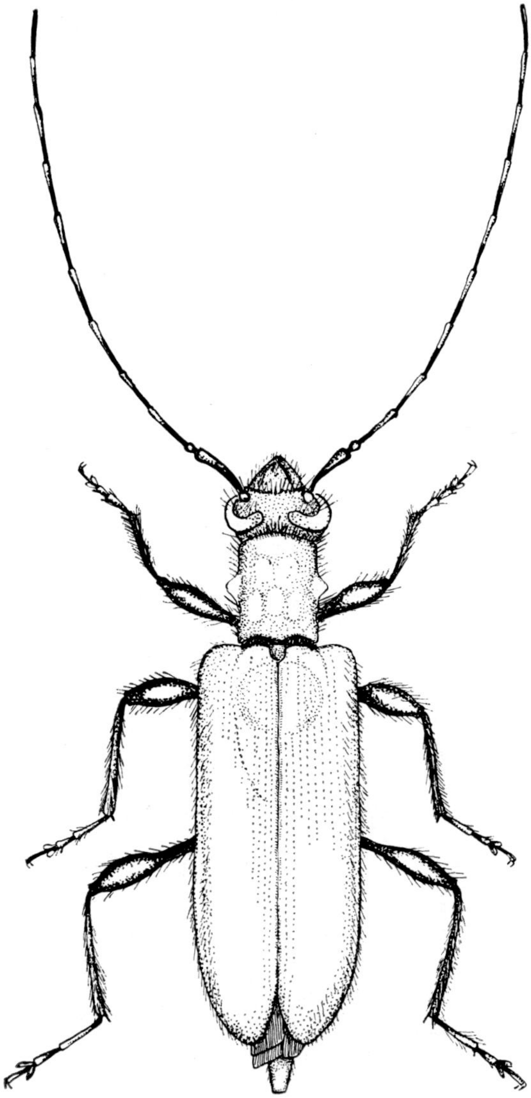

Fig. 138. *Obrium cantharinum* (L.). Length: 8·5 mm.

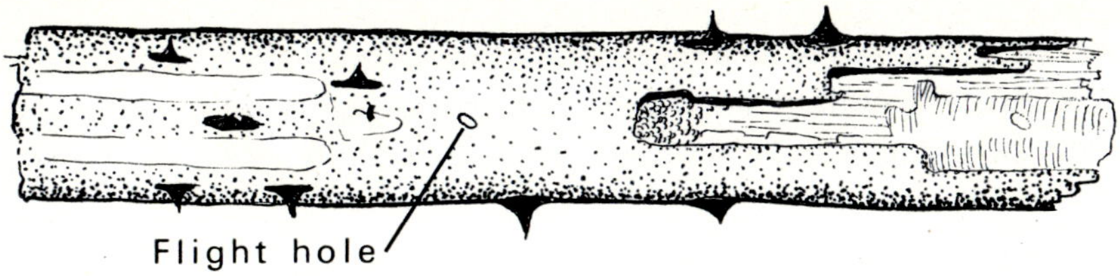

Fig. 139. Piece of old bramble stem showing a flight hole of *Gracilia minuta* (F.). The stem is 21 mm in diameter.

Romaleum rufulum Haldeman

A common cerambycid throughout the Eastern United States where it is known as the red oak borer, the host timber being *Quercus*. It is up to 25 mm in length and is brown with yellowish pubescence, the latter being denser on the head and elytra. The apices of the elytra are produced into two small spines but there are no lateral spines on the prothorax. Great damage is done by this insect to mature living trees by the larvae which, after feeding in the cambial layer for a year, bore their galleries deep into the heartwood. The life cycle takes two to three years and often good trees are killed by concentrated attacks by this beetle. It is occasionally imported into Britain in North American oak.

Obrium cantharinum (L.) (Fig. 138).

A very rare British species found only in southern England. It is light shining reddish-yellow with rows of small punctures in the elytra and varies from 6 to 10 mm in length. Joy says that it is found chiefly in fruit trees but *Populus tremula* and *Quercus* are recorded as host trees on the Continent. It has an annual life cycle and the larvae feed under the bark, only entering the sapwood to pupate. The adult beetles are usually found in June and July and are attracted to the flowers of UMBELLIFERAE. Another extremely rare species in this genus is *Obrium brunneum* (Fabricius) but it is very doubtful whether this is an indigenous insect. In contrast to the previous species it is found only in coniferous timbers, *Abies excelsa, Pinus* and *Picea* being recorded; the larvae are found under dead bark. The adult beetle may be found from May to July.

Gracilia minuta (Fabricius) (Colour Plate, I; Fig. 139).

Host plants of the larvae are *Rubus, Rosa canina, Corylus, Salix, Aesculus, Betula, Quercus, Crataegus, Rhamnus alternus* and *Ceratonia siliqua*. Usually the dead dry twigs and small branches are infested and the larvae make their galleries just under the bark. They pupate in the sapwood or in the pith channel. This is a local species, but important economically on account of the damage it causes to wickerwork, rustic work and other craft work. As in attack by *Leptideella brevipennis*, the infested wood is completely destroyed.

The House Longhorn Beetle

Hylotrupes bajulus (Linnaeus) (Colour Plate, B; Fig. 140)

In Britain this insect is known as the 'House Longhorn Beetle', in Germany 'Hausbock', in Spain 'Cerambicido de la madere labrada', in France 'Capricorne des maisons' and in the United States 'Old House Borer'. It has been remarked that *Hylotrupes bajulus* has been the subject of more research in Germany in connection with its biology and destruction than has been the case in any other insect species.

Status in Britain

Hylotrupes bajulus has been known in Britain since 1795. Kaufmann summarized the forty or so published records of this insect in Britain from 1830 to 1946 and he gave his opinion that it was a 'casual importation', but Hickin in 1947 suggested that it was an 'established introduction'. Fisher

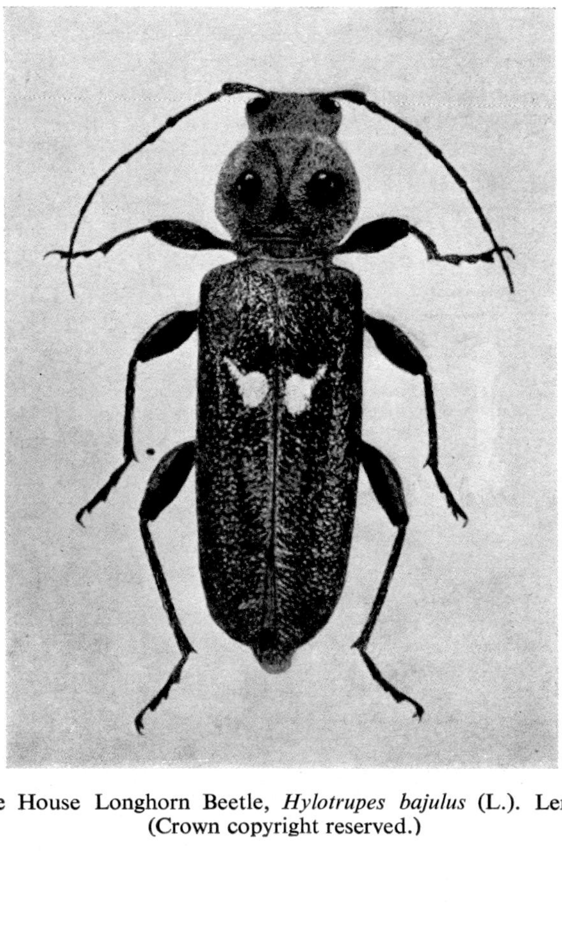

Fig. 140. The House Longhorn Beetle, *Hylotrupes bajulus* (L.). Length: 15 mm. (Crown copyright reserved.)

and Harris in 1949 and White in 1954 and 1959 have also discussed the past records, present distribution, and economic status of *H. bajulus* in Britain. From the evidence of old larval damage in old buildings now apparently inactive and old specimens in collections, it seems probable that it was widespread 50 to 100 years ago. It is now extremely rare to find this species wild—outside buildings, so that it must be assumed that the indigenous race died out somewhere round the turn of the century. However it has occurred in pest proportions in a limited area of north-west Surrey, probably from the early nineteen-thirties when Parkin recorded the first of three instances of this new phase of Longhorn attack in Britain. In 1954 Tame and Snelling recorded a survey of sixty houses in Byfleet, Surrey, when twenty-seven were found to be infested, and judging from the frequency that commercial control operations have occurred in some areas of Camberley, Weybridge and Addlestone and the surrounding area something like this distribution is to be expected. The largest company reports about thirty House Longhorn control operations annually.

Although only isolated cases are known of this insect being imported in building timber, instances of living larvae being found in packing cases, orange boxes, etc., are numerous, most frequently originating from Spain. This seems to be the most likely source of the recent sporadic attacks.

World distribution

Hylotrupes bajulus is distributed from central Norway to North Africa and from Portugal to Siberia. In addition it has been introduced and has established itself in many other countries including Australia, South Africa and the United States. Cymorek remarks that in central Europe and in some South European countries *H. bajulus* is the most important softwood-destroying insect. He also states that in many territories it has disappeared as a 'wild' species occurring in the open, but that its characteristics of being able to live and reproduce in dry barkless timber are the 'causes and conditions for its extensive domestic occurrence'.

This view is also held by Rasmussen who further suggests that the original habitat disappeared with the Mediterranean primaeval forests when they were destroyed by felling and sheep-grazing in ancient times. 'Only the simultaneous presence of a replacing biotope, timbers in houses, saved the species from extinction'.

Host timbers

In Britain *Hylotrupes bajulus* is confined to the genera *Pinus*, *Picea* and *Abies*, but in continental Europe it has been recorded on *Alnus*, *Corylus*, *Populus*, sometimes on *Quercus* and surprisingly on *Genista scorpius*; in America it has occurred on *Conium maculatum*. The sapwood of the dry timber is attacked and not before the major part of this is destroyed does the larva bore into the heartwood. In recent years in Britain it has only been found once under what can be termed natural conditions, in a pine stump, even after an extensive search in the pine woodlands which are a common landscape feature in much of the area in which the insect abounds.

The first timber attacked in a building is usually that in the roof space and there have been various theories put forward to account for this, the most quoted being that this situation is very hot in summer; but perhaps the alternation of cold in winter is of equal importance. The great spread of *Hylotrupes bajulus* in Europe in recent years has been explained by the replacement of thatch by slates, tiles and bituminized felt, all the latter absorbing heat whilst the former had a heat-insulating effect. Bletchly has pointed out that the area in Britain where the species is established is also an area of high summer temperature, believed to be the highest in Britain. The dilemma exists as to whether the insect is indeed limited to its present distribution in Surrey by climatic conditions or whether it can extend its distribution far wider in the British Isles if means for its dispersal were at hand. Several authors have referred to the activity of the insect in a hot summer. A series of such summers could well lead to many other foci of infestations occurring, probably each originating not from dispersal from our own established population but from the orange boxes which White has shown to be a constant means of introduction from abroad.

Biology

Identification

The adult beetle is variable in size, very small males being only about 7 mm in length, whilst the largest females are up to 25 mm. It is somewhat compressed dorsoventrally and the pronotum is almost circular. The integument is greyish-brown to black and covered with greyish pubescence. On the pronotum there are two shining black eye-like areas divided by a central longitudinal band which, with the surrounding oval areas of pubescence, give it an owl-like appearance (Fig. 141). On each elytron

Fig. 141. Pronotum of *Hylotrupes bajulus* (L.).

there are four greyish or whitish areas arranged in two transverse sloping bands, which may more or less coalesce. The males are mostly smaller than the females and have longer antennae. The female has a pronounced elongated pygidium.

Courtship and copulation

Preliminary courtship lasts about a minute and a half and copulation about the same time. During courtship legs and antennae are sometimes bitten off. Both males and females are able to copulate several times even in the course of one day. It is not necessary for the female to copulate more than once but it has a stimulating effect on the female to recommence egg-laying.

Oviposition

The female begins her search for a suitable egg-laying site immediately after copulation and although it is usual for oviposition to commence as soon as a suitable site has been found, on occasions this might be delayed for several days. The female is influenced in site selection in a general way by smell (the terpenes, α and β pinene seem to be attractive) and light, but nearer the point of actual egg laying the tactile sense is brought increasingly into play to the exclusion of the former senses. The female can be persuaded at this stage to lay on such unpromising materials as cement or wax, provided suitable-sized cracks are available. As Cymorek states 'at the decisive moment the females do not have the capacity to decide whether the material is suitable for the existence of their progeny'. The particular mechanical requirement of the female for the oviposition site is a crack of 0·25 to 0·6 mm in width extending to 2 or 3 cm into the wood. This is the sort of crack caused by drying or shrinkage. If the crack is wider than that stated then the beetle will get inside it or lay in the lateral extensions. Each egg is placed carefully, the position being determined by the vaginal palps. The average duration of oviposition is 12 days and during this time from two to eight clutches are laid, later clutches being smaller than the earlier. There is great variation in the egg number. the smallest complete oviposition recorded was 86 and the largest 582, whereas the average appears to lie somewhere between 140 and 200. The eggs are dull white and spindle-shaped and are 1·2–2 mm in length and 0·5 mm broad.

Relative humidity and temperature play a large part in determining the length of the egg stage. Steiner found that at 26·3°C and between 70 and 80 per cent relative humidity the eggs hatch in 9 to 10 days, but at 31·5°C and between 90–95 per cent relative humidity it is reduced to 5·5–5·6 days. There is, however, as much as 24 hours difference in the hatching time shown by different members of the same clutch. If the surface of the wood crack is rough and conditions are otherwise good the young larvae can burrow into the wood within 12 hours.

Larvae (Fig. 142)

Two detailed studies have been made, that by Duffy in 1953 and by

Dürr in 1957. The larva is typically cerambycid in form; although generally cylindrical it is somewhat flattened, broadest at the thoracic region and gradually tapering towards the hinder end, but the abdominal segments prior to the apical two enlarge somewhat before tapering again. The colour is a shiny ivory white and it has a fleshy appearance; the intersegmental grooves are very distinct. The body has a sparse covering of long, thin, yellow hairs. The head, of which the hinder part is invaginated into the prothorax, is very small but the jaws which are heavily pigmented are very powerful. The larva is up to 24 mm in length and up to 7·5 mm

Fig. 142. Larva of a House Longhorn Beetle exposed in its tunnel after the surface veneer of sound wood was cut away.

broad at the prothorax (maximum breadth). The mandibles are broad with a deep longitudinal median depression and at their base close to the antennae are three ocelli (Fig. 143). In the majority of cerambycid larvae there is only a single ocellus on each side. Duffy states however that in *Hylotrupes bajulus* there are in addition to the three pigmented pairs of ocelli, a pair of rather widely spaced unpigmented ocelli on either side.

The larva of *Hylotrupes bajulus* bears small, although well defined, legs. The tarsus is long, roughened and tapering. The legs are probably of little value to the insect, their function having been taken over by the ambulatory ampullae, both dorsal and ventral, which the larva uses for moving along its tunnel. Three anal lobes are present, each being triangular, and they are contractile into the tenth abdominal segment.

A prepupal stage in which the head assumes a hypognathous position has been described by Dürr.

Larval development

The optimum conditions for larval growth lie between 28 and 30°C and from 26 to nearly 50 per cent moisture content in the wood. The young larvae grow at a faster rate compared with the older larvae and at first keep just underneath the outer skin of the sapwood. This is the area of highest nutrient value. The larval galleries, which are oval in cross-section, are filled with rejected wood particles and faecal pellets. Becker has stated that the less nutritive value there is in the wood for the larvae the less is the proportion of droppings compared with the unconsumed gnawings in the frass. The frass is rarely ejected by the larva but it can be detected under the surface as long blister-like protrusions and it sometimes happens

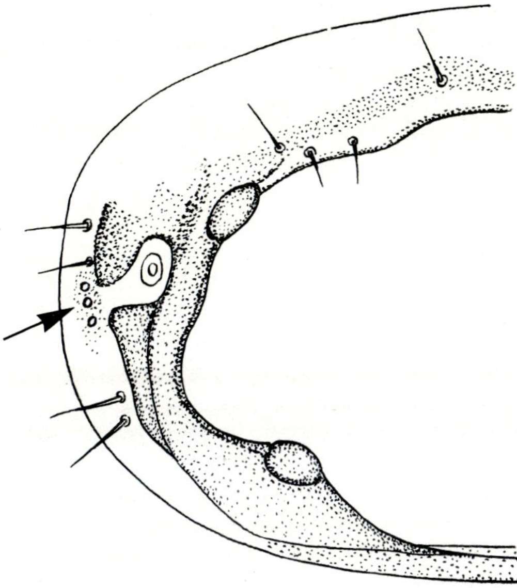

Fig. 143. Right half of mouth frame of *Hylotrupes bajulus* larva from the front, showing the three ocelli (arrowed) at base of antenna.

when the wood substance has been largely consumed, that the greater volume of the frass causes the outer veneer of unconsumed wood to split, usually along the grain. The faecal pellets are cylindrical but on dessication they split into two parts. The larvae can clearly be heard gnawing the wood, especially in warm weather when they are most active. The length of larval development is usually from 3 to 6 years but 2 years and up to 10 years are not uncommon and there is one record of 32 years. At a weight of 40 mg pupation is possible but only very small males will result, usually the pupation weight lies between 150 and 500 mg. The lighter larvae will usually only produce males. A weight of 860 mg has been recorded for larvae subjected to additional albumen feeding.

It is recorded that lead and other metals have been perforated by larvae and adults of this species.

The nutrition of the larva has been studied by a number of investigators. Apparently no symbiotic organisms are present, but Mansour (1934) considered that the larvae secrete the enzyme cellulase and are thus able to digest cellulose. This is confirmed by Parkin (1940) who places this species in his group iii—those wood-boring larvae able to utilize cell contents and the carbohydrates of the cell wall including cellulose. Falck (1930), who compared the analysis of the frass with that of the unconsumed wood, considered that the larvae removed about 21 per cent of the cellulose and hemicellulose content of the latter. Becker (1943) showed that the larva cannot thrive on a sole diet of cellulose, and that if the oils, resins and fats were removed from the wood it has a beneficial effect on the larva, but if the concentration of these wood extractives was increased even by a small amount the effect was detrimental to the larva. Becker also found that even very small amounts of albumen in the wood were a necessity for larval development. Dürr (1957) showed that the minimum larval plus pupal period was only one year when the protein and ash (giving an indication of mineral content) percentage was high. When it was low this period was four years. Dürr also showed that 25 per cent of the protein is depleted by the larvae from the wood.

Pupae

The pupa is white and varies from 14 to 25 mm in length according to Duffy, but his specimens could not include pupae producing the diminutive males. The mandibles resemble those of the adult in being pointed. The first and second pairs of legs lie ventral to the wings but the third pair lie dorsally. The length of the pupal stage has been recorded in Europe as two to three weeks, but in South Africa, Dürr has observed that it lasts from 29 days to more than 44 days. Cymorek confirmed the information given by Becker for the duration of the pupal stage at 70–80 per cent relative humidity. At 21°C it was 22·5 days, at 24°C 17 days, at 25°C 15 days, at 26°C 14 days and at 28°C 11 days; several pupae being used in each case.

Dürr also states that the winter is passed in the larval stage and pupation occurs during the spring months or early summer, and although pupae are generally found near the surface of the wood they are sometimes found at a depth of 10–15 cm. Duffy states that pupation usually takes place about May though sometimes it occurs in autumn or even winter.

When the eclosion of the adult has taken place, several days may elapse before it emerges, during which time the integument hardens; it removes a bung or stopper made by the larva prior to pupation and bites out the oval flight hole.

The adult beetles emerge during July to September (in Germany from mid-June) usually when the weather is hot. Although more females than males are normally observed, investigation of 5,239 reared beetles gave 50·7 per cent males, 49·3 per cent females. The duration of the adult stage was found, in the case of females, to average 25 days.

Parasites and predators

On the European mainland the clerid beetle *Opilo domesticus* L. is an important predator of this and other wood-boring insects. It does not occur in Britain. Perhaps this would be a beneficial introduction. Hymenopterous parasites are *Doryctes leucogaster* (Nees), *Rhoptrocentrus piceus* Marsh., *Sclerodermus domesticus* Klug, *Caenocryptus minator* (Grav.), *Ephialtes imperator* Kreich. (*manifestator* auctt. nec L.) and *E. tuberculatus* Fourc.

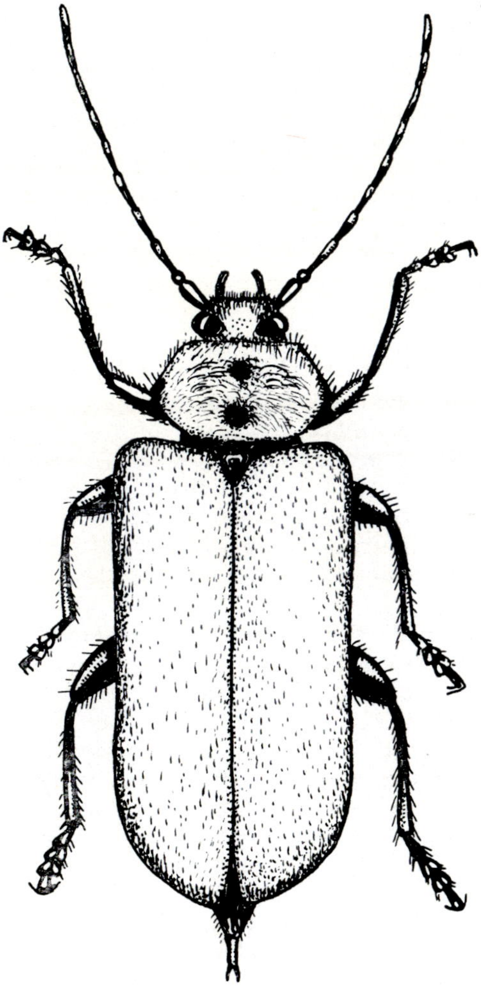

Fig. 144. *Pyrrhidium sanguineum* (L.). Length: 11 mm.

Pyrrhidium sanguineum (Linnaeus) (Fig. 144)

This very beautiful red longhorn appears somewhat flattened in appearance. The legs, antennae and body are black, the red coloration being due to the presence of a fine, thick pubescence of that colour. There is a superficial similarity to the Cardinal beetles (the two species *Pyrochroa coccinea* and *P. serraticornis*) both in colour and size. In the two *Pyrochroa* species, however, the antennae are pectinate whilst in *P. sanguineum* they are filiform.

The beetle is very rare as an indigenous insect being confined to a locality in Herefordshire (Allen, 1973). It does, however, get imported into our timber yards from the Continent and notice is taken of it on account of its bright coloration. Although it is most commonly met with on *Quercus*, it is recorded from *Aesculus*, *Carpinus*, *Castanea* and *Fagus* and once only on *Pinus*. Attacks occur in the freshly cut branches or standing trees that have been killed. The tightly packed powdery frass in the galleries in the cambial layer often loosens the bark. Duffy states that the mature larva tunnels into the sapwood in the autumn and there excavates an oval cell 2 cm by 1 cm; pupation takes place in March or April, the adults emerging in May or June. The adult beetle is recorded as boring through metal sheeting obviously in attempting to emerge from the pupal chamber, but there do not appear to be records of its emerging from woodwork indoors in Great Britain.

Callidium violaceum (Linnaeus) (Fig. 145).

This beetle is a deep blue-black in colour and the elytral colouring throws up most attractively a number of points of violet and mauve colour. Although local in its distribution it is found in England, Scotland and Ireland. It infests dry seasoned wood used out of doors, such as posts and rustic work, where the bark is still adhering to the wood. Joy says 'especially in new pine palings'. Softwoods are normally infested, *Larix*, *Picea* and *Pinus* having been recorded and more rarely *Alnus*, *Salix* and *Quercus*. The greater part of the larval gallery is excavated in the sapwood, the inner bark just being scored. The older larva does not construct a narrow gallery but feeds in a large flat oval area; before pupation it burrows into the wood curving round until it is parallel with the grain. It then pupates facing the entrance. The complete life cycle takes 2 years but one of 15 years has been recorded. The adults are found in May and June. With regard to its economic importance Duffy states that although the actual damage is superficial the accumulation of water and fungal spores accelerates decay. He knows of no instance of a standing tree being attacked.

Phymatodes

Phymatodes lividus (Rossi) (Fig. 146). This species is given by Duffy as being imported occasionally from the Continent but Joy although stating that it is very rare gives Berkshire as a locality. The two specimens in the author's collection are labelled Reading and one bears the name of a wine stores. Host timbers are given as *Quercus*, *Castanea*, *Aesculus* and *Salix* and the significance of the wine stores mentioned above is that this species

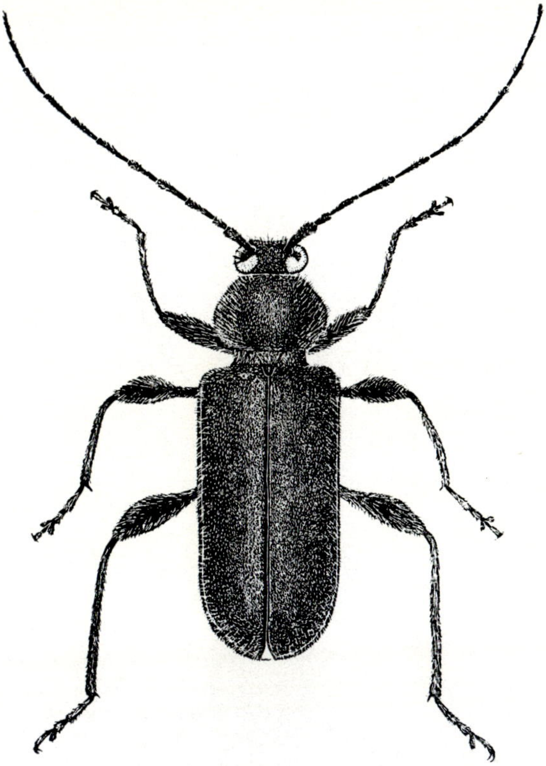

Fig. 145. *Callidium violaceum* (L.). Length: 12 mm.

was formerly of economic importance in that it attacked the chestnut hoops formerly used around wine casks. Ultimately the hoops failed and the casks burst with the loss of the contents.

The larvae feed under the bark of recently cut or dead trees and Duffy states that the larval habits are similar to those of *Pyrrhidium sanguineum*.

Phymatodes testaceus (Linnaeus) (Fig. 147). This insect is very variable in colour, from yellowish-brown through reddish to a deep blue-black. Joy in 1932 said that this species was local in southern England, extending as far north as Nottinghamshire. Today it appears to be much more common, probably on account of the large stocks of oak held at the end of the war and for some years afterwards and stored in an unbarked condition. Although *Quercus* is the preferred host timber it has been recorded from *Fagus*, *Salix*, *Prunus avium*, *Castanea*, *Carya*, *Picea*, *Tsuga* and *Fraxinus*. The eggs are laid under the bark of recently cut trunks or of recently dead standing trees. The larval galleries are tightly packed with frass and as they are situated in the inner bark adjacent to the cambium they sometimes loosen the bark from the wood in heavy infestations.

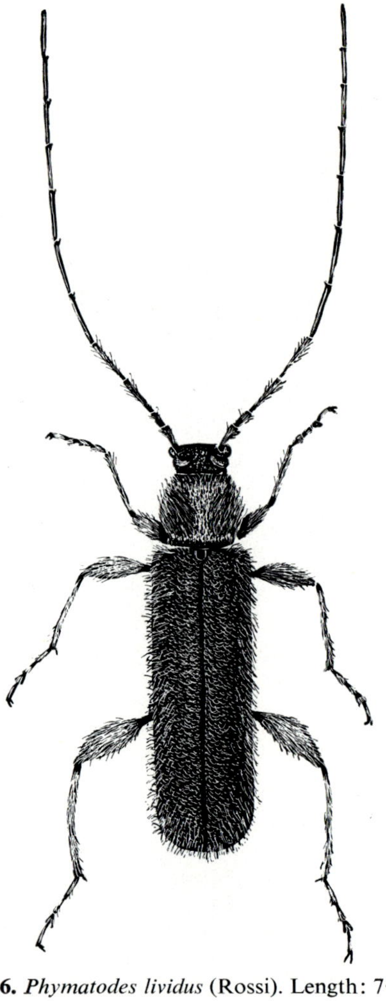

Fig. 146. *Phymatodes lividus* (Rossi). Length: 7·5 mm.

Duffy says that if the bark is thin the larva constructs a pupal cell in the sapwood but if the bark is thick a shallow oval pupal cell is made either in the bark or just beneath it. Larval development normally takes about two years and the pupal stage lasts about three weeks. Adults emerge from May to July.

Often this beetle is of considerable economic importance as the sapwood of oak boles can be almost completely destroyed. Although Duffy says that the damage is comparatively superficial and does not seriously detract from the utility of the timber, there have occurred a large number of cases where *Phymatodes testaceus* adults have emerged from oak block flooring. For some time after the last war certain types of the latter were

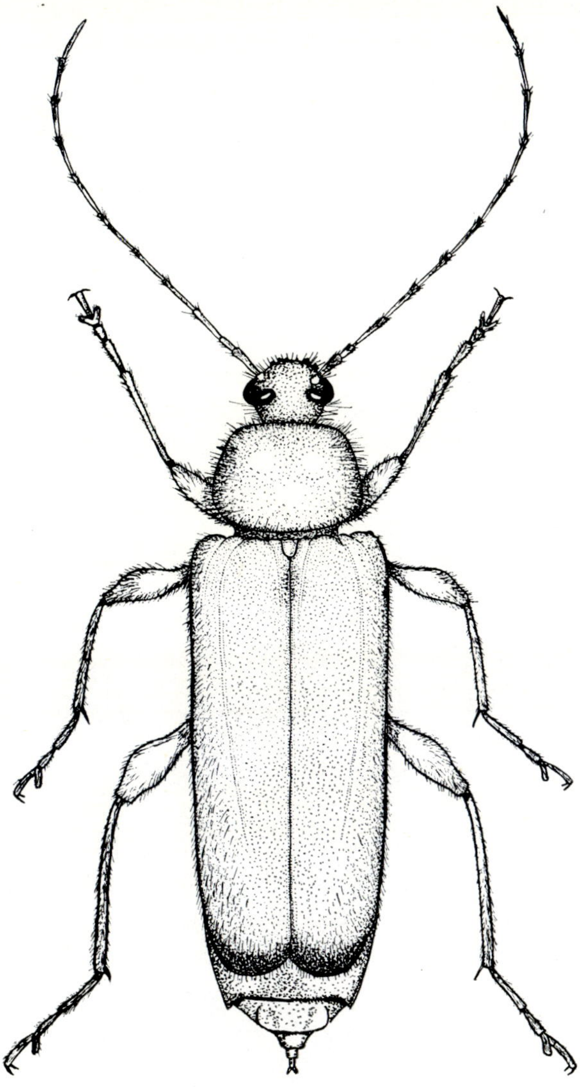

Fig. 147. *Phymatodes testaceus* (L.). Length: 11 mm.

of low grade, containing a high proportion of sapwood and as *P. testaceus* is able to complete its development in converted and fabricated timber there were many instances of these fairly large beetles emerging from such floors, much to the consternation of the householder. The beetles, however, do not lay eggs on such woodwork. Craighead in America draws attention to the damage caused by this species to the bark of oak and hemlock when used for tanbark. In addition to damage caused by indigenous examples of this species, imported timber from Europe often contains them.

Poecilium alni (Linnaeus) (Fig. 148).

Kloet and Hincks considered that this local species should be placed in the genus *Phymatodes* (Mulsant 1839) and questioned the validity of Fairmaire's genus *Poecilium* (1863). Recent workers however appear to

Fig. 148. *Poecilium alni* (L.). Length: 5·4 mm.

have reverted to the use of this latter genus. The host timbers are given as *Quercus, Alnus, Ulmus, Rosa, Castanea* and *Fraxinus*, the first two mentioned being preferred. Duffy states that the larvae feed under the bark of recently dead or decaying slender twigs but that they are also found in recently erected posts and lopped branches. They make their tunnels straight and parallel with the grain of the wood, boring in the sapwood or in the pith canal in order to construct a pupal chamber. This occurs in March and April and the beetles emerge in April and May, the life cycle being an annual one.

Trinophylum cribratum Bates (Fig. 149)

This species was first discovered in England in 1946 and at first was thought to be a new species until it was recognized as having been described by Bates in India. Duffy has given an account of it and its economic importance in timber yards and has remarked on the great damage which can be done in a relatively short time. It is of moderate size varying from

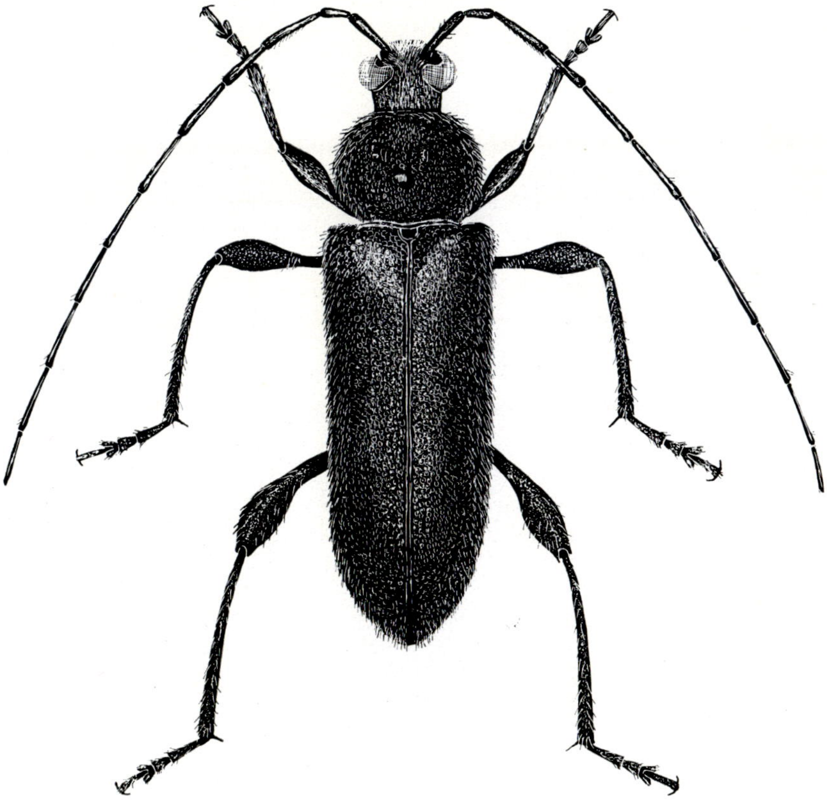

Fig. 149. *Trinophylum cribratum* Bates. Length: 16 mm.

10·5 to 17 mm and is of dark chestnut colour covered with sparse brown setae on the upper surface. The pronotum is globular, covered with setae except for three small areas, the largest in the centre of the disc and two smaller anterior to it. The egg is 1·6 mm in length and 0·6 mm wide, rounded at one end, the other end slightly attenuated. It is white when first laid, later turning pale yellowish white. The larva is similar to that of *Phymatodes testaceus* with which it is often found at least when discovered in British timber yards. It may be separated from the latter species however by the shape and texture of the dorsal ampulla: in *T. cribratum* the hinder transverse fold is distinct and the central area is tuberculate whereas in *P. testaceus* it is not present or indistinct and the whole area is granulate. Detailed figures are given by Duffy enabling the larvae of these two species to be separated in the field with the assistance of a ×15 lens. The host timbers are given as *Quercus*, *Fagus*, *Carpinus*, *Juglans*, *Larix*, *Pinus*, *Betula*, *Platanus* and *Fraxinus* in Britain although Beeson only recorded *Quercus dilatata* and *Q. incana* as the host timbers in India.

The adults which emerge in May and June are most active from dusk until midnight and are inactive during daylight. They can fly strongly and are attracted to light. The eggs are deposited singly in bark crevices and the total number of eggs laid is upwards of forty-four. The young larvae feed on the inner face of the bark but then also score the surface of the sapwood to a depth of about 3 mm. The galleries are usually parallel to the grain, tightly packed with frass and fairly straight. They are about 20 mm in width when the larva is mature. The pupal cell is constructed in the sapwood, the entrance to which is oblique and oval. The pupal cell entrance gallery is usually at least 50 mm in length and 40 mm deep in the sapwood. It is plugged with one or two wads of coarse fibres. The eventual oval flight-holes are conspicuous and measure 5–8 mm by 2·8–3·8 mm.

The total length of the life cycle is stated by Duffy to be one year but Moore considers that two years may be required on occasion.

Plagionotus

Three species of this genus are frequently imported into Britain. They are wasp-like and agile, causing them to be readily observed in timber yards.

Plagionotus arcuatus (Linnaeus). This beetle is listed by Kloet and Hincks as a British indigenous species but there are grave doubts concerning this. The recorded host timbers are *Quercus*, *Prunus* and *Salix* but Duffy thinks the last named is questionable and importations of the insect are usually in association with *Quercus*. Fraser states that the larval galleries are frequently about 20 mm wide and cut into the sapwood to a depth of about 3 mm. The frass is powdery and tightly packed. Prior to pupation the larva constructs the final escape gallery usually to within 1mm from the surface of the bark, sometimes even puncturing the surface, before turning to burrow into the sapwood where the pupal chamber is sited. The entrance to the pupal chamber is plugged with fibrous shavings.

Plagionotus detritus (Linnaeus) (Fig. 150). The biology and economic importance of this specis is similar to the last named, indeed the two are frequently found together. Although *Quercus* is the usual host timber, *Castanea* has also been recorded.

Plagionotus speciosus (Say) (Fig. 151). The beetle illustrated was found in a timber yard in Bolton, Lancs., in 1948. It had been imported in some N. American pine.

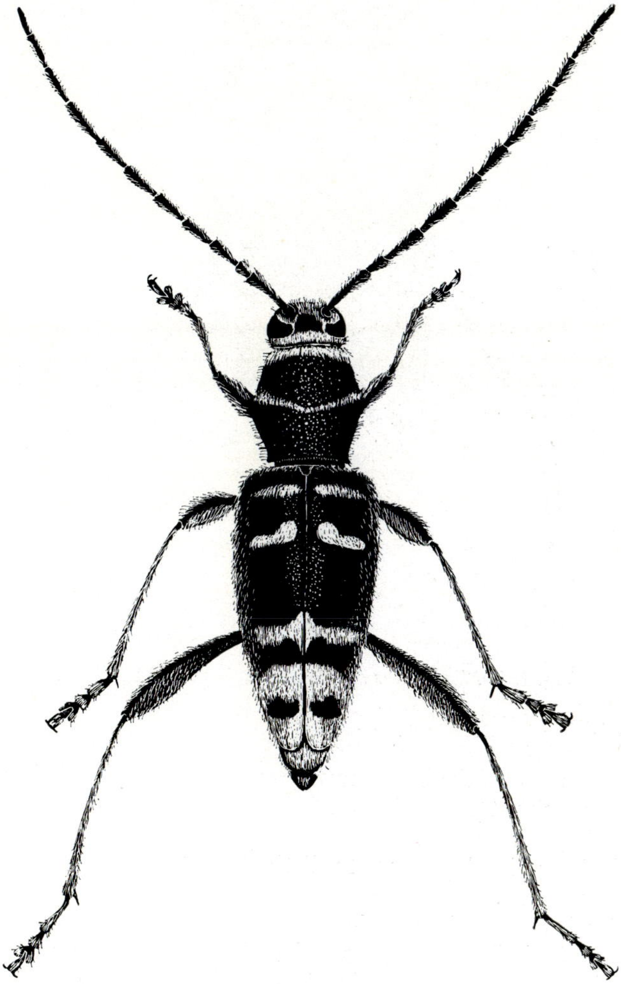

Fig. 150. *Plagionotus detritus* (L.). Length: 17 mm.

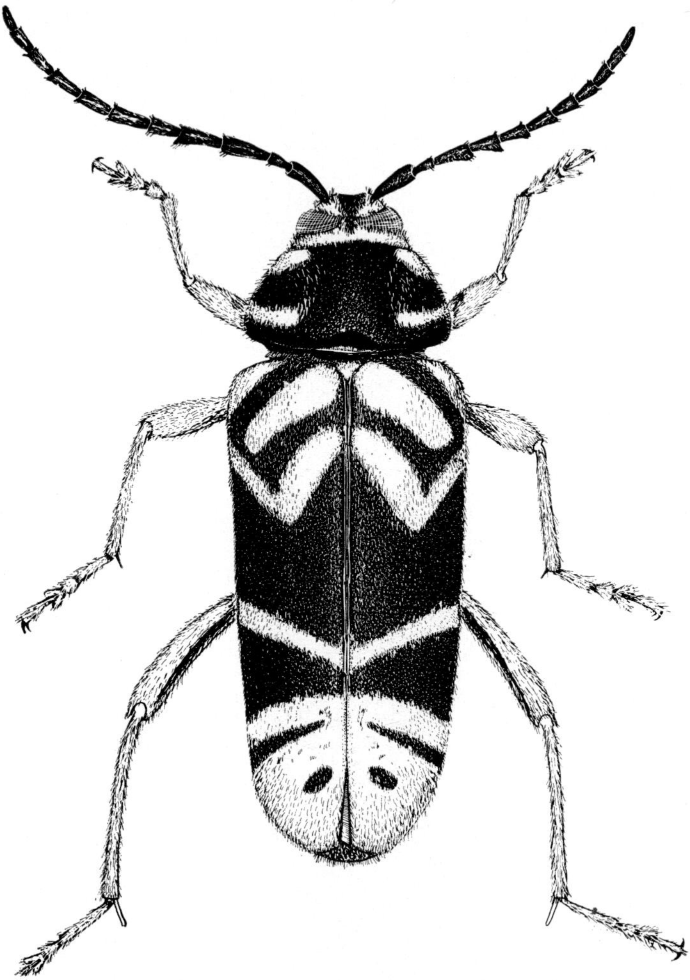

Fig. 151. *Plagionotus speciosus* (Say). Length: 23 mm.

Clytus arietis (Linnaeus) (Fig. 152).

This is perhaps one of the best known cerambycid beetles, being generally common throughout the British Isles. Its transverse black and yellow banding and quick wasp-like movements make it easily observed in the garden. The long hind legs serve to distinguish it from the very much rarer *Plagionotus arcuatus*.

The following host timbers have been recorded, *Quercus*, *Castanea*, *Rubus*, *Acer pseudoplatanus*, *Pyrus malus*, *Ficus*, *Prunus cerasus*, *Crataegus*, *Rosa*, *Corylus*, *Fagus*, *Betula*, *Salix*, *Laurus nobilis* and *Wistaria*. Until the larva is about half grown it feeds entirely just under the bark, but thereafter

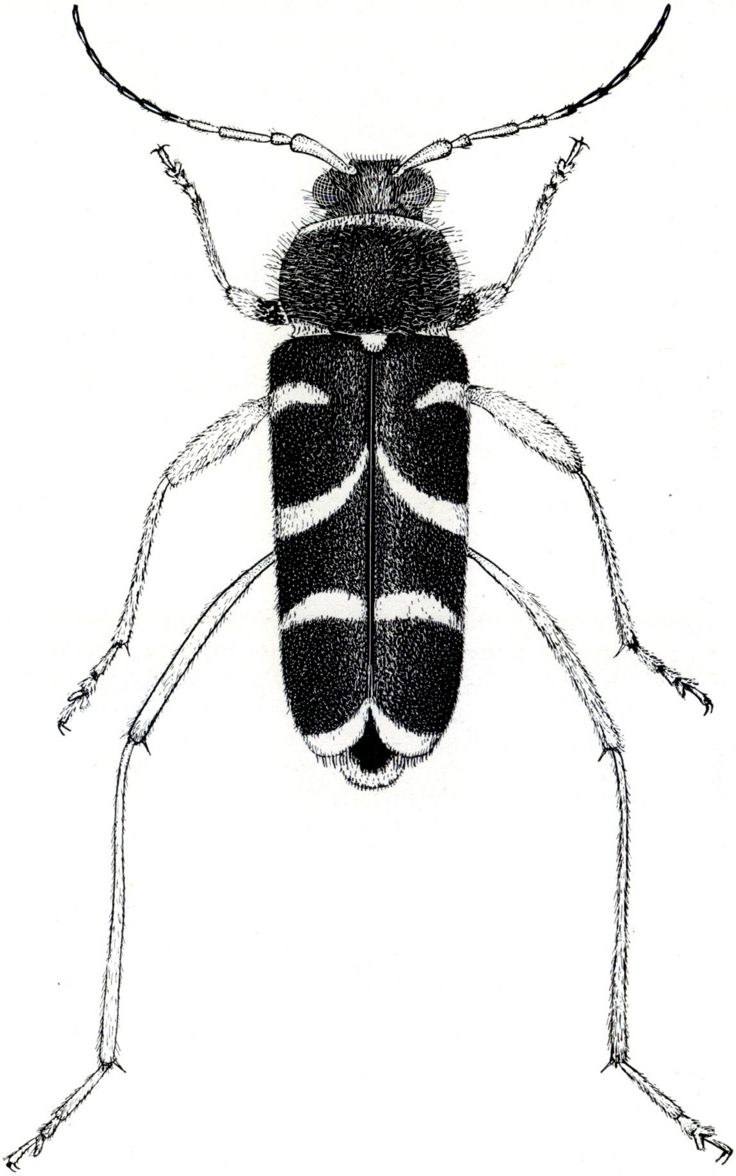

Fig. 152. *Clytus arietis* (L.). Length: 13 mm.

it tunnels into the sapwood where, finally, the pupal cell is constructed parallel with the grain. Usually only dead trees, dry, fallen branches, garden rustic work or dead parts of living trees are attacked but four cases are known to the writer, of this insect emerging from beech furniture, in one instance of which legal proceedings were instituted. It is recorded also as emerging from a dry branch, previously heated, in a museum exhibit three years after installation. Linscott has also recorded it as emerging from a wooden stopper of a cosmetic jar of only 14·6 g weight (including a plastic screw cap). Duffy states that pupation takes place either in August or September or in March or April and that the adults emerge in May or June. The life cycle is usually of two years' duration.

Anaglyptus

Two species of this genus may be encountered in Britain. *Anaglyptus mysticus* (Linnaeus) is a British native found locally in southern England extending as far north as Yorkshire. It has been recorded from *Quercus*, *Prunus*, *Cornus*, *Pyrus*, *Diospyros*, *Betula*, *Benzion*, *Acer*, *Fagus*, and *Crataegus*, the latter being considered by Duffy to be the preferred host. This author states that the larvae are found under the bark and in the wood of very dry dead boles and branches, and those trees that have been scorched by fire are particularly prone to attack. The larval galleries (Fig. 153) are similar to those of *Clytus arietis* but Duffy thinks that they tend to enter the heartwood more frequently. The frass is very fine and tightly packed and the life cycle is usually completed in two years. This is a species in which delayed development of the larva can take place in very dry conditions, Hickin having recorded the discovery of a larva at least 13 years old in the wood of a dressing table.

Anaglyptus verrucosus Olivier is imported in timber from North America. The recorded host timbers are *Quercus*, *Prunus*, *Cornus*, *Pyrus*,

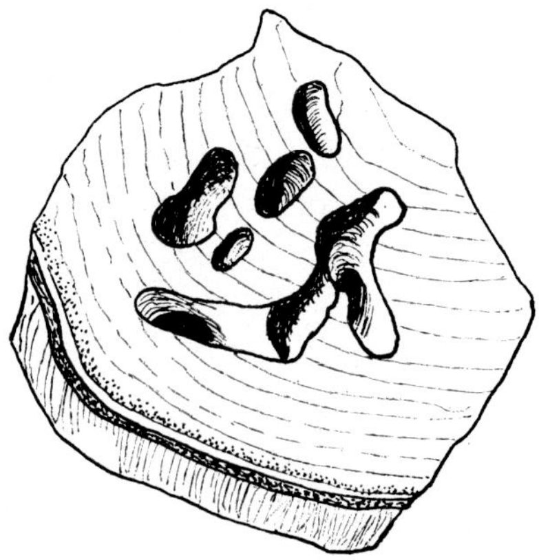

Fig. 153. Larval tunnels of *Anaglyptus mysticus* (L.) shown in a section of hawthorn stem about 30 mm across.

Diospyros, *Betula*, *Benzion*, *Acer*, *Cydonia vulgaris*, *Castanea*, *Fagus*, *Tilia*, *Carya*, *Juglans*, *Robinia* and *Vitis vinifera*. Craighead states that *Quercus* is the preferred host timber and that the larvae are found in solid, dead wood sometimes being associated with the larvae of *Neoclytus longipes* Kirby. The life cycle is completed in one year.

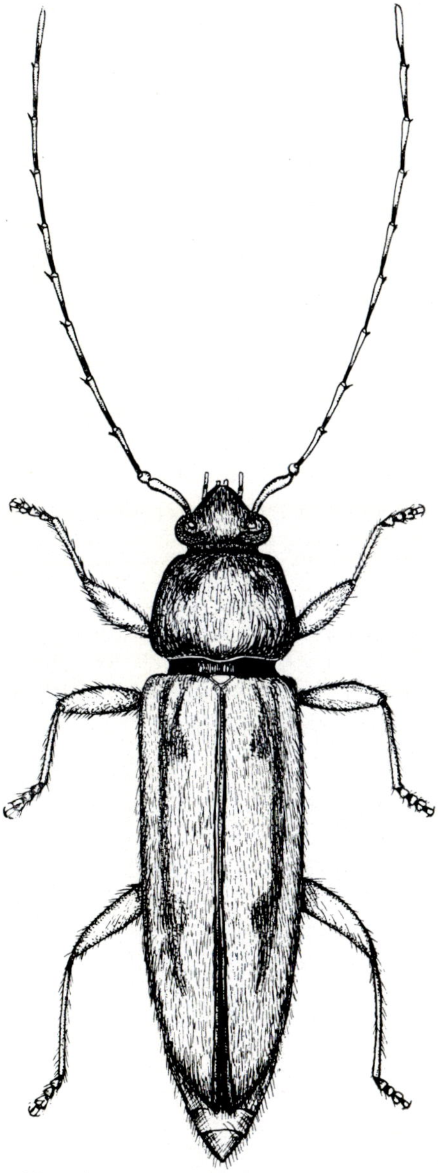

Fig. 154. *Elaphidion nanum* (F.). Length: 11 mm.

Elaphidion

Three species of *Elaphidion* are listed by Duffy as occasionally being imported into Britain in timber. *Elaphidion nanum* (Fabricius) (Fig. 154) comes into this country from Jamaica and South America in *Guaiacum officinale* but *Conocarpus* is also recorded as a host timber. Duffy states that the larvae make meandering galleries approximately two-thirds in the bark and one-third in the sapwood. Pupation often occurs near the origin of the larval gallery and the beetles emerge in September and October.

Elaphidion mucronatum (Fabricius) has been imported from North America in *Tilia*. This species is destructive to rustic furniture in North America where it is known as the Spined Bark Borer. It has been recorded from *Quercus*, *Chamaerops palmetto*, *Celtis*, *Vitis*, *Cercis canadensis*, *C. japonica*, *Fagus*, *Acer*, *Rhus*, *Morus*, *Castanea*, *Populus*, *Liriodendron*, *Asimina*, *Pyrus*, *Sassafras* and *Taxodium*.

Elaphidion spinicorne (Drury) is also occasionally imported from Jamaica and South America but little information is available concerning this species.

Megacyllene guttata Chevrolat (Fig. 155).

This beetle has been recorded emerging from a log of *Bulnesia arborea* in 1956 in London.

Phoracantha

A number of species of this genus have been recorded in Britain, having been accidentally imported with timber. *P. semipunctata* (Fabricius) and *P. recurva* Newman are native to Australia and New Zealand, *Eucalyptus* being the host timber in both cases. In the case of *P. punctata* Donovan the host timber is *Acacia homalophylla*.

Purpuricenus hochleri (Linnaeus).

This handsome red and black insect is occasionally imported into Britain from Europe. The host timbers are recorded as *Fagus, Picea, Pinus, Pyrus, Salix, Populus, Crataegus* and *Juglans*. The larvae can feed in sound wood and indeed they are known to emerge from structural timber in buildings and from furniture. They do seem to prefer, however, timber in which there is some decay.

LAMIINAE

The fifth and last of the subfamilies is the LAMIINAE. Duffy's keys have again been used.

1. Prothorax with a single pair of lateral and/or dorso-lateral tubercles or spines 2

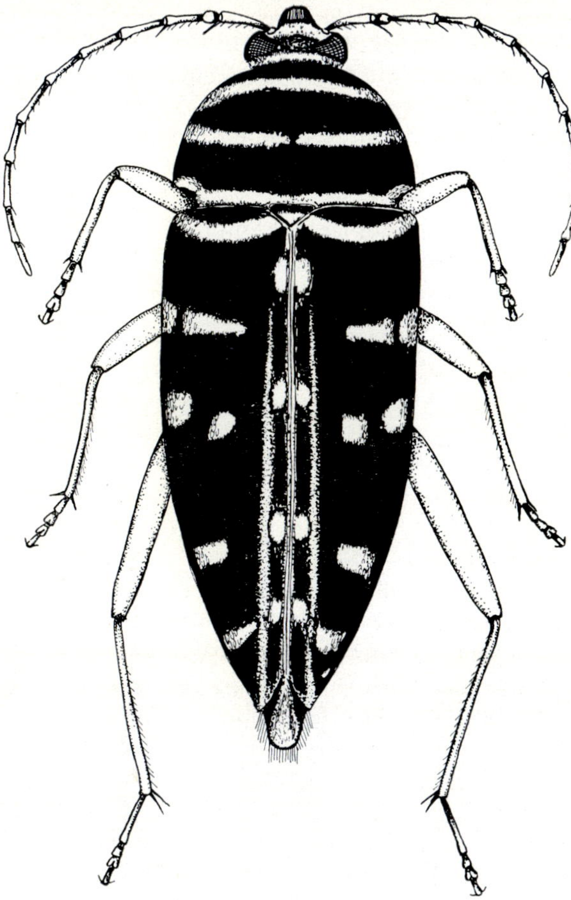

Fig. 155. *Megacyllene guttata* Chevrolat. Length: 13 mm.

– Prothorax without a single pair of lateral or dorso-lateral spines, but sometimes (*Coptops*) with two pairs of small tubercles one above the other 8

2. Elytra each with two conspicuous thorn-like tubercles on basal third, and with a strongly protuberant blade-like carina extending beyond the middle. Prothorax with a pair of paramedian conical tubercles as well as a pair of dorso-lateral tubercles. Elytra covered with light brown or greenish pubescence. Length 16–35 mm. African. From *Triplochiton* . *Ancylonotus tribulus* (Fabricius)

– Elytra without thorn-like tubercles 3

3. Shoulders of elytra strongly protuberant and very coarsely punctured like a honey-comb. Elytra covered with light brown pubescence and each with a large triangular area of dark brown pubescence on outer half beyond the middle. Length 24–35 mm. African. From *Chlorophora* and *Morus* . . . . *Phryneta leprosa* (Fabricius)

– Shoulders of elytra not or scarcely protuberant and never so coarsely punctured. Elytra without triangular areas of dark pubescence 4

4. Femora not clavate 5

– Femora strongly clavate 6

5. Antennae extending well beyond apex of elytra; first segment much shorter than third segment. Legs slender. From coniferous timber . . *Monochamus*. 15

– Antennae not extending nearly as far as apices of elytra; first segment as long as third segment. Legs stout. Length 14–30 mm . . *Lamia textor* (Linnaeus)

6. Antennae extending well beyond apices of elytra (often considerably) and covered with decumbent pubescence. Elytra without raised carinae . . . 7

– Antennae not or scarcely extending beyond apices of elytra, and bearing numerous long erect setae. Elytra each with two or three longitudinal carinae. Chiefly from dead twigs of fruit trees . . . . . *Pogonocherus*. 20

7. Antennae in the male about four times as long as the body; in the female about twice as long. Pygidium of female strongly produced into a tubular process. Elytra with scattered reddish, bead-like granules showing through pubescence. Length 13–19 mm. Northern species, in *Pinus* *Acanthocinus aedilis* (Linnaeus)

– Antennae less than twice as long as body in both sexes. Pygidium of female not produced. Elytra with deep punctures showing through pubescence. Length 6–10 mm. Southern species. In deciduous trees, especially *Quercus*. *Leiopus nebulosus* (Linnaeus)

8. Antennal segments unicolorous. Tarsal claw bifid, split or toothed basally. Anterior coxae subcontiguous 12

– Antennal segments white, yellow or grey basally, and black or brown apically. Tarsal claws not bifid, split, or toothed basally. Anterior coxae distinctly separated 9

9. Outer side of first antennal segment near apex with a thin transverse black carina (often partly obscured by pubescence). Elytra short, less than twice as long as combined basal width. Mesosternum convex, protuberant between middle coxae 10

– Outer side of first antennal segment without a black carina. Elytra elongate, two and one-half to three times as long as combined basal width. Mesosternum flat, not protruding between the middle coxae 11

10. Prothorax with two pairs of lateral, short, blunt tubercles, one above the other. Elytra with scattered greyish-white patches of pubescence, especially at the base; shoulders very strongly prominent. Length 12–20 mm. African *Coptops aedificator* (Fabricius)

– Prothorax without lateral tubercles. Elytra with greyish white patches confined to middle; shoulders slightly prominent. Length 10–15 mm. Chiefly from *Quercus* . . . *Mesosa nebulosa* (Fabricius)

11. Antennae twelve-segmented. Hind tarsi as long as hind tibiae. Length 12–17 mm. From herbaceous plants, especially *Carduus* and *Heracleum* *Agapanthia villosoviridescens* (Degeer)

– Antennae eleven-segmented. Hind tarsi distinctly shorter than hind tibae. From deciduous trees . *Saperda*. 22

12. Elytra yellow with apices black. Eyes completely divided into two parts. Prothorax with a very deep transverse impression on each side near base. Length 3–5 mm . . *Tetrops praeusta* (Linnaeus)

– Elytra completely grey, dark or black. Eyes not completely divided into two parts. Prothorax without a deep transverse impression near base . . . 13

13. Prothorax and abdomen yellow. Elytra strongly emarginate at apex. Hind tibiae not nearly reaching apices of elytra. Length 16–20 mm. From *Salix* . . . . *Oberea oculata* (Linnaeus)

– Prothorax and abdomen black. Elytra not or scarcely emarginate at apex. Hind tibiae reaching apices of elytra 14

14. Elytra each with one or two slightly raised longitudinal ridges and very gradually tapering posteriorly. Front femora and tibiae yellow. Length 6–10 mm. From UMBELLIFERAE *Phytoecia cylindrica* (Linnaeus)

– Elytra without ridges and slightly broadened towards apices. Front femora and tibiae black. Length 9–12 mm. Chiefly from *Tilia* . *Stenostola ferrea* (Schrank)

15. Apices of elytra produced at suture into a short spine. Elytra light reddish-brown with small scattered white markings. Length 23–30 mm. *M. titillator* (Fabricius)

– Apices of elytra rounded, without spines. Elytra black, usually with white or yellow markings 16

16. Antennae reddish-brown; segments usually obscurely ringed with white in the female 17

– Antennae black; segments distinctly ringed with white in the female 19

17. Elytra with apical third very densely and evenly covered with greenish-yellow pubescence. Scutellum completely and densely covered with yellow pubescence. Length 20–35 mm. European *M. rosenmulleri* (Cederhjelm)

– Elytra with apical third not covered with greenish-yellow pubescence; either with small white patches or with sparse brown, patchy pubescence. Scutellum not completely covered with pubescence 18

18. Pubescence of scutellum narrowly but distinctly interrupted by a longitudinal median black line. Elytra with a slightly bronzed lustre. Length 13–23 mm. North American . . . *M. scutellatus* Say

– Pubescence of scutellum broadly interrupted at base by a semicircular glabrous area, so that the pubescent area is U-shaped. Elytra without a bronze lustre. Length 13–23 mm. European *M. galloprovincialis* (Olivier)

19. Elytra (viewed laterally) with a very shallow, rather broad transverse depression just before the middle. Pubescent area completely covering scutellum. Length 26–32 mm. European . *M. sartor* (Fabricius)

– Elytra without transverse depression. Pubescent area of scutellum narrowly but completely divided longitudinally. Length 18–25 mm. European. . . . . . . *M. sutor* (Linnaeus)

20. Elytra subtruncate at apex; not dentate or mucronate. Length 5–7 mm. Northern. From coniferous trees . . . . *P. fasciculatus* (Degeer)
– Elytra emarginate at apex and dentate or mucronate at external angles. Southern. In deciduous trees, especially *Pyrus* 21
21. Elytra each bidentate at apex, the sutural angle being produced into a short blunt tooth or spine. Scutellum covered with white pubescence. Length 6–7 mm . . . *P. hispidulus* (Piller & Mitterpacher)
– Elytra each dentate at apex, the sutural angles not produced. Scutellum dark brown or black. Length 4–6 mm . . . . . *P. hispidus* (Linnaeus)
22. Prothorax concave between bases of antennae. Elytra without bands or markings of a contrasting colour; covered with scattered coarse punctures and yellow or greyish pubescence; apices acutely pointed. Length 20–28 mm. Chiefly from *Populus* . . . *S. carcharias* (Linnaeus)
– Prothorax convex or flat between bases of antennae. Elytra with distinct bands or markings of a lighter colour; apices rounded 23
23. Prothorax broadly and transversely depressed just before middle and hind margin. Elytra depressed, black and with a broad bright yellowish-green longitudinal band of pubescence extending along suture and branching at intervals, and with a few scattered spots of similar pubescence. Length 14–18 mm. Northern. Chiefly from *Prunus*, *Pyrus*, *Quercus* . . *S. scalaris* (Linnaeus)
– Prothorax scarcely or not transversely depressed. Elytra slightly convex with rather scanty yellowish pubescence and with a few scattered patches of similar but denser pubescence forming conspicuous patches. Length 9–14 mm. Southern. From *Populus* and *Salix* . . *S. populnea* (Linnaeus)

Notes on the species

Ancylonotus tribulus (Fabricius). This beetle is occasionally imported in the timber *Triplochiton scleroxylon* from West Africa. It is also recorded as damaging *Coffea*.

Phryneta

Two species of this genus are occasional immigrants into Britain in timber from Africa. *P. leprosa* (Fabricius) attacks *Chlorophora excelsa* and

Morus whilst *Phryneta spinator* (Fabricius) is recorded from *Cupressus horizontalis*, *Cupressus lusitanica*, *Ficus*, *Funtumia*, *Melia azedarach*, *Prunus persica*, *Pyrus communis*, *P. malus*, *Salix* and *Vitis vinifera*. The larva of *P. spinator* ejects frass through a series of holes made in the bark and the average length of the life cycle is three years. The usual length of the larval life is two and a half years.

Monochamus

Joy does not list any species of this genus as British. Kloet and Hincks however give *rosenmulleri*, *sartor*, *sutor* and *galloprovincialis* as British. Duffy gives all these as imported, except *sutor* which he considers to be doubtfully indigenous. This latter author lists seven of the species as imported.

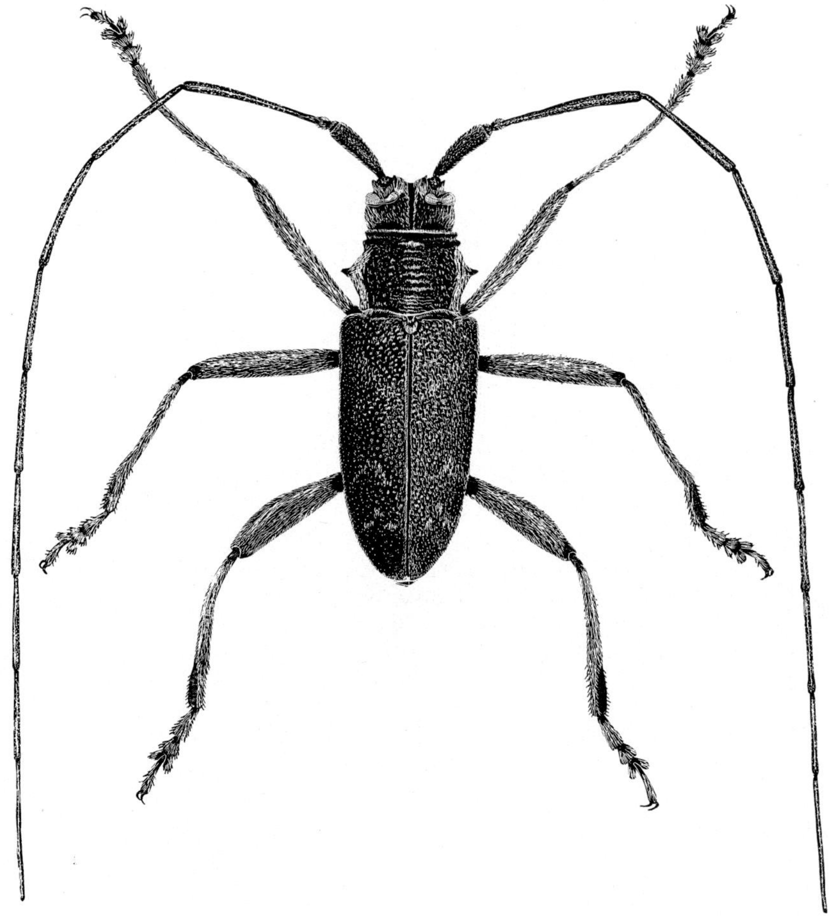

Fig. 156. *Monochamus galloprovincialis* (Ol.). Length: 21 mm.

Monochamus titillator (Fabricius). This insect is a native of North America. The male has exceptionally long antennae, one specimen in the writer's collection having antennae 7·5 cm in length. The elytra are brown with scattered patches of white pubescence. The tarsal segments are exceptionally prominent and provided with pads of long, fine, black setae on the under surfaces. The host trees are *Pinus*, *Picea* and *Abies*; felled or injured trees mainly being attacked. The female gnaws a pit in the bark extending to the bast in which nine eggs are laid. When the larva is mature it bores right into the heartwood and then returns to the outer sapwood. The flight holes are round, about 10 mm in diameter and the usual length of the life cycle is two years.

Monochamus rosenmulleri (Cederhjelm). Occasionally imported in timber of *Abies* and *Pinus* from Scandinavian and other continental countries. The adults cause damage to the young shoots of *Abies*.

Monochamus scutellatus Say. This species is also from North America and its biology is similar to that of *M. titillator*.

Monochamus galloprovincialis (Olivier) (Fig. 156). Imported occasionally from continental Europe. Host timbers are *Pinus* and *Picea*.

Monochamus sartor (Fabricius). Also imported in timber of *Pinus* and *Picea* from continental countries. The larva is said to make a creaking noise.

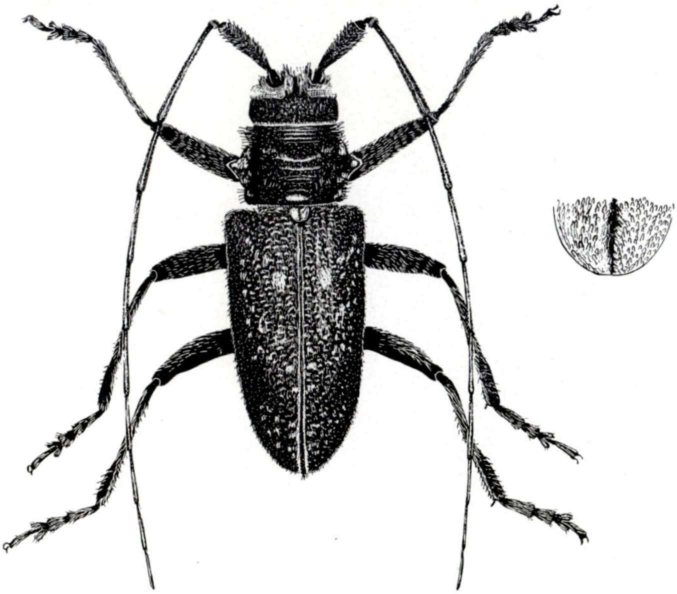

Fig. 157. *Monochamus sutor* (L.). Length: 18 mm. Right, enlargement of scutellum.

Monochamus sutor (Linnaeus) (Fig. 157). This species enters Britain from Scandinavian and other Continental Countries in timber of *Picea* and *Pinus*.

Monochamus griseoplagiatus Thompson. Duffy also lists this West African beetle as being very occasionally imported into Britain, but no further information is available.

Monochamus maculosus Haldeman. This is another North American species similar in its biology to *M. notatus*.

Monochamus marmorator Kirby. Another North American insect. It is said that the larval galleries are more transverse than in other species, which cause a more rapid girdling of the trunk.

Monochamus notatus Drury. A record exists of an adult of this species emerging from a piece of pine furniture 15 years after manufacture; it must have spent this time mostly as a larva. Host timbers are *Pinus* and *Abies*.

Fig. 158. *Lamia textor* (L.). Length: 27 mm. Right, enlargement of tarsal claw.

Monochamus ruspator Fabricius. This is a West African beetle whose host is *Citrus*. Nothing is known of its biology.

Lamia textor (Linnaeus) (Fig. 158).

This is a large stout beetle, dull pitchy-black in colour but under a lens numerous shining tubercles are revealed. Duffy gives the length as 14–20 mm and Joy quotes 18–28 mm; the specimens in the author's collection measure up to 30 mm. There is a group of stiff bristles under the strongly arched tarsal claws.

The recorded host timbers are *Populus nigra*, *Populus tremula* and *Salix*. The specimens in the author's collection are recorded from *Betula alba*. The life cycle takes place in the branches and the roots of healthy

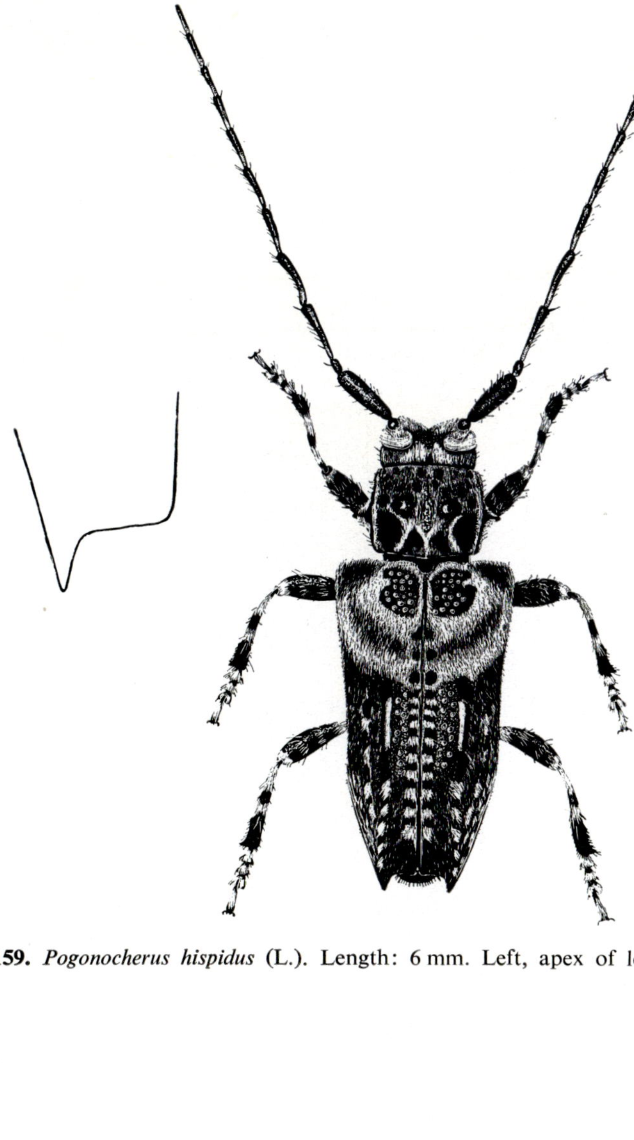

Fig. 159. *Pogonocherus hispidus* (L.). Length: 6 mm. Left, apex of left elytron.

trees; usually the larval gallery follows the pith channel. Sometimes the adults emerge from the ground but it is thought that in these cases only the roots have been attacked. The adults are found during June and July. It is a rare insect in Britain.

Pogonocherus

The three species of *Pogonocherus* are small insects with variable patterns of whitish pubescence with patches of black, erect, spine-like setae on pronotum and elytra. The femora and tibiae are banded. They are easily separated by the shape of the apices of the elytra.

Pogonocherus hispidus (Linnaeus) (Fig. 159). The larvae of this insect

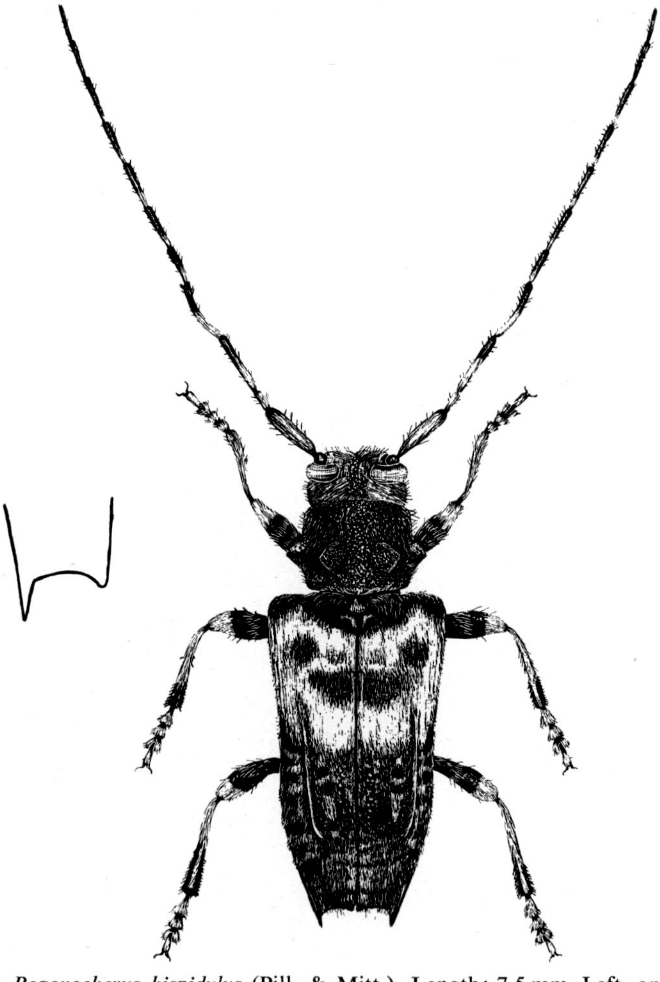

Fig. 160. *Pogonocherus hispidulus* (Pill. & Mitt.). Length: 7·5 mm. Left, apex of left elytron.

are found in dead or decaying twigs of *Alnus*, *Castanea*, *Corylus*, *Fagus*, *Ficus*, *Hedera*, *Ilex*, *Laurus nobilis*, *Lonicera*, *Prunus*, *Pyrus communis*, *P. malus*, *Quercus*, *Rubus*, *Pyrus aucuparia*, *Tilia*, *Cotoneaster*, *Ulmus*, *Viscum album*.

Eggs are laid singly in the bark and the resulting larvae feed in the cambium. This causes the thin bark of the twigs to crack and curl and expose the lighter coloured sapwood in a characteristic manner. The mature larvae bore into the sapwood, in which the pupal cell is constructed, and the entrance is plugged with frass. Adults emerge during April and May but exceptionally they may appear during the previous autumn. It is found locally throughout the British Isles.

Pogonocherus hispidulus (Piller & Mitterpacher) (Fig. 160). The life history of this species is similar to that of *P. hispidus* but the recorded

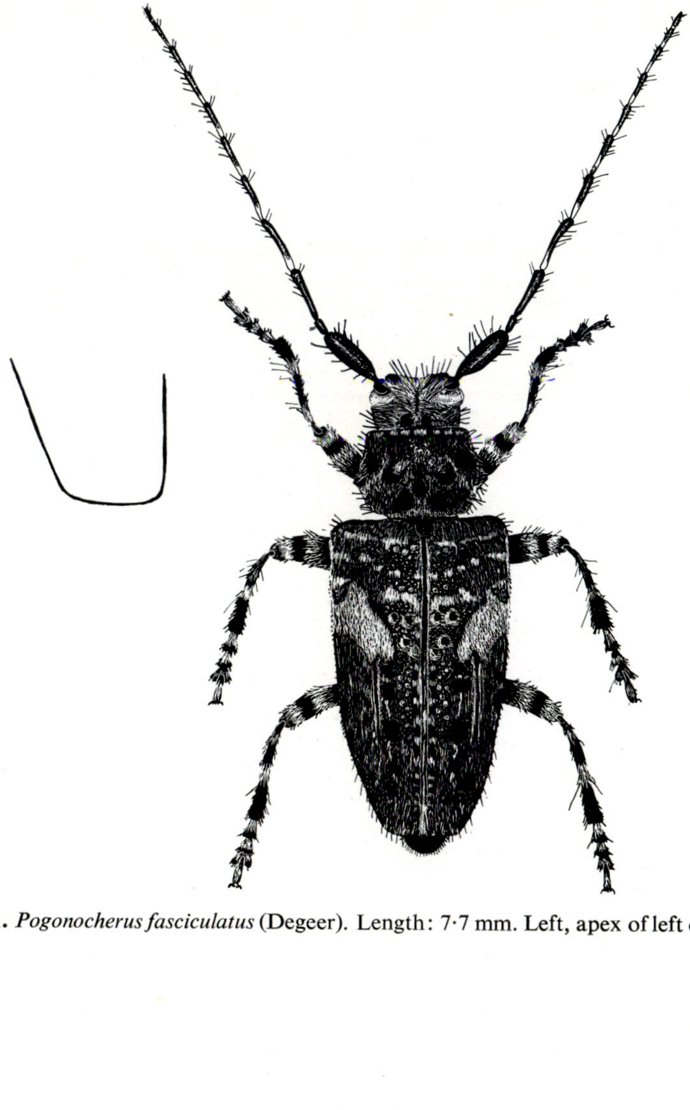

Fig. 161. *Pogonocherus fasciculatus* (Degeer). Length: 7·7 mm. Left, apex of left elytron.

host timbers and plants are *Alnus*, *Pyrus communis*, *P. malus*, *Viscum album*, *Hedera*, *Corylus avellana*, *Carpinus betulus*, *Sambucus nigra*. Duffy considers that *Pyrus malus* is the preferred host. It is a local insect but recorded from England, Scotland and Ireland.

Pogonocherus fasciculatus (Degeer) (Fig. 161). Again the life history of this beetle is similar to that of *P. hispidus*. It is however a rare insect only known from coniferous trees, being recorded from the twigs of *Pinus* and *Picea* in Lincolnshire and Scotland.

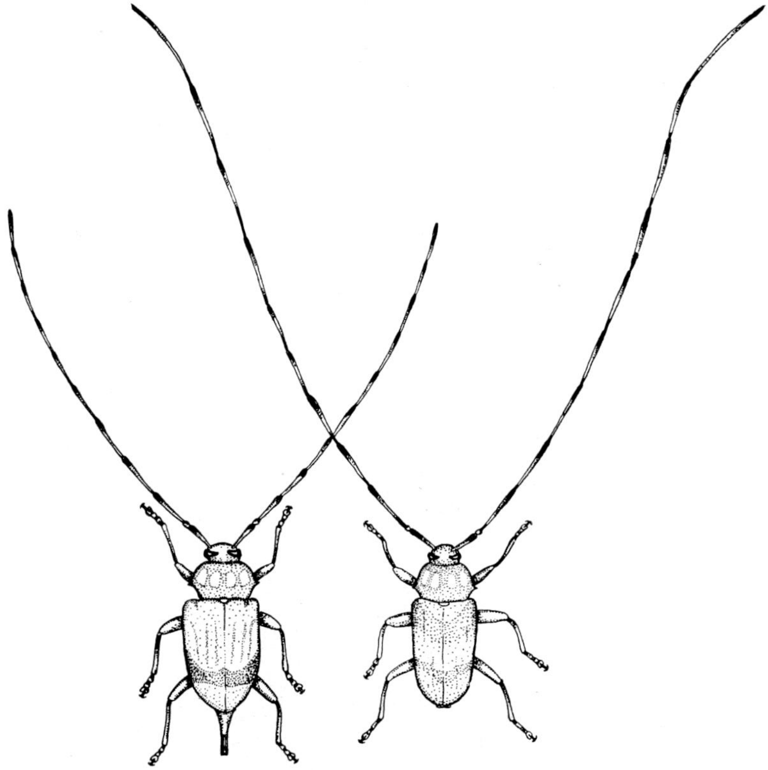

Fig. 162. *Acanthocinus aedilis* (L.). Left, female. Right, male, length: 17 mm.

Acanthocinus aedilis (Linnaeus) (Figs. 162 and 163).

Formerly this insect, which has been given the common name of 'The Timberman', was found only in Scotland and Northern Ireland but since the Second World War, it has been found in several localities in England. It is frequently found in imported timber of *Pinus* and *Picea* especially in pit props in coal mines. The damage caused, however, can never be looked upon as serious. The adults are characterized by the extreme length of the antennae—the males having much the longer, and the female possessing a long ovipositor. The elytra are covered with greyish pubescence through which numerous chestnut-coloured tubercles protrude. The general ground colour of the elytra is sometimes mauve or purplish.

Recently dead or freshly felled trees are attacked. The female gnaws a pit for receiving the eggs and in addition uses the blade-like lower valve of the tip of the ovipositor for excavating the pit even deeper. The larvae feed only in the bast and do not even score the sapwood. If the bark is thin the pupal cell is constructed in the outer sapwood, if thick it is found

Fig. 163. Right elytron and ovipositor of *Acanthocinus aedilis*.

in the bast. The main emergence period is during August and September but overwintering of the adult in the pupal cell has frequently been observed. The flight-hole is circular.

Leiopus nebulosus (Linnaeus) (Fig. 164).

This indigenous British insect is local in its distribution. The head, pronotum and elytra bear a grey and whitish pubescence in a characteristic pattern. The whitish transverse bar across the centre of the elytra and the area of the apices of the elytra are spotted with black. The pubescence at the top of the head is fernlike and distinctive.

This species attacks recently dead or dying branches, or sometimes slender trunks of the following recorded species: *Quercus*, *Castanea*, *Prunus armeniaca*, *P. persica*, *Ulmus*, *Alnus*, *Robinia*, *Corylus*, *Pyrus malus*, *P. aucuparia*, *P. communis*, *Rosa*, *Baccharis halimifolia*, *Carpinus*, *Fagus sylvatica*, *Acer*, *Betula*, *Cydonia vulgaris*, *Juglans*, *Pinus*, *Salix*, *Crataegus*, *Larix* and *Picea*, of which the first two named are preferred. The larvae feed in the cambial layer and if the bark is thick (as in *Quercus*) the pupal cell is hollowed out in the bast and the outer sapwood. If, however, the

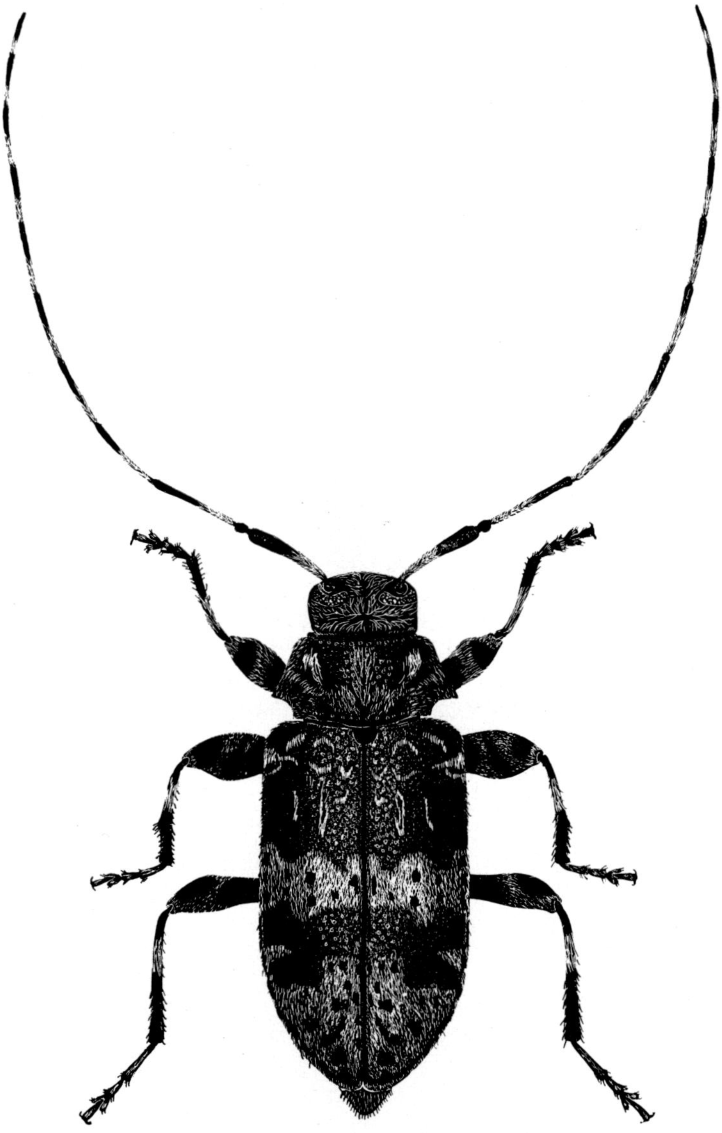

Fig. 164. *Leiopus nebulosus* (L.). Length: 9 mm.

bark is thin (as in *Castanea*) the pupal cell is constructed within the sapwood or in the heartwood or may even be in the pith canal. If the pupal cell lies in the cambium it is surrounded by a low wall of frass, but if in the wood a plug of frass seals the entrance. The adults emerge in June and July.

Coptops aedificator (Fabricius) (Fig. 165).

This is a stout beetle of moderate size and bears a complicated pattern of greyish-white pubescence on the elytra, pronotum and head. There are spots and bars on the femora and tibiae.

This species is imported from time to time in timber from Africa and India and the recorded host timbers are: *Acacia arabica*, *Acacia* spp., *Acrocarpus fraxinifolius*, *Aegle marmelos*, *Afzelia* sp., *Albizia lebbek*, *Artocarpus* sp., *Bauhinia purpurea*, *B. retusa*, *B. vahlii*, *B. variegata*, *Bombax malabaricum*, *Buchanania latifolia*, *Butea frondosa*, *Coffea arabica*, *Careya arborea*, *Cassia siamea*, *Cudrania javanensis*, *Dalbergia paniculata*, *Excaecaria agallocha*, *Ficus glomerata*, *F. religiosa*, *F.* sp.,

Fig. 165. *Coptops aedificator* (F.). Length: 17 mm.

Funtumia, Garuga pinnata, Hevea braziliensis, Holoptelea integrifolia, Lannea grandis, Manihot glaziovii, Millettia auriculata, Mucuna imbricata, Pavetta indica, Pongamia glabra, Pterocarpus marsupium, Pueraria tuberosa, Semecarpus anacardium, Shorea robusta, Spatholobus roxburghii, Terminalia belerica, T. tomentosa, Wrightia tinctoria.

The larvae are found in the cambial layer or entirely in the inner bark, and Beeson and Bhatia state that even the pupal cell is constructed in the latter so that damage to sapwood is only superficial. Fraser observed during rearing experiments that if the wood dried out the larvae would construct a tunnel into the sapwood in order to pupate. A circular but ragged flight-hole is made, and June is the usual month of emergence in India, although emergence can take place throughout the year. It has an annual life cycle.

Mesosa nebulosa (Fabricius) (Fig. 166).

The pitchy-black ground colour of this insect is almost completely covered with a pattern of greyish, reddish-yellow and white pubescence. The latter is situated in a transverse band between angular black bars near the middle of the elytra. It is a rare insect found in southern and eastern England. The preferred host timber is *Quercus* but *Acacia, Alnus, Betula, Carpinus, Castanea, Corylus, Fagus, Ilex aquifolium, Populus, Pyrus malus, Robinia, Salix* and *Tilia* have all been recorded as being attacked. Duffy states that oviposition takes place usually on the uppermost branches of dead or unhealthy trees. The larva feeds in the cambial layer for a short time only, before boring into the sapwood. In the case of small diameter branches the larva enters and bores along the pith canal. When fully grown it constructs an oval, curved pupal chamber in the outer sapwood parallel to the wood grain. The adult eclodes in the late autumn but remains in the pupal cell until it finally emerges during the following July.

Agapanthia villosoviridescens (Degeer) (Fig. 167).

This species is moderately large and elongate. The colour of the integument is black, but central and lateral longitudinal bands of yellowish pubescence on the head and pronotum give it a characteristic appearance. The pubescence is denser at the hinder end of the elytra, and in the central area it occurs in small transverse patches.

This beetle is never a timber-borer but attacks only certain herbaceous plants, *Heracleum, Carduus*, being the chief food plants in Britain. In addition *Aconitum* spp., *Senecio aquaticus, Eupatorium cannabinum, Artemisia vulgaris* and *Urtica dioica* are sometimes infested. The eggs are deposited on healthy plants, usually two only on each plant, the female first biting a cavity in the stem wall to contain them. The young larvae, on hatching, bite their way through into the pith canal then make their way along to the base of the stem and sometimes into the root stock. In appearance the larva is most uncerambycid-like, being very elongate. It can move rapidly up and down the stem by means of rows of tubercles

Fig. 166. *Mesosa nebulosa* (F.). Length: 12 mm.

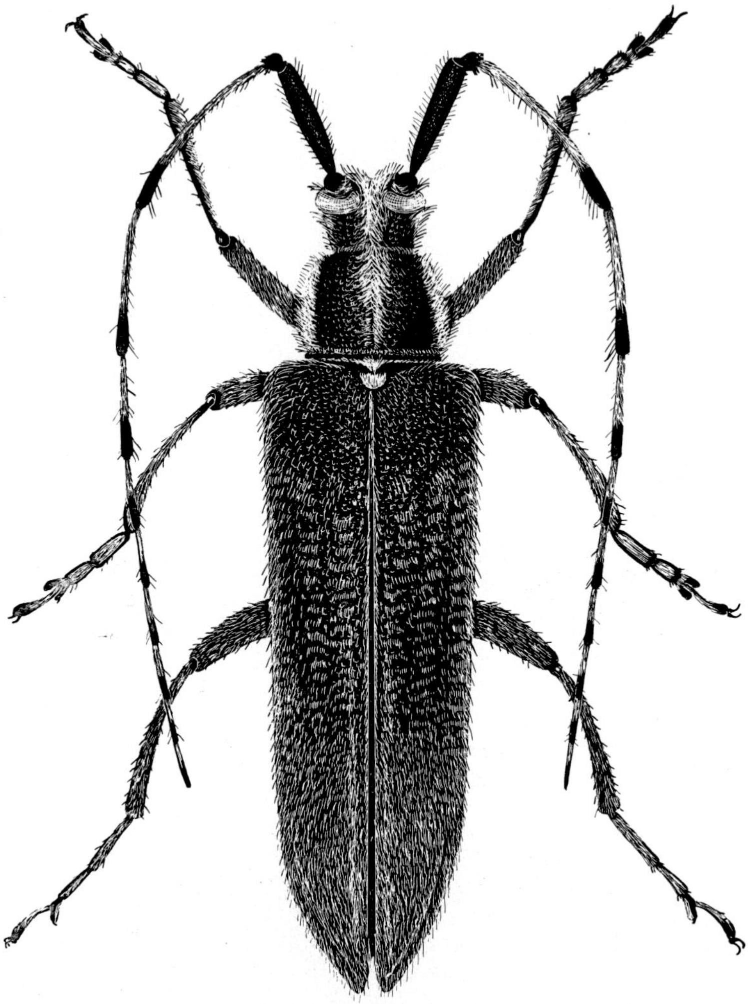

Fig. 167. *Agapanthia villosoviridescens* (Degeer). Length: 16 mm.

on the dorsal ampullae and this activity is reminiscent of a lepidopterous larva. The pupal cell is hollowed out in the stem with a loose plug of fibres at the base, and often much higher in the stem there is another plug. The pupa is furnished with groups of sharp spines on the abdominal tergites, by means of which the pupa also can move up and down the stem within a few inches of the pupal cell. The adult beetles emerge during June and July, having completed the life cycle in one year. It is very rare in Britain, found in southern England only as far north as Lincolnshire.

Saperda

In this genus there are three indigenous British insects, and in addition two species are listed by Duffy as being occasionally imported into Britain from North America.

Saperda carcharias (Linnaeus). This rare British insect is more or less unicolorous being covered with yellowish or yellowish-grey pubescence except for black tubercles. There is only the slightest elytral pattern—faint lunar transverse bands near the middle of the elytra being present. Various species of *Populus* and *Salix* are given as the host trees although *Populus tremula* is preferred. Before egg laying commences the adults gnaw the leaves of the tree, often causing considerable damage. The female then bites an oviposition pit in the bark of the main trunk, often only about 30 cm above ground, or a larger branch of young tree. One egg is laid in each pit which is covered with a sticky secretion. The eggs hatch in about one month or they may 'over-winter', not hatching until the following June. The young larva first burrows into the sapwood to a depth of 25 mm or so when it turns downwards and inwards, constructing a gallery into the pith canal. It follows the pith canal upwards and then bores into the outer sapwood and constructs what will be the eventual flight hole, which is plugged with wood particles. The larva then returns to the vertical gallery where it burrows upwards until it is fully fed when it constructs a pupal chamber by plugging beneath it with wood fibres. The life cycle is completed in four years and the adults emerge in July or August. This species is of considerable economic importance especially in plantations of commercial species of *Populus*.

Saperda populnea (Linnaeus) (Fig. 168). This is a local British insect characterized by five or sometimes six round patches of yellowish pubescence on each elytron. There is also a thick yellowish pubescence on lateral margins of the pronotum but only a sparse central longitudinal band. The ground colour is black, and the elytra are covered with sculptures and small discreet patches of pubescence. Various species of *Populus* and *Salix* are given as the host timbers with *Populus tremula* preferred, but Duffy has recorded it as attacking saplings of *Betula*. The female selects a slender twig for oviposition, usually not more than 10 mm in diameter. She first of all gnaws a horseshoe-shaped abrasion in the bark in the centre of which is a very small receptacle for the single egg. The larva causes the twig to swell and form a gall—the only British cerambycid to exhibit this phenomenon (Fig. 169). It is of interest to note that somewhat similar galls are produced by a Clearwing Moth (*Conopia flaviventris* Staud.) and by a weevil (*Cryptorhynchus lapathi* (L.)). The swellings produced by the former, however, are smaller and those by the latter larger, than those produced by *Saperda populnea*. The *populnea* gall is packed with frass of a light orange colour whilst on the outside there is a patch of disintegrating or partially decayed bark. There is an annual life cycle and the adults emerge in May and June.

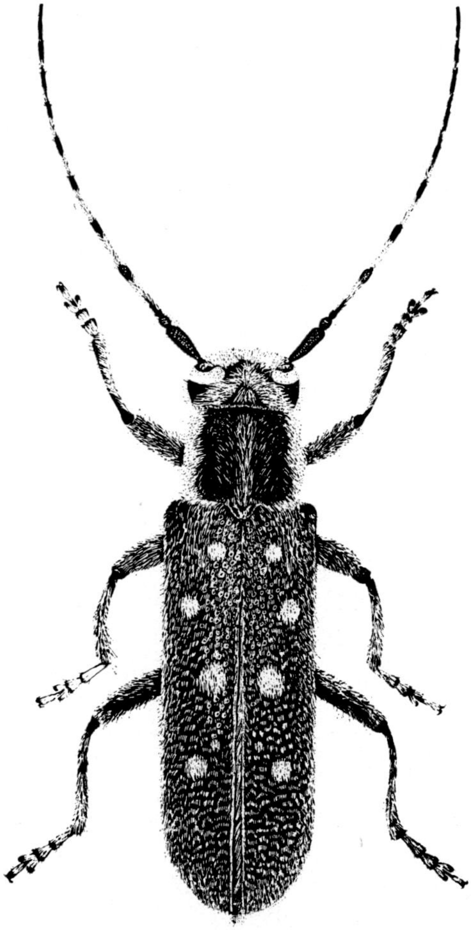

Fig. 168. *Saperda populnea* (L.). Length: 14 mm.

Saperda scalaris (Linnaeus). This rare species is one of the most attractive of British insects. The elytra are characterized by a pattern of black bars and ellipses which, however, do not reach the suture, and by yellowish-green or apple green pubescence. The pattern is ladder-like (hence *scalaris*). The host trees are given as: *Acer, Alnus*, *Betula*, *Castanea*, *Corylus*, *Fagus*, *Ilex*, *Juglans*, *Populus alba*, *Prunus avium*, *P. cerasus*, *P. padus*, *P. spinosa*, *Pyrus aucuparia*, *P. communis*, *P. malus*, *Quercus*, *Salix*, *Sambucus nigra*, and *Ulmus*.

Fraser has given an account of the biology of this beetle from which the following is taken. The female makes an incision in the bark of dead wood and then lays a single egg in it which cannot be detected from the

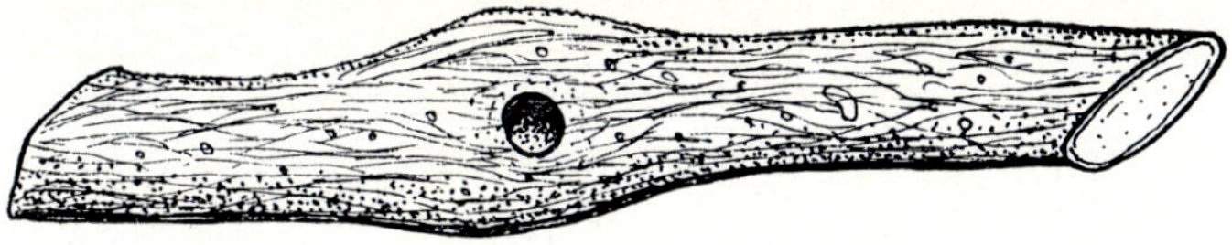

Fig. 169. Gall and flight hole of *Saperda populnea* (L.) in *Populus tremula*. The piece of twig is 70 mm long.

surface. The larva feeds in the cambial layer, the gallery being mainly straight and the frass fibrous and dry. Larvae have been found in the same wood (fence posts) as *Leiopus nebulosus* (L.), *Clytus arietis* (L.), *Rhagium bifasciatum* F. and *R. mordax* (Degeer). In the autumn of the second year the larvae are full grown and then they burrow shallowly into the sapwood to form a pupal chamber which is plugged with loose dry shavings. The adults feed on leaves, gnawing ragged holes along the veins (Fig. 170).

Saperda discoidea Fabricius. This species is occasionally imported into Britain from North America in timber of its host trees, *Carya*, *Juglans*, *Caryocar*, and *Ulmus*. Only dying or recently dead trees are attacked; especially susceptible to attack are those killed by the beetle *Scolytus quadrispinosus* Say.

Saperda calcarata Say. The host tree of this North American species is *Populus*. The young larvae often completely girdle the tree, thus killing it, they then enter the heartwood. Craighead states that the trees are sometimes so honeycombed with larval galleries that they break off in the wind and piles of frass several inches deep are noticed under infested trees! It is thus not surprising that this insect has totally ruined many poplar plantations in North America.

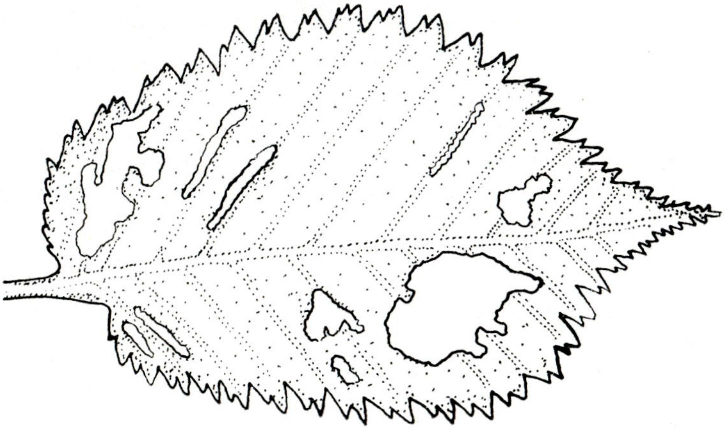

Fig. 170. Elm leaf damaged by adult of *Saperda scalaris* (L.).

Tetrops praeusta (Linnaeus) (Fig. 171).

This beetle is very small, seldom reaching 5 mm in length. It is found locally in England. The fore legs are much lighter in colour than the middle and hind legs. There are transverse whitish bands across the elytra near the middle. The complete division of the eyes is sometimes difficult to observe even with a microscope, owing to their very dark colour. There is

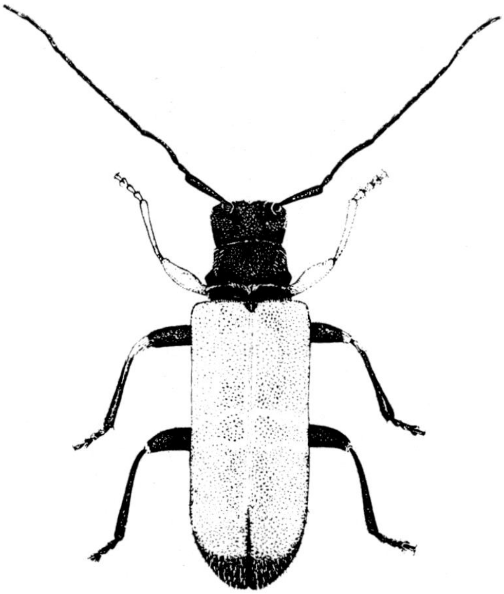

Fig. 171. *Tetrops praeusta* (L.). Length: 4·5 mm.

some confusion concerning the host timbers attacked on account of the great similarity in the life history to that of *Pogonocherus hispidus*. Larvae of the two species often occur together so that Duffy recommends that the recorded list of host timbers is treated with reserve. They are however given as *Fraxinus, Pyrus malus, Pyrus communis, Prunus domestica, P. spinosa, P. persica, P. armeniaca, Rosa, Tilia, Quercus, Ulmus* and *Crataegus*. Duffy gives *Pyrus malus* as the preferred host.

Oberea oculata (Linnaeus) (Fig. 172).

This is another rare insect in Britain being found very locally in the eastern counties and Kent. In appearance it is very distinctive. The head, antennae and elytra are black, the latter being covered by white pubescence through which rows of black tubercles emerge. The pronotum is brownish-orange and bears two or four black spots.

The host timber is given as *Salix babylonica* and other species of *Salix*. Eggs are laid on the smooth bark of healthy twigs.

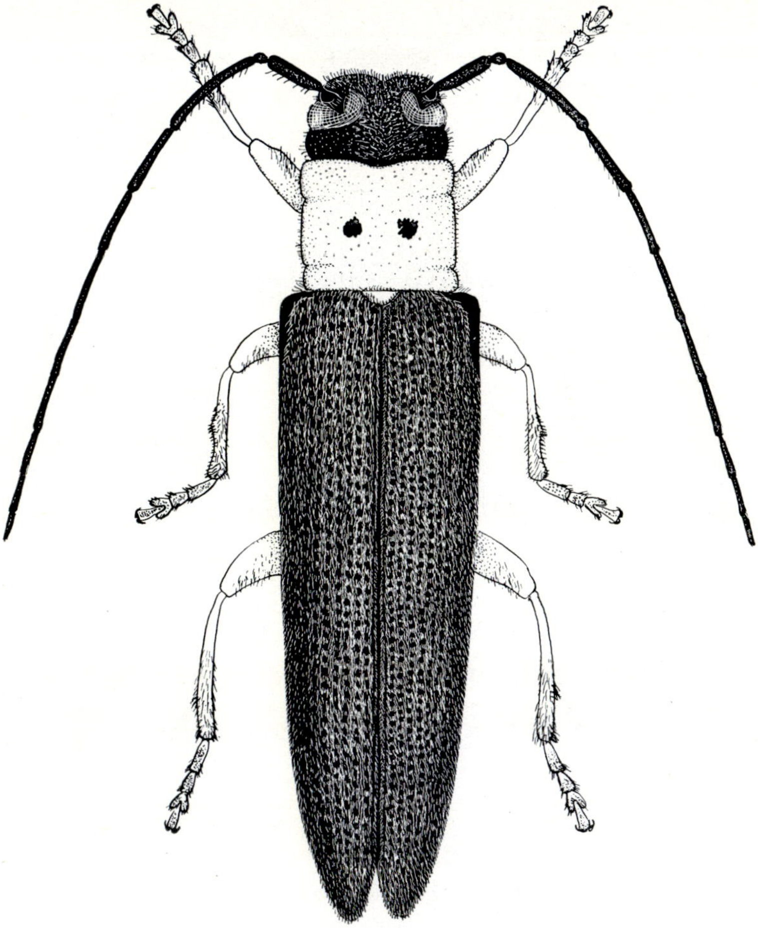

Fig. 172. *Oberea oculata* (L.). Length: 17 mm.

The young larva bores straight into the pith canal in which it spends the rest of the larval stage, extending the gallery sometimes longer than 30 cm. The larva is very active and from time to time ejects the frass which can give an indication of its presence. Emergence of adults takes place during July and August.

Phytoecia cylindrica (Linnaeus) (Fig. 173).

This species is dull black in colour with the exception of the distal femora, tibiae and tarsi of the front legs which are bright brownish-yellow. It is rare in England being found only in the South and Midlands, not being found north of Derby. It is not a wood-borer but attacks certain species of

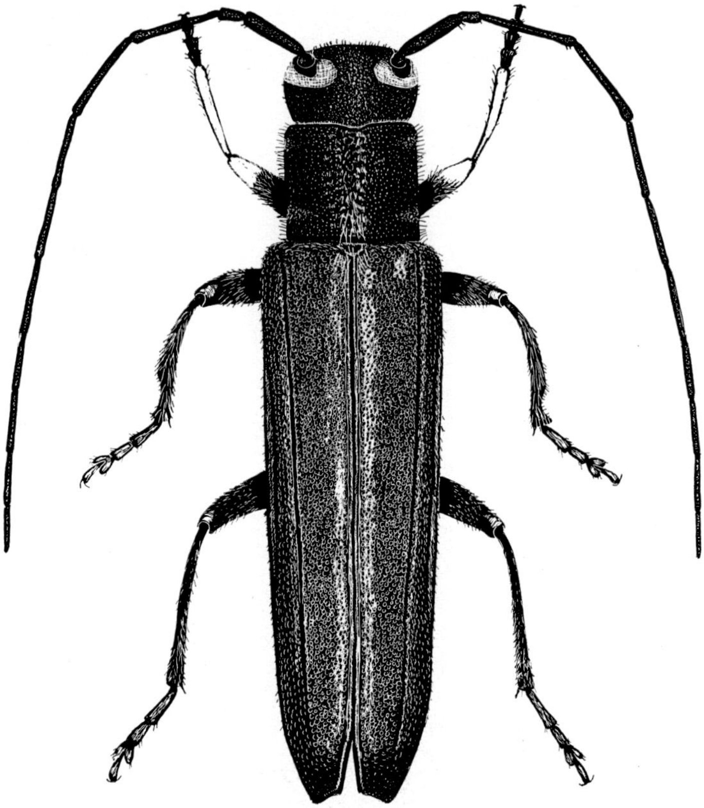

Fig. 173. *Phytoecia cylindrica* (L.). Length: 10 mm.

UMBELLIFERAE, *Daucus carota, Chaerophyllum temulum, Heracleum sphondylium,* and *Anthriscus sylvestris*, having been recorded. A single egg is laid in the stem just below the inflorescence and the resulting larva bores into the pith canal and thence hollows out the whole of the stem. The duration of the larval stage is no more than three months as the adults eclode in the autumn. They overwinter in this stage, not emerging from the stem until the following May or June.

Stenostola ferrea (Schrank) (Fig. 174).

This rare insect is black with a bluish or greenish-blue sheen. It is covered with greyish-white pubescence and in the female there are patches of longer white pubescence to a greater or lesser extent in two longitudinal bands on the pronotum, in the centre of the anterior margin of the pronotum, and around the eyes.

The host timbers are recorded as *Tilia*, *Salix*, *Salix caprea*, *Populus tremula*, the first named being preferred. Duffy states that attack by this species takes place in recently dead or freshly cut branches, generally not more than 25 mm in diameter, and usually when branches are lying on the ground decaying slightly. Until the time the larva is about half grown it feeds in the cambial layer, but it then bores into the sapwood and if the branch is thin it will continue along the pith canal and eventually construct

Fig. 174. *Stenostola ferrea* (Schrank). Length: 11 mm.

its slightly curved pupal chamber within it. April or May are the usual months for pupation, and emergence of the adult beetle takes place in May or June. The beetles often feed on the leaves of *Tilia*, perforating the leaf in the same manner as *Saperda carcharias*.

Niphona hookeri Gahan (Fig. 175).

This species is imported from time to time from the Orient in bamboo. Two adults emerged from a piece of bamboo 120 cm in length and only 1 cm in diameter. The flight hole is oval, 7 mm long, 5 mm wide. The adults are reported as feeding on the young bark of fruit trees such as *Citrus*. The beetle is densely covered with pale yellowish-brown pubescence with a central longitudinal band of darker colour in which black turbercles show through, each of which being surrounded by a ring of white pubescence.

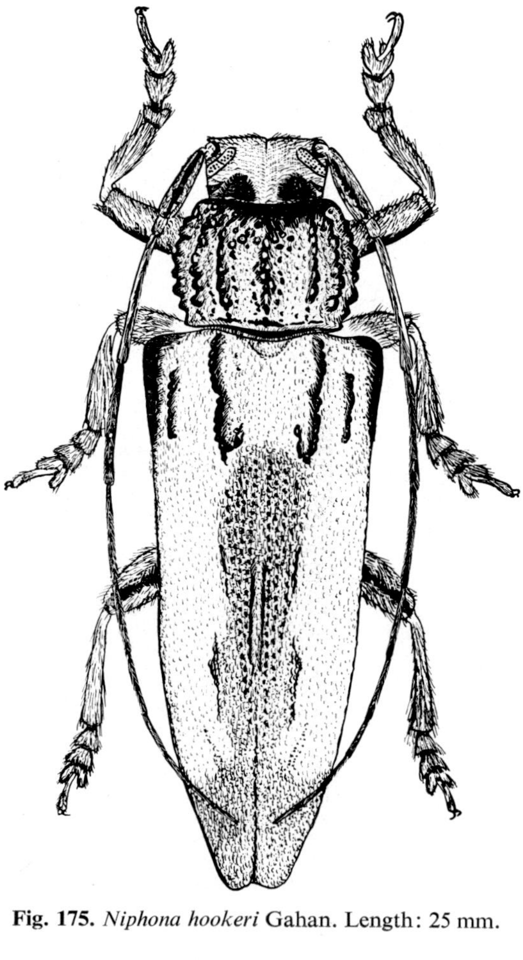

Fig. 175. *Niphona hookeri* Gahan. Length: 25 mm.

REFERENCES

ALLEN, A. A. 1973. The British status of *Pyrrhidium sanguineum*. *Ent. mon. Mag.* **109**: 98.

ANON. 1946 and 1952. The House Longhorn Beetle. F.P.R.L. Leaflet No. 14, D.S.I.R. London.

BECKER, G. 1938. Zur Ernährungsphysiologie der Hausbockkäferlarven (*Hylotrupes bajulus* L.). *Naturwissenschaften* **26**: 462–3.

BECKER, G. 1942. Zur Sinnesphysiologie des Hausbockkäfers. *Naturwissenschaften* **30**: 253–6.

BECKER, G. 1943. Zur Oekologie und Physiologie Holzzerstörender Käfer. *Z. angew. Ent.* **30**: 104–18.

BECKER, G. 1944. Der natürliche Schutz des Laubholzes gegen Hausbockkäferlarven und seine Ursache. *Z. angew. Ent.* **30**: 391–417.

BEESON, C. F. C. 1941. *Ecology and Control of the Forest Insects of India.* Dehra Dun.

BENOIT, J. and JACQUIOT, C. 1953. The Main Problems of Wood Preservation in France. *Record of the Annual Convention of the British Wood Preserving Association. London.*

BLAIR, K. G. 1941. *Rhagium bifasciatum* Fab., and its Varieties in Britain. *Proc. S. Lond. ent. nat. Hist. Soc.* **1940–41**: 39–42.

CRAIGHEAD, F. C. 1923. North American Cerambycid larvae. *Bull. U.S. Dept. Agric.* **787**: 1–12.

CRAWSHAY, G. A. 1907. The Life History of *Tetropium gabrieli* Weise. *Trans. ent. Soc. Lond.* **69**: 183–212.

CYMOREK, S. 1961. *Hylotrupes bajulus* (CERAMBYCIDAE) Entwicklungszyklus. *Encyclopaedia Cinematographica* E374/1961 Institut für den Wissenschaftichen Film. Göttingen.

DUFFY, E. A. J. 1945. The Coleopterous Fauna of the Hants–Surrey border. *Ent. mon. Mag.* **81**: 169–97.

DUFFY, E. A. J. 1946. A Contribution towards the Biology of *Prionus coriarius* L. *Trans. R. ent. Soc. Lond.* **97**: 419–42.

DUFFY, E. A. J. 1951. Dorsal Prolegs and Extreme Cephalic Modification in the First Instar of *Agapanthia villosoviridescens* Deg. (COL. CERAMBYCIDAE). *Ent. mon. Mag.* **87**: 313–81.

DUFFY, E. A. J. 1952. *Handbooks for the Identification of British Insects*, **5** (12), COLEOPTERA, CERAMBYCIDAE. Royal Entomological Society of London.

DUFFY, E. A. J. 1953. *Trinophylum cribratum* Bates. (COL. CERAMBYCIDAE). A New Indian Pest of British Timber Yards. *Entomologist's Gaz.* **4**: 254–64.

DUFFY, E. A. J. 1953. *A Monograph of the Immature Stages of British and Imported Timber Beetles* (CERAMBYCIDAE). British Museum (Natural History). London.

DUFFY, E. A. J. 1957. *A Monograph of the Immature Stages of African Timber Beetles* (CERAMBYCIDAE). British Museum (Natural History). London.

DUFFY, E. A. J. 1960. *A Monograph of the Immature Stages of Neotropical Timber Beetles* (CERAMBYCIDAE). British Museum (Natural History). London.

DUFFY, E. A. J. 1963. *A Monograph of the Immature Stages of Australasian Timber Beetles* (CERAMBYCIDAE). British Museum (Natural History). London.

DURR, H. J. R. 1957. The morphology and bionomics of the European Houseborer, *Hylotrupes bajulus* (COLEOPTERA CERAMBYCIDAE). *Entomology Memoirs* **1** (4): 1–136. Dept. of Agric. Pretoria. South Africa.

ECKSTEIN, K. 1936. Holzzerstörende Bockkäferlarven *Ergates faber* L., der Mulmbock, *Leptura rubra* L., der Rothalsbock und *Hylotrupes bajulus* L., der Hausbock. *Z. angew. Ent.* **23**: 281–92.

FISHER, R. C. and HARRIS, E. C. 1948. A Note on the Status of the House Longhorn Beetle *Hylotrupes bajulus* L. in Great Britain. *Proc. VIII International Congress of Entomology*. Stockholm, 1948.

FISHER, R. C. 1951. House Longhorn Beetle—an Enemy in the Roof. *Municipal Journal*, pp. 1790–1, 1797.

FISHER, R. C. 1954. A Longhorn Beetle New to Britain. *Timber Technology*. **62**: 75

FOWLER, W. W. 1889. *The Coleoptera of the British Islands*. **4.** London.

FRASER, M. G. 1948. *Ent. mon. Mag.* **84**: 47.

FRASER, M. G. 1948. *Saperda scalaris* L. (COL. CERAMBYCIDAE) in Cheshire. *Ent. mon. Mag.* **84**: 120.

FRASER, M. G. 1949. Notes upon the biology of certain species of exotic CERAMBYCIDAE (COL.) occasionally imported into Britain. *Ent. mon. Mag.* **85**: 97–9.

FRASER, M. G. 1950. Notes on the biology of *Saperda scalaris* L. (COL. CERAMBYCIDAE). *Ent. mon. Mag.* **86**: 33–6.

FRASER, M. G. 1950. Further notes on the biology of *Saperda scalaris* L. (COL. CERAMBYCIDAE). *Ent. mon. Mag.* **86**: 73.

GILMOUR, E. F. 1948. *Trinophylum cribratum* Bates (COL. CERAMBYCIDAE). New to Britain. *Ent. mon. Mag.* **84**: 12–16.

GUNN, D. 1919. The Fig and Willow Borer (*Phryneta spinator*). *Bull. Dept. Agric. S. Afr.* **6**: 1–22.

HARDOWIN, R. 1945. Le cycle evolutif de *Clytus arietis* L. sans le rosier. *Bull. Soc. ent. Fr.* 1945: 58–60.

HARWOOD, P. 1946. Status of *Obrium brunneum* Fab. (COL. LONGICORNIA) in Britain. *Ent. mon. Mag.* **82**: 303.

HICKIN, N. E. 1947. The Status of *Hylotrupes bajulus* L. (COL. CERAMBYCIDAE) in Surrey. *Ent. mon. Mag.* **83**: 132.

HICKIN, N. E. 1957. House Longhorn Beetle (*Hylotrupes bajulus* L.) in Ireland. *Ent. mon. Mag.* **93**: 96.

HICKIN, N. E. 1959. Alien Insects Found Alive in Britain, 7–8. *Entomologist's Gaz.* **10**: 17.

JOY, N. H. 1932. *A Practical Handbook of British Beetles*, 2 vols. London.

KAUFMANN, R. R. U. 1944. Notes on *Rhagium bifasciatum* (F.), (COL. CERAMBYCIDAE) and its Varieties. *Ent. mon. Mag.* **80**: 262–3.

KAUFMANN, R. R. U. 1945. Further notes on the Colour Forms of *Rhagium bifasciatum* Fab. (COL. CERAMBYCIDAE). *Ent. mon. Mag.* **81**: 186–7.

KAUFMANN, R. R. U. 1946. The Distribution of *Rhagium bifasciatum* Fab., and its Aberrations in the British Isles. *Naturalist*, **1946**: 193–246.

KAUFMANN, R. R. U. 1947. Notes on *Grammoptera holomelina* Poole, and the other forms of *G. ruficornis* F. (COL. CERAMBYCIDAE). *Ent. mon. Mag.* **83**: 10–12.

KAUFMANN, R. R. U. 1947. Notes on the occurrence of *Hylotrupes bajulus* L. (COL. CERAMBYCIDAE) in the British Isles. *Ent. mon. Mag.* **83**: 37–40.

KAUFMANN, R. R. U. 1948. Notes on the Distribution of the British Longhorn COLEOPTERA. *Ent. mon. Mag.* **84**: 66–85.

KLOET, G. S. and HINCKS, W. D. 1945. *A Check List of British Insects*. Stockport.

KORTING, A. 1960. Durch Hausbockbefall bedingte Gewichtsminderungen des Holzes als Kriterium für die Stärke der Frasstätigkeit. *Holz Roh–u. Werkst.* **18**: 167–70.

LAING, F. 1919. Insects Damaging Lead. *Ent. mon. Mag.* **55**: 278–9.

LEPESME, P. 1948. Cerambyciares nouveaux ou peu connus de l'Afrique tropicale. *Revue Zool. Bot. afr.* **40**: 256–7.

LLOYD, R. W. 1950. *Pyrrhidium sanguineum* L. (COL. CERAMBYCIDAE) in Herefordshire. *Ent. mon. Mag.* **86**: 36.

MENDIZABAL, M. 1943. Cerambidisos de interes agricola. *Bol. Pat. veg. Ent. agric. Madr.* **11**: 387–410.

MOORE, B. P. 1958. *Trinophylum cribratum* Bates (COL. CERAMBYCIDAE). Reared from the egg. *Ent. mon. Mag.* **94**: 16.

PARKIN, E. A. 1934. Occurrence of the Longhorn Beetle, *Hylotrupes bajulus* L. in England. *Forestry*, **8**: 150–4.

PERRIS, E. 1847. Observations sur les larves du *Clytus arietis* de la *Saperda punctata* et de la *Grammoptera ruficornis*, pour servir a l'histoire des longicornes. *Ann. Soc. ent. Fr.* (2) **5**: 547–54.

RASMUSSEN, S. 1967. *Hylotrupes* (Col., Cerambycidae) in dead trees on Fårön, a Swedish island. *Ent. Meddr.* **35**: 223–6.

REITTER, E. 1912. *Fauna Germanica. Die Käfer des Deutsches Reiches*, **4.** Stuttgart.

RITCHIE, E. 1920. The Structure, Bionomics and Economic Importance of *Saperda carcharias* L. 'The Large Poplar Longhorn'. *Ann. appl. Biol.* **7**: 299–343.

TOOKE, F. G. C. and SCOTT, M. H. 1944. Wood-boring Beetles in South Africa: Preventive and Remedial Measures. *Dept. agric. Union S. Afr. Bull.* **247** (*Ent. Ser.* 14): 28–36.

TOOKE, F. G. C. 1949. Beetles injurious to timber in South Africa. *Sci. Bull. Dept. agric. S. Afr.* **293**: 1–95.

TORRE-BUENO, J. R. DE LA. 1950. *A Glossary of Entomology*. Brooklyn Entomological Society, Brooklyn, New York.

VITÉ, J. P. 1952. *Die Holzzerstörenden Insekten Mitteleuropas*. Göttingen.

WHITE, M. G. 1959. The Status of the House Longhorn Beetle. *Timber Technology*. **67**: 406–11.

XAMBEU, V. 1898–1902. Moeurs et métamorphoses d'insectes. *Echange* **14–18**: 1–220 (spec. pag.).

CHAPTER EIGHT

Ambrosia Beetles and Bark Beetles

The families PLATYPODIDAE and SCOLYTIDAE

Ambrosia Beetles (Pinhole Borers)

Ambrosia beetles are so called because Schmidberger in 1836 found that the larvae were not feeding on the wood but were grazing on a sticky lining in the tunnels. This he likened to 'Ambrosia' or 'food of the Gods'. Hartig in 1844 recognized that this lining was fungal growth. Ambrosia beetles are also called Pinhole Borers, perhaps more widely than the former name, in the United Kingdom. This is on account of the appearance of the infested timber containing long, straight but narrow tunnels.

Ambrosia beetles are classified in the two families PLATYPODIDAE and SCOLYTIDAE, and there are more than a thousand different species already described and named. They are found throughout the world where there are forests, but in Britain there is only one indigenous platypodid—*Platypus cylindrus* (F.) and fifty-seven scolytids, of which a few are probably immigrants. Only a small proportion, however, of the British SCOLYTIDAE (six in fact) are Ambrosia beetles, the greater proportion being commonly referred to as 'Bark beetles' which are described later.

It is extremely rare for the life cycle of Ambrosia beetles and Bark beetles to be continued in building timber already utilized in buildings. In almost all cases Ambrosia beetle attack occurs in the woodland or forest when a tree falls by wind or storm or is felled by man. Sometimes however an attack by these beetles proceeds in a standing tree but it is thought that in such cases, as is so often the case in wood-boring insect attack on living trees, the latter possess diminished vigour due to one or another causes. Bark beetles too are usually only the concern of the forester, and if they occur indoors they are either fortuitous intruders or have been brought indoors in baskets of logs for burning. If they happen to be at an advanced stage of development some species are able to metamorphose and, gnawing their way out of the wood, suddenly appear indoors in large numbers. Neither platypodids nor scolytids are able to reinfest dry seasoned wood indoors.

The Timber Infestation Surveyor, however, must be able to recognize the presence of these beetles in timber, even though the attack will not proceed further. This is because of the very widespread nature of Ambrosia and Bark beetles' attack, sometimes in pestiliential proportions in tropical forests, and converted timber is often imported and utilized containing the evidence of this attack. It must therefore be possible to identify the cause of the wood damage and to distinguish it from other infestations which are likely to develop further.

An account of the biology of British Ambrosia beetles has been given by Baker (1956). Roberts (1960) has described the Ghana species, *Trachyostus ghanaensis* Schedl.

As an example of the life history of an Ambrosia beetle that of *Platypus cylindrus* is given below, but it must be borne in mind that there is wide variation within the group and it must not be assumed that the details given necessarily apply to all species. The reader is referred to the considerable literature which is given at the end of Bletchly's 1960 paper.

Platypus cylindrus (Fabricius). The Oak Pinhole Borer (Fig. 176)

This is the sole indigenous species of the PLATYPODIDAE. It is easily distinguished from all the scolytids by being elongate, the head very prominent and the first tarsal segment unusually long. It is pitchy-brown, sometimes pitchy-red and varies in length from 5 to 5·5 mm. The antennae are short but possess large flat clubs. In Britain it is a southern species distributed mainly from Kent to the New Forest, Oxfordshire and the

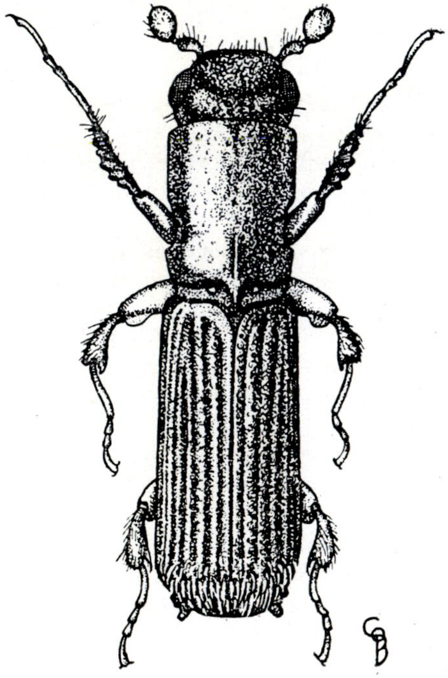

Fig. 176. *Platypus cylindrus* (F.). Length: 5 mm. (Royal Entomological Society of London.)

Forest of Dean. It is only of minor importance as a forest pest, the attack usually taking place in stumps but sometimes also in freshly felled logs in the timber yard. It usually attacks oak but has been recorded also in beech, ash, elm, and sweet chestnut.

Life cycle

This description starts with the adult beetles finding a suitable stump or freshly felled log. Baker found that the beetles, which have a strong quick flight, arrive on the cut surface, the nearby bark, or the ground without any preliminary circling and are obviously strongly attracted to the smell of fermenting sap. This has also been observed by Browne (1952). As a point of interest in this regard there are records of playpodids flying on to rum casks and boring into the oak staves along the interfaces where there was a leak of rum and thus causing an even greater leak! The present author has also seen the same phenomenon in oak casks containing lime juice just shipped from Dominica.

The state of fermentation of the sap appears to be the main factor in the selection by the adults of a boring site. The male appears first on the stump and running over the cut surface finds a crack or a crevice in the bark and then commences to bore. The female on the other hand runs or flies around the stump until a newly constructed tunnel is observed by the pile of bore-dust at the mouth. She then enters and if acceptable to the male they both emerge from the tunnel. Mating takes place on the surface. They then re-enter the tunnel but the female now enters first followed by the male, and thereafter it is the female who carries out the boring, the male merely removes and ejects the bore-dust. It should be noted that this bore-dust is white and splintery and has not travelled through the gut of the beetle.

The female lays her white glistening eggs in the main tunnels, in small batches most often of three to four, commencing in early autumn and continuing into winter. Baker found that the number of ripe eggs in the ovaries of a female varied from one to eight and that after the pair have entered the wood no further copulation takes place, the eggs being fertilized from sperm stored in a small sac-like spermatheca. The young, legless, and flattened larvae move along the tunnel with a 'ripple-like' motion. This stage probably represents two instars. Then the larvae become cylindrical for a further two instars and almost take up the full diameter of the tunnel. Baker thought that the first three instars were undergone in one month, by far the greater length of the larval life being taken up by the last instar. It has been observed that some larvae require two years to develop although usually only one year is spent in this stage. There is great variation in the complexity of the tunnel system, varying from a relatively short unbranched tunnel 8 cm or more in length, to a complicated branch tunnel system, recorded by Parkin to be over 2 metres in length.

The larvae graze on the lining of the tunnel which is composed of small fragments of wood on which the fungal growth occurs. The wood fragments are successively passed and repassed through the larval gut. The larvae, however, do occasionally bore to a small extent, the wood fragments being carried away by the adults.

Baker investigated the fungi associated with *Platypus cylindrus* and consistently found three species. It seems probable that all these fungi play some part in the constitution of the Ambrosia and thus in the nutrition

Fig. 177. Obeche timber showing pinhole borer damage.

of the insect. Baker found first, a species of *Graphium* which was usually more evident in the entrance shaft and tunnels in which there were no larvae present. Secondly a species of *Cephalosporium* which was more evident in the fresh tunnels in which larvae were feeding. These two species are actually conidial stages of *Ceratostomella* a genus of the ASCOMYCETES. *Cephalosporium* grown on sterile oak produces a brown stain very similar to that usually produced in the wood surrounding *Platypus cylindrus* tunnels (Figs. 177 and 178).

The third fungus present in the Ambrosia is a yeast which is always present in the fresh tunnels. It is interesting to note that several successful attempts have been made to rear Ambrosia beetles entirely on a specific fungus isolated from their tunnels. The species *Xyleborus germanus* has been cultured on *Ceratostomella* and *Xyleborus fornicatus* has been reared throughout its complete life cycle on *Monacrosporium ambrosium*.

Identification of Ambrosia beetle damage

It is important that anyone examining timber in buildings should be able to distinguish between the effects of insect damage caused by those species

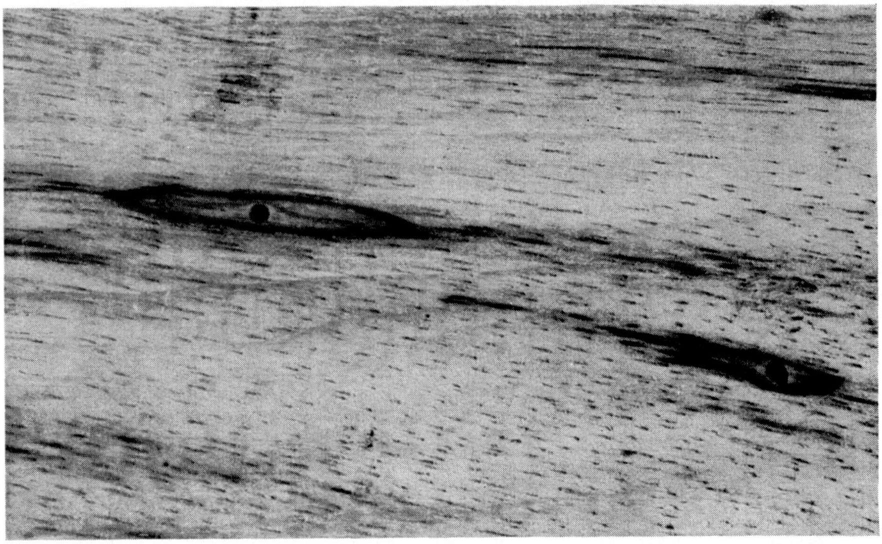

Fig. 178. Timber showing shot-hole borer damage with associated flame-shaped staining.

which can complete their development and possibly reinfest the dry wood (either the same piece of timber in which the larval and pupal life took place or some other wood in the building), and the effects of insect damage caused by those species which infest the dying tree or freshly felled tree but always die when the moisture content drops to that found in wood in buildings (approximately 8–20 per cent). In the latter eventuality the insect damage will not worsen, its only defect will be visual. This is the case with damage to wood by Ambrosia beetle, whereas Bark beetle is purely superficial, tunnels seldom penetrating the sapwood more than a fraction of an inch. In Ambrosia beetle attack the tunnels frequently penetrate to the centre of the heartwood.

Although in the timber yard it is sufficient only to separate Ambrosia beetle attack from that of *Lyctus* yet frequently timber is observed in older buildings which show Ambrosia beetle attack which took place when it was a standing tree or freshly felled log. In the table the features of Ambrosia beetle attack are compared with all other forms of wood-boring insect attack found indoors.

Notes on the species

The British Ambrosia beetles in the family SCOLYTIDAE are grouped in the subfamily IPINAE. Short descriptions are given below.

Anisandrus dispar (Fabricius) has the common name of 'shot-hole borer' although many other species are given this title on account of the similarity of their borings to shot-holes (Fig. 178). The female is from 3 to 3·5 mm in

Ambrosia Beetle Attack [Pinhole Damage]	Damage by Other Wood-boring Insects Found Indoors.
Hardwoods most frequently attacked. Softwoods less frequently. In sapwood and heartwood.	Hardwoods less frequently attacked. Softwoods more frequently. Confined to sapwood, or heartwood damaged only after sapwood consumed.
Wood dust absent from tunnels.	Wood dust present in tunnels, either wholly or in part.
Wood never powdered or tunnels never run into each other.	Sapwood often completely powdered when attack severe or tunnels run into each other until wood is coarsely porous.
Holes and tunnels usually darkly stained; often surrounded by a flame-shaped streak of discoloured wood.	Holes and wood not darkly stained, discoloured wood absent.
Tunnels across grain of wood.	Tunnels may run in all directions but tend to be along grain in early stages of attack.
Holes circular varying from 0·5 mm to 3 mm in diameter, according to the species of beetle. Sawn surfaces may reveal all sections, from longitudinal to transverse, of the brood tunnels and sometimes showing larval cradles.	Holes circular from 1·5 to 3 mm or oval 6 mm. Longitudinal sections of galleries never revealed except by abrasive wear.
Damage ceases when timber is seasoned.	Damage progressively increasing.

length, stout and cylindrical, but the scarcer male is flatter and smaller, only 2–2·5 mm in length, poorly developed and without wings. This beetle attacks oak and holly and occurs also in old orchards where it attacks plum and apple. It is found rarely in southern England.

Xyleborus. There are two very local species in this genus found only in southern England, *X. xylographus* (Say) and *X. dryographus* (Ratzeburg). The former beetle (Fig. 179) is found attacking oak, beech, sweet chestnut and a variety of other hardwood species and has been recorded on pine (Forest of Dean). It is common on the European Continent. The brood tunnel does not follow the annual rings and consists of a simple narrow tunnel in the sapwood at the end of which is a chamber in which larvae, pupae and adults occur. In North America, species of this genus are of considerable economic importance especially in the south and on the Pacific coast. *X. dryographus* is rarer than *X. xylographus*, occurring in oak and beech in Kent, Sussex, Surrey, Oxfordshire and Gwent.

Trypodendron. There are two species in this genus which are characterized by the transverse pronotum and the very strongly clubbed antennae.

Trypodendron domesticum (Linnaeus) is found in a wide variety of

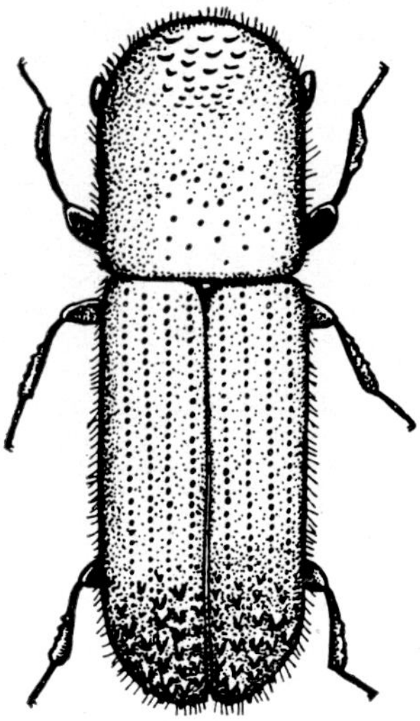

Fig. 179. *Xyleborus xylographus* (Say). Length: 2·2 mm.

broad-leaved trees and shrubs, but although widely distributed, is rarely of economic importance. The larvae are situated in 'cradles' (short tunnels just sufficiently large to accommodate the larvae) above and below the main tunnel.

Trypodendron lineatum (Olivier) resembles the preceding species with regard to its egg and brood tunnels, but in contrast is found only on pine, spruce and larch. From time to time this insect causes serious damage to wind-blown timber in Scotland.

Xyloterus quercus (Eichhoff) is rarer than the preceding species probably having a distribution only in southern England, and is found attacking oak and beech. It is sometimes found infesting the same tree as *T. domesticum*.

Bark Beetles

Munro has given a good general account of the SCOLYTIDAE of Britain. For the reasons given above only a brief summary is given here. As is suggested by the popular name of this group most species spend the greater part of their life cycle in the bark or in the cambial layer, only rarely penetrating into the sapwood. Some species are of great economic importance in forestry, more on account of the fungal diseases which can more easily infect the tree later, rather than the slight damage caused by the beetles and larvae. The Elm Bark Beetles *Scolytus scolytus* and *S. multistriatus* are pests causing serious loss, whilst the adult stage of the Pine Shoot Beetle *Blastophagus piniperda* burrows into the leading shoots of pine trees, causing them to be crippled or possibly to die.

There are two rather distinctive life cycles according to whether the male is monogamous or polygamous. In the former the female bites out a small chamber in the bark of the tree in which she has chosen to lay her eggs. She then flies off and returns with a male with which she mates, either in the nuptial chamber or in its proximity. The female then bores a long straight tunnel in the cambium depositing her eggs alternately on each side as she progresses. When the larvae hatch out they commence to bore outwards, the galleries thus formed usually making a constant angle with the 'mother gallery'. This angle has been found to differ in the different species and is of importance in identifying the damage. As the galleries extend outwards they increase in size and as they are all in the same plane, of necessity they get closer together. They are usually filled with the bore-dust. If a large number of beetles excavate their burrows a large area of the cambium is destroyed and the bark is loosened. This reveals the characteristic patterns of galleries radiating from the 'mother' or 'brood' channel (Fig. 180). When mature, the larvae pupate and after the final metamorphosis the adults bore their way out through the bark, the pattern of flight holes again often being a characteristic design. Woodpeckers are adept at finding the brood galleries and, picking off the loosening bark, they consume the circle of developing larvae.

The Two-toothed Bark Beetle, *Pityogenes bidentatus* (Herbst), is an example of one of the polygamous species. The male, in this case, gnaws out a circular chamber in the bark and in this nuptial chamber he pairs with from four to six females. The latter then construct their brood galleries radiating out from the nuptial chamber. Larvae hatching from the eggs construct radiating galleries so that a very complicated tunnel pattern results, especially where the attack is heavy and the larval tunnels overlap at the extremities.

Some of the more important species of the SCOLYTIDAE in Britain are as follows:

Scolytus

There are six British species all characterized by the front tibiae possessing a hook-shaped claw.

Scolytus scolytus (Fabricius) found from April to August, is the Common Elm Bark Borer in England. There are two generations in the year and the immature adults damage small branches and twigs. The adults are 3–6 mm in length.

Scolytus multistriatus (Marsham), the Small Elm Bark Beetle, is 2·0–3·3 mm in length. The brood tunnels usually occur on the smaller branches of the tree.

Scolytus ratzeburgi Janson is the Birch Bark Beetle and is found only locally in Scotland. It is the largest species in the genus being up to 6·5 mm in length.

Fig. 180. Sapwood surface of elm showing the pattern of attack by *Scolytus scolytus* (F.).

Scolytus mali (Bechstein), the Large Fruit Tree Bark Beetle, attacks several fruit trees as well as whitebeam (*Pyrus aria*) and *Laurus*. It measures from 3·5 to 4·5 mm in length. It is very local in southern England.

Scolytus rugulosus (Müller), the Small Fruit Tree Bark Beetle, is widely distributed in England and Wales. It infests fruit trees and *Crataegus*. There may be three broods annually. This is the smallest *Scolytus* measuring from 2 to 2·5 mm in length.

Scolytus intricatus (Ratzeburg) attacks the base of young shoots of oak. It is the only *Scolytus* whose mother gallery is horizontal or transverse, those of all other species being vertical or longitudinal. It is mainly distributed in southern England but has been recorded in some northern counties.

Leperesinus

One species, *L. fraxini* (Panzer), occurs in Britain. It is known as the Common Ash Bark Beetle and is probably the best known of the Bark beetles, on account of ash being a popular wood for logs. It is about 3 mm in length and the mother galleries are horizontal. There are usually two broods annually but probably only one in the north. The mature larvae burrow to a depth of about 25 mm and adults often emerge from dry wood.

Hylesinus

There are two British species, both of which attack only ash (*Fraxinus*) and rarely some allied trees.

Hylesinus crenatus (Fabricius) is known as the Large Ash Bark Beetle. It measures from 4 to 6 mm in length and occurs mainly on dying ash trees. The larval tunnels are long and irregular and the mother gallery is horizontal. It is a common insect.

Hylesinus oleiperda (Fabricius) is known as the Lesser Ash Bark Beetle. It is from 2·5 to 3·5 mm in length and the larvae are to be found in the branches and twigs of ash. It is common in England and Wales, but has not been found north of the Tweed. In Europe it causes damage to olive trees.

Blastophagus

These are the Pine Shoot Beetles.

Blastophagus minor (Hartig), the Lesser Pine Shoot Beetle, is restricted to certain districts of Scotland. The egg tunnel is transverse.

Blastophagus piniperda (Linnaeus), the Pine Shoot Beetle, on the other hand has a longitudinal egg tunnel and it is much more widely distributed. It attacks Scots pine, spruce and larch. There are two broods in the year. In very favourable conditions they may invade and kill vigorous trees.

Hylurgops

Contains the single British species *palliatus* (Gyllenhal) which attacks pine, spruce and larch. In thicker bark there are often collective tunnels.

Hylastes

This genus contains five British species, the commonest of which is *H. ater* (Paykull). The egg tunnels are cut in the stumps or roots of pine, spruce and larch. Young trees are often girdled and killed, especially where replanting has occurred after felling.

Ips

Ips sexdentatus (Boerner) is frequently found on imported timber (pine). It is established in the Forest of Dean and occurs also in Scotland. From time to time outbreaks occur in other parts of the country. The tenebrionid beetle *Hypophloeus fraxini* Kug. is predacious on this insect.

Ips typographus (Linnaeus). Although not established in Britain this European species attacking spruce is frequently imported. It is of great economic importance in Europe and is potentially dangerous to our spruce plantations.

Onthotomicus

Onthotomicus erosus (Wollaston) is a Mediterranean species which occurs in the Forest of Dean, where it was thought to be introduced in pit-prop timber.

Onthotomicus laricis (Fabricius), in spite of its name, only rarely attacks larch and is usually found on pine. It is a common beetle in England.

REFERENCES

ANON. 1921, revised 1946. The Black Pine Beetle (*Hylastes ater*) and other closely allied beetles. Forestry Commission Leaflet. No. 4. London.

ANON. 1941. Pinhole or Powder-post Damage. Forest Products Research Leaflet No. 17. D.S.I.R., H.M.S.O. London.

BAKER, J. M. 1956. Investigations on the Oak Pinhole Borer *Platypus cylindrus* Fab. A progress report. *British Wood Preserving Association Convention Record.*

BASTIN, H. 1948. Bark Beetles. *Illustrated London News*, 644. 4th Dec. 1948.

BATRA, L. R. 1966. Ambrosia fungi: Extent of specificity to ambrosia beetles. *Science, N.Y.* **153**: 193–5.

BLETCHLY, J. D. 1960. A Review of Factors affecting Ambrosia Beetle Attack in Trees and Felled Logs. *XI Int. Congr. Ent.* Vienna. 1960. This paper contains an important list of 53 references to the literature up to 1959.

BOOCOCK, D. 1960. The Behaviour and Treatment of the Oak Pinhole Borer. *Timber Technology* **68**: 73–4.

DUFFY, E. A. J. 1953. *Handbooks for the Identification of British Insects,* **5** (15), COLEOPTERA, SCOLYTIDAE and PLATYPODIDAE. Royal Entomological Society of London.

FISHER, R. C., THOMPSON, G. H. and WEBB, W. E. 1953. Ambrosia Beetles in Forest and Sawmill. *Forestry Abstracts*, part 1, **14**: 381–9. Part 2, **15**: 3–15 (111 references are given in this part).

JONES, T. 1959. Ambrosia beetles (SCOLYTOIDEA), their Biology and Control in West Africa. *West African Timber Borer Research Unit. Tech. Bull.* **2**: 1–14.

MUNRO, J. W. 1926. British Bark Beetles. *For. Comm. Bull.* No. 8. H.M.S.O. London.

ROBERTS, H. 1960. *Trachyostus ghanaensis* Schedl. (COL. PLATYPODIDAE). An Ambrosia Beetle attacking Wawa (*Triplochiton scleroxylon* K. Schum.) *West African Timber Borer Research Unit. Tech. Bull.* **3**: 1–17.

CHAPTER NINE

Wasps, Ants, Bees, Moths and Mayflies

The Hymenoptera, Lepidoptera and Ephemeroptera

The insects commonly known as Wood Wasps are placed in the order of insects known as HYMENOPTERA containing the ants, bees, wasps and ichneumon flies.

The more important characters distinguishing hymenopterous insects are the possession of two pairs of membraneous wings, the hinder pair smaller than the fore pair and with the wings of each side interlocked with rows or groups of hooklets (Fig. 181). The mouthparts of the adult are adapted for biting but may also be adapted for lapping or sucking. An ovipositor is present which may be modified for sawing, piercing, or stinging.

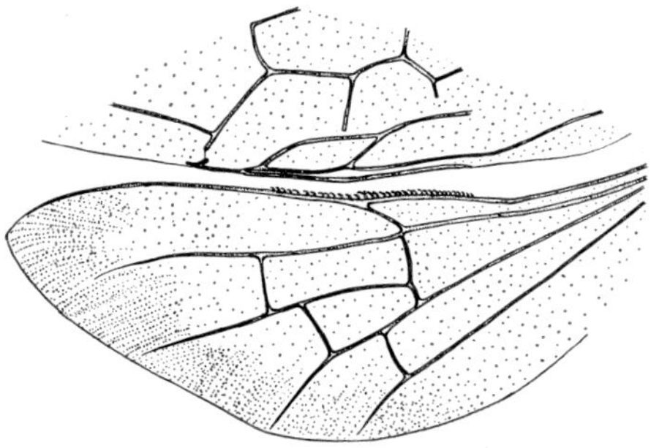

Fig. 181. Hooking mechanism of wings of *Sirex cyaneus* F.

The HYMENOPTERA is a very large order, over 100,000 species having already been described and it is thought that many thousands are still to be discovered. But much more remarkable is the very wide range of form and behaviour found in this group. The order is divided into two suborders, firstly, the SYMPHYTA containing morphologically the more primitive species in which there is no marked constriction between first and second abdominal segments (Fig. 182), a feature which is such a marked character of the second suborder, the APOCRITA (Fig. 183). The SYMPHYTA, with the exception of one genus *Orussus*, are phytophagous and contain no parasitic species. (It is of interest to note that *Orussus* is parasitic on BUPRESTIDAE.) It is in the suborder SYMPHYTA

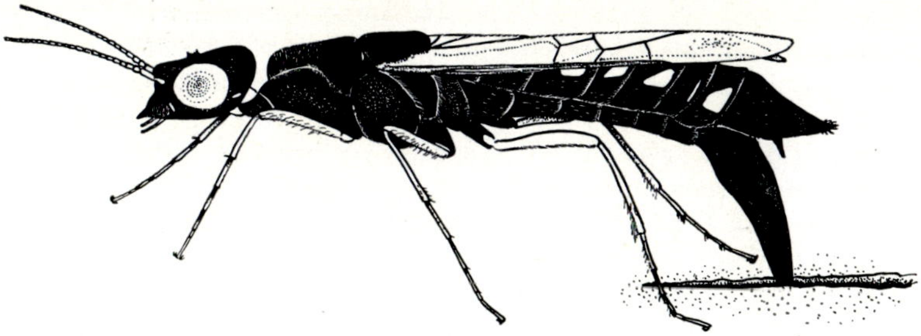

Fig. 182. Side view of female *Xiphydria camelus* (L.) ovipositing. Length: 12 mm.

that the Wood Wasps are classified in which they are joined with the saw-flies and stem saw-flies. It is worth mentioning that in the APOCRITA are found insects living in social communities, the ants, wasps of the family VESPIDAE and bees of the family APIDAE and in addition a number of large groups whose species are parasitic on other insects.

The Wood Wasps

There are six superfamilies in the SYMPHYTA, the one that concerns us as containing those families with wood-boring larvae is the SIRICOIDEA. This consists of three families, the SIRICIDAE, the XIPHYDRIIDAE, and the SYNTEXIDAE, but as the last-named family contains only a single North American species not found in Britain it will not concern us here.

World distribution

The family SIRICIDAE is a fairly small one divided into two subfamilies and eight genera and only about seventy species and subspecies. They are indigenous to the Holarctic and Oriental regions but have been imported in timber into many countries of the world, in some of which they are viewed as serious forestry pests. In Europe there occur three genera containing

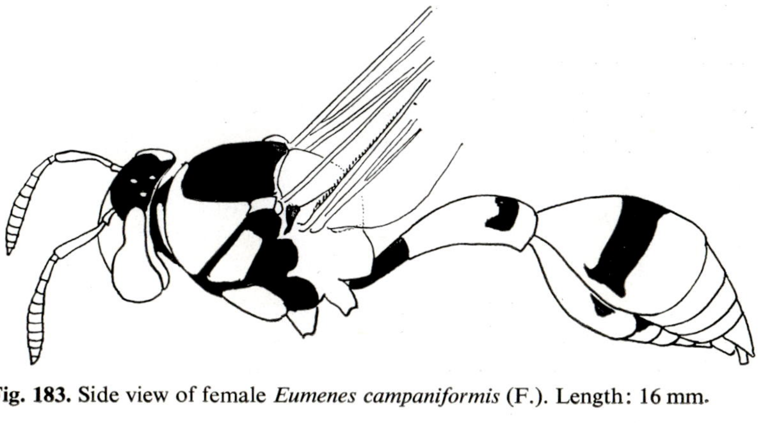

Fig. 183. Side view of female *Eumenes campaniformis* (F.). Length: 16 mm.

eleven species and subspecies. There are some difficulties in the way of establishing which species are endemic to Britain but Benson considered that the form of *Urocerus gigas* known as *taiganus*, and possibly *Sirex juvencus*, were probably native to the Caledonian forest in North Britain. The other form of *S. gigas* and two species of *Sirex* are established aliens in Britain. Four other forms of *Urocerus gigas*, one species of *Sirex*, one species of *Xeris* and one species of *Tremex* are found rarely or occasionally in imported timber in Britain.

The family XIPHYDRIIDAE likewise is a fairly small family of about seventy species. It is, however, world-wide in distribution. Two genera containing six species occur in Europe whilst in Britain there are two species of only one genus—*Xiphydria*.

The XIPHYDRIIDAE may be separated from the SIRICIDAE by the possession of a transverse furrow across the mesonotum and an exceptionally long neck. The females do not possess a horn-like projection on the apical segment of the abdomen.

SIRICIDAE

The British SIRICIDAE may be separated by means of the following keys which are those of Benson.

1. Antennae filiform and long (longer than C+ stigma of forewing) and set close together (so that the distance between them is only about one and a half times as much as the distance of one from the nearest eye-margin). Eyes not more than one and a half times as broad as long. Labial palps three-segmented. Cerci present. Anal cell of forewing contracted from about the middle. In Coniferous trees (PINACEAE). . SIRICINAE. 2

– Antennae slightly swollen in middle and short (shorter than C+ stigma of forewing) and set far apart (three times as great as the distance from one of them to the nearest eye-margin). Eyes at least twice as broad as long. Labial palps two-segmented. Anal cell of forewing contracted in basal third. Angiosperm trees. TREMECINAE: *Tremex*

2. Head above with a white or yellow spot present behind each eye (Fig. 184). Cornus, at apex of ♀ abdomen, constricted towards base and broadened apically 3

– Head above without a pale spot behind each eye (Fig. 185). Cornus of ♀ may be shouldered, but it is never constricted towards the base and then broadened apically. . . . . . . *Sirex*. 4

3. Head without a genal carina present on lateral hind margin. Pronotum entirely black, and, in dorsal view, is truncate in front with medial length less than the length of the head behind the eyes. Hind tibia with two apical spurs. Third antennal segment shorter than fourth. ♀ ovipositor shorter than abdomen + cornus and not longer than a forewing . . . . *Urocerus*. 11

– Head with a genal carina present on lateral hind margin. Pronotum with a pale dorsal stripe each side and in dorsal view, the front is strongly emarginate, projecting forward at the angles, with its medial length at least as long as the length of the head behind an eye. Hind tibia with only one apical spur. Third antennal segment longer than fourth. ♀ ovipositor longer than abdomen + cornus and about one and a third times as long as a forewing . . . . *Xeris*

4. Females 5

– Males 8

5. Legs mostly reddish yellow (except that the coxae and apical tarsal segment may be piceous). Wings at most only slightly infuscate. Ovipositor shorter than forewing 6

– Legs mostly bluish-black (except that the apical half of the hind coxae may be yellow). Wings dark violaceous to slightly smoky. Ovipositor about as long as a forewing . . . *areolatus* (Cresson)

6. Apical tarsal segment of all legs yellow. Sawsheath about as long as oblong plate or longer; ovipositor would reach beyond base of cell 3R1 if stretched along front margin of forewing from base. Mesopleura blue-green and in the middle with shining interspaces larger than the punctures. Middle tergites less densely rugulose and shining laterally. Antenna may be brown at base . . . 7

– Apical tarsal segment of all legs piceous. Sawsheath clearly shorter than oblong plate and ovipositor would reach from the base of wing only to base of cell 3R1. Mesopleura violet and so densely punctured that in the middle the interspaces between the punctures are smaller than the punctures. Middle tergites very densely rugulose and not shining laterally. Antennae always entirely black (Fig. 187, top) . . *noctilio* (Fabricius)

7. Sawsheath longer than oblong plate; ovipositor would reach almost to apex of radial cell if stretched along front margin of forewing from base. Antenna all black (Fig. 187, bottom) . *cyaneus* (Fabricius)

– Sawsheath as long as oblong plate; ovipositor would reach only to the middle of radial cell from base of forewing. Antenna often with brown basal segments (Fig. 187, centre) . . *juvencus* (L.)

8. Head and thorax black or blue-black. 9

– Head and thorax metallic green. Posterior or sometimes all the legs, except their coxae, rufous. Abdomen, except one or two basal segments, red . . . . *areolatus* (Cresson)

9. Apical tarsal segment of all legs yellow. Mesopleura in the middle with shining interspaces larger than the punctures 10

– Apical tarsal segment of all legs piceous. Mesopleura so densely punctured in the middle that the shining interspaces are smaller than the punctures. Abdomen with apical sternites and tergites as well as two or three basal tergites black . . . . *noctilio* (Fabricius)

10. Antennae with basal segments of flagellum red or brown; if infuscate, then with apical sternites and tergites black . . . . . . *juvencus* (L.)

Fig. 184. Head of *Urocerus gigas* (L.).

Fig. 185. Head of *Sirex noctilio* (F.).

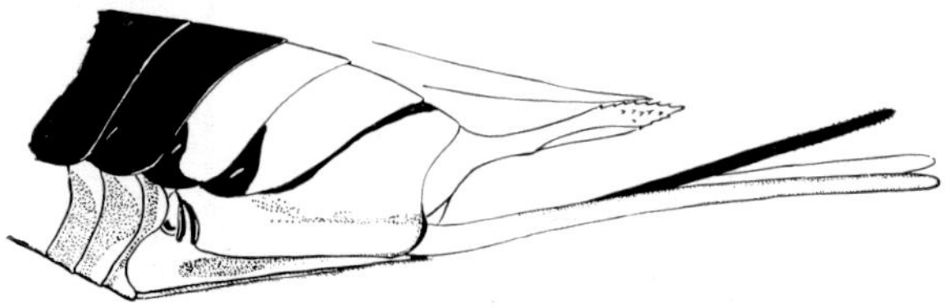

Fig. 186. Apical segments of abdomen of *Urocerus gigas* (L.), female.

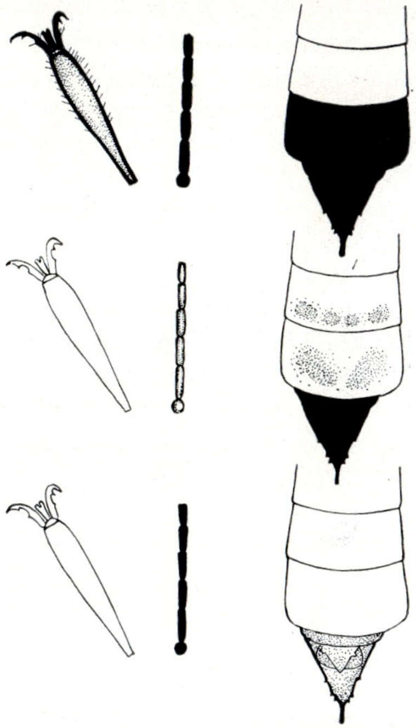

Fig. 187. Apical tarsal segment, basal antennal segments and apical abdominal segments of: top, *Sirex noctilio* (F.); centre, *Sirex juvencus* (L.); bottom, *Sirex cyaneus* (F.).

– Antenna entirely black and abdomen not infuscate at apex . . . . . *cyaneus* (Fabricius)

11. Females 12

– Males 18

12. Abdomen entirely black, or with at most white cornus and white flecks at the sides of some of the tergites. Legs black except for white bases of tibiae and basitarsus 13

– Abdomen banded with yellow or brown 14

13. Abdomen entirely dull black. Wings yellow with yellow venation. Antennal flagellum yellow entirely . . . . . . *californicus* (Norton)

– Abdomen black with white cornus and white flecks at the sides of some of the tergites. Wings more or less smoky with black venation. Antennal flagellum white except for the black apex and two or three basal segments . . . . *albicornis* F.

14. Head behind the eyes yellow except at most for

the infuscate medial furrow of the post-ocellar area 17

Head with the yellow behind the eyes divided into two widely separated spots by the broad black post-ocellar area. Hind tibia with at most the extreme apex infuscate. Stigma at least slightly darker than the costa. Ovipositor clearly shorter than a forewing 15

15. Ovipositor sheath black and shorter (would reach only to about the middle of the cell 3R1 if stretched along from margin of forewing from the base). Ninth tergite at least black at base 16

– Ovipositor sheath brown and longer (would reach to about apex of cell 3R1 of forewing from base). Ninth tergite at most black only in its basal groove (Fig. 186) . . . . . *gigas gigas* (L.)

16. Ninth tergite at least partly yellow above and eighth entirely yellow . . . . . *gigas taiganus* Benson

– Ninth tergite entirely black and eighth mostly black . . . . *gigas flavicornis* (F.)

17. Hind tibia with basal two-thirds black. Abdomen with at least tergites 3–7 and 9 banded with black above. Claws with large subapical tooth longer than its basal breadth. Ovipositor as long as forewing . . . . . *fantoma fantoma* Fabricius

– Hind tibia all yellow with at most the extreme apex brown. Abdomen mostly yellow above with only tergites 6 and 7 banded apically with black, and sometimes 4 and 5 with lateral spots. Claws with a minute subapical tooth not longer than its basal breadth. Ovipositor only about two-thirds as long as forewing . . . . *tardigradus* Cederhjelm

18. Abdomen with at least apex infuscate 19

– Abdomen entirely reddish-yellow . . . *californicus* (Norton)

19. Head with yellow behind the eyes divided into two widely separated spots by the broad black post-ocellar area 20

– Head entirely yellow behind the eyes except for the infuscate medial furrow to the post-ocellar area . . . . *fantoma fantoma* Fabricius

20. Abdomen with seventh tergite black. Antenna generally infuscate apically from about the seventh segment 21

– Abdomen with seventh tergite yellow. Antenna with flagellum entirely yellow . . . . *gigas gigas* (L.)

21. Hind basitarsus 6·5 to 8 times as long as broad . *albicornis* F.

– Hind basitarsus 4 to 5·5 times as long as broad *gigas flavicornis* (F.) *gigas taiganus* Benson

Biology of the SIRICIDAE

Our knowledge of the biology of the SIRICIDAE in Britain is mainly due to the papers of Chrystal, Chrystal and Myers, and Parkin. As British insects, Wood Wasps are first mentioned by Moses Harris in the *Exposition of British Insects* of 1792. Although in that work an insect is figured and described as *Sirex torvus*, almost certainly *S. cyaneus* is referred to. Donovan in 1806 reports the capture of *S. juvencus* in a house in London but the excellent illustration is again that of *S. cyaneus*. Curtis, in his *British Entomology* of 1829, records *S. gigas* and *S. juvencus* and Westwood in the *Introduction to the Modern Classification of Insects* in 1836 refers to the occurrence of *S. juvencus* in Wyre Forest in Worcestershire.

Flight period

Urocerus gigas is usually seen on the wing from June until early October and *Sirex cyaneus* from the end of July to the beginning of October. A loud buzzing noise is made by the insects when they fly in the hot sun. In German literature the noise is known as *schwirren*. Early writers referred to the adult wood wasps feeding on other insects which they captured, but later authors consider that this is erroneous.

Mating

This is thought to take place at the tops of trees. All the evidence seems to point to a preponderance of females over males at something like two to one.

The ovipositor (Fig. 188)

This consists of three separate segments, a dorsal valve and two ventral valves. The dorsal valve, which is thinner, extends around almost three-quarters of the ovipositor tube, and represents the inner gonapophyses of abdominal segment 9 fused together. The ventral valves which represent the gonapophyses of segment 8 are much thicker and fit into longitudinal grooves at the lateral extremities of the dorsal valve. When an egg is passing down the ovipositor the ventral valves are extended and the egg is visible through the slit.

Oviposition

Chrystal says that this takes place with greater vigour in warm sunny weather than on cloudy wet days. Before commencing to bore, the female *Sirex* makes a close examination of the bark with her antennae. A few

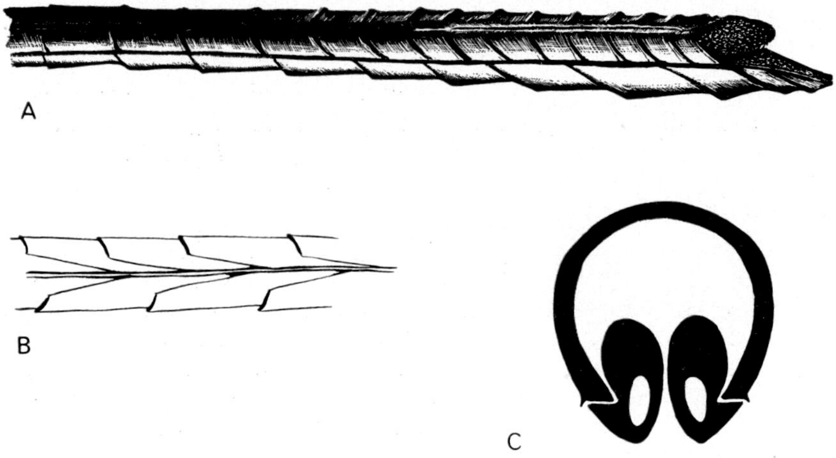

Fig. 188. A, Apical 3 mm of ovipositor of *Urocerus gigas* (L.); B, Basal part of ovipositor of *U. gigas* (ventral view); C, Transverse section of ovipositor of *Sirex cyaneus* (F.) (after Chrystal).

trial borings are made before oviposition starts. Egg-laying is accomplished by a complex pattern of movements. The insect first lifts itself by forcing the ovipositor sheaths against the bark. This exposes the flexible and very elastic ovipositor which is then forced into the bark. The ovipositor moves rapidly up and down with a sawing movement, and the abdomen moves rhythmically during the operation.

When the ovipositor is fully inserted, an egg is laid at the bottom of the shaft. The ovipositor is slowly withdrawn a short distance and another egg is laid. The eggs and tunnel wall are coated with a white, glistening, glue-like secretion which apparently lubricates the egg-passage. This secretion may contain the oidia of symbiotic fungi (see below).

Egg number

The number of eggs laid per tunnel is three to seven. By dissection, Scheidter found an average of over 1,000 eggs in the ovaries of *Sirex augur* but in *S. noctilio* the number was only 400. Chrystal found 300–400 eggs in ovaries of large female *Urocerus gigas*. Sometimes the female dies with the ovipositor stuck fast in the wood and in some cases this may be due to oviposition being attempted in wood with too vigorous a sap flow.

Eggs

In *Urocerus gigas* and *Sirex cyaneus* the eggs are cigar-shaped, with one end more pointed than the other. They vary in length between 1·25 and 1·50 mm. The young larvae can be seen through the translucent chorion, the mandibles and the terminal spine being very dark in colour. The eggs hatch in about three to four weeks.

Larvae

The young larva generally bores at right angles to the egg shaft and keeps within the outer sapwood until it is about 8–9 mm in length, it then turns inwards to the heartwood usually boring in a vertical direction. The tunnel becomes tightly packed with frass in which the moulted skins of the larva can be found. After a time the larva commences to direct its gallery towards the outside of the wood and the pupation chamber is most usually constructed at about 12–20 mm from the wood surface. The direction of the gallery, however, is often influenced by the position of knots and cankerous growths, which the larva will usually but not always avoid. The length of the larval gallery has been given by Hartig in the case of *Sirex juvencus* as being 60–75 cm long. Hanson measured a tunnel of *Urocerus gigas* in silver fir which was 38 cm. Chrystal measured a number of tunnels of *Sirex cyaneus* in a larch log 18 cm in diameter, the shortest was 15·6, the longest 26 cm. There is some suggestion that male pupae are found more superficially than the females, a phenomenon well known in other wood-boring insects, notably in the COLEOPTERA.

The siricid larva (Fig. 189a) is cylindrical and whitish in colour. The head is ovoid, yellowish-brown and directed downwards. It is generally well developed; the antennae are three-jointed; the labrum is rather strongly asymmetrical with the anterior margin spinous in the young larvae and heavily sclerotized when mature. The mandibles are asymmetrical, the right lies above the left and bears four teeth. The left bears only three

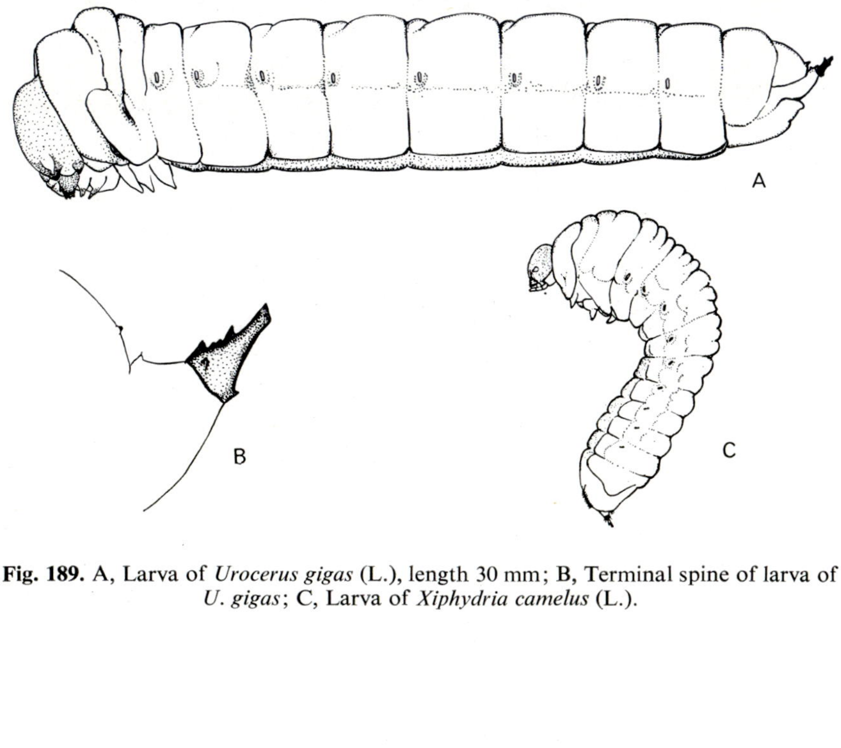

Fig. 189. A, Larva of *Urocerus gigas* (L.), length 30 mm; B, Terminal spine of larva of *U. gigas*; C, Larva of *Xiphydria camelus* (L.).

teeth. The palps of the maxilla and labium are blunt and fleshy. The pronotum is large and hood-like and there are three pairs of mamma-like thoracic legs.

Parkin has found that between the first and the second abdominal segments, perhaps of only female larvae, there lie a pair of deep folds in which are situated epidermal structures overlying columnar epithelium tissue. The cuticle is pitted in a characteristic fashion and in the pits there occur tangled aggregates of fungal oidia. Parkin calls these the hypopleural organs (Fig. 190). In *Urocerus gigas* the organ is shorter and broader and

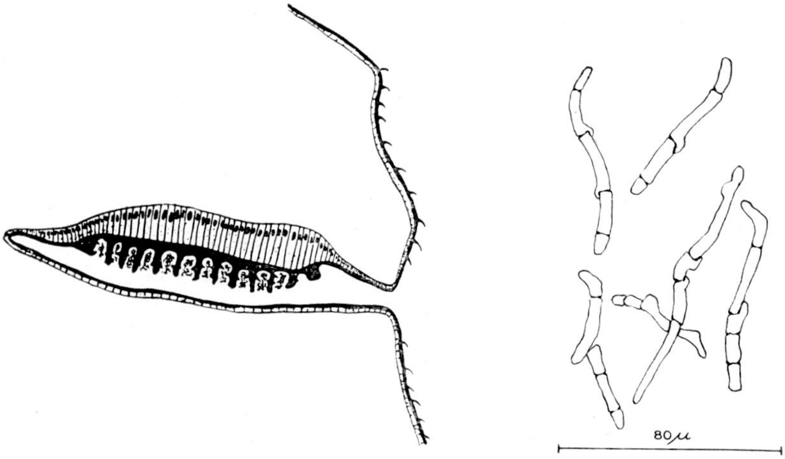

Fig. 190. Left, longitudinal section of hypopleural organ of larva of *Sirex cyaneus* (F.); length of pit, 0·6 mm (after Parkin). Right, oidia of *Stereum sanguinolentum* from *Urocerus gigas* (L.) (after Müller).

the rows of pits are more numerous than in *Sirex cyaneus*. This can serve to distinguish the female larvae of these two species.

On the dorsal surface of the last abdominal segment there is a deep longitudinal groove and two small tubercles, below which there is an upwards projecting terminal spine of characteristic shape which is heavily sclerotized (Fig. 189b). The terminal spine is thought to serve in packing the frass tightly in the gallery and also to serve as a support, as it is often driven into the tunnel side for this purpose.

Length of life cycle

In his studies on *Sirex cyaneus*, Chrystal found that the minimum length of the life cycle was two years with an average of two and a half years and with a maximum under natural forest conditions of three years. Those individuals overwintering in the adult stage in the trees will be in the latter category.

Parasites

One of the remarkable features of the biology of the SIRICIDAE is the phenomenon of parasitism. The two more important species concerned are HYMENOPTERA: *Rhyssa persuasoria* (L.) and *Ibalia leucospoides* (Hochenw.). The former is a large slender insect in the PIMPLINAE group of the ICHNEUMONIDAE, whilst the latter is a cynipid of the subfamily IBALIINAE. Chrystal and Myers have contributed in great measure to our knowledge of the biology of these two remarkable insects.

Rhyssa persuasoria (L.)

The female *Rhyssa persuasoria* thoroughly explores the tree bark with her antennae which are continually palpated and curved downwards and sometimes even backwards, so that their dorsal surfaces are in contact with the bark. When deciding to bore, the insect elevates itself to the fullest extent appearing almost to stand on its head and the ovipositor is held or guided between the hind coxae. The ovipositor sheath is looped at first but is finally held in line with the long axis of the body. When the ovipositor has cut a shaft near to its fullest extent an egg may be deposited —passing down the ovipositor canal which is so fine that a human hair cannot pass through. It is very surprising that a large number of borings, at least 90 per cent, are made that are nowhere in the vicinity of a *Sirex* larva. It is not known what senses are employed by the *Rhyssa* in seeking out the host but it is believed that sight plays an initial part in selecting the tree likely to be infested, but smell must influence the final selection of the boring site. The egg of *Rhyssa* is at least 12 mm in length but a narrow pedicel takes up 9 mm. An account of the passage of the egg down the ovipositor does not appear to exist but it seems virtually certain that the substance of the egg must flow into the pedicel during the operation and then be extruded back again when the egg emerges from the ovipositor. Three eggs are known to have been laid on a single larva. The larva feeds as an ectoparasite on the *Sirex*, whilst lying on top of it. Chrystal and Myers reared a larva which reached a length of 16 mm before spinning a diaphanous silken cocoon. In the pupal stage, the ovipositor is curled forwards over the back and reaches the head. On eclosion the adult bites its way through the remaining wood and bark, making a circular flight hole somewhat smaller than that of *Sirex*. The complete life cycle usually occupies a year. Adults are found in spring and early summer and the feeding stage of the larva is thought to take about five weeks, but thereafter it remains in a resting stage until the following early spring before it pupates. It is thought possible however in some circumstances that a second year could be occupied by the larval resting stage. Only *Sirex* larvae nearly full fed are successfully parasitized: not only are they nearest to the outside of the wood but younger larvae would not contain sufficient nutriment to enable the *Rhyssa* to complete its development, although a male of only 8·4 mm length is recorded. In contrast to the latter a female of 35 mm from head to tip of abdomen (the ovipositor being an additional 45 mm) is before the present author.

Ibalia leucospoides (Hochenw.)

Ibalia are most active immediately after the main activity period of *Sirex*. The female makes a very detailed search of the trunks where *Sirex* has been egg-laying until she finds the actual egg shaft. She uncurls her ovipositor and pushes it down the narrow shaft to the *Sirex* eggs, in each of which an egg is laid. The egg is very similar in shape to that of *Rhyssa* but whereas the latter is laid pedicel first, in *Ibalia* it is laid egg body first. When it first hatches the young larva possesses relatively enormous sickle-shaped jaws but as it develops they become proportionately smaller. The average size when full grown is 12–15 mm. They are well segmented and the head is strongly retractile. During the whole of the larval life it is endoparasitic and although two larvae have been observed in one host larva, never more than one has been known to emerge as adult. The pupation chamber lies parallel to and about 1 cm from the wood surface. The length of the life cycle is thought to be at least two years.

Symbiotic fungi

Another remarkable phenomenon exhibited by the SIRICIDAE is their symbiosis with a species of fungus. In the abdomen of the female adults of most species, except *Xeris spectrum*, there lies a pair of pear-shaped glands near the base of the ovipositor. These glands contain the oidia of a fungus, now known from the work of Cartwright to be the species *Stereum sanguinolentum* (A. & S.) Fr. The oidia consist of small lengths of from one to four cells with clamp connections visible (Fig. 190). This fungus is known commonly as causing a rot of many species of coniferous trees but as Parkin has pointed out, whereas the *Stereum* most often lives in the absence of the siricid, the latter cannot exist without the *Stereum*. When the eggs of *Sirex* are laid the oidia are forced into the ovipositor by muscular action and thus the two symbionts are drawn together. When the oidia reach the wood they develop rapidly into hyphae which precede the boring larva. It appears that the fungal hyphae play an important part in the nutrition of the larva, although it is not clear how this is effected. The method of transference of the fungus from larva to pupa and from pupa to adult has not yet been finally elucidated, but Parkin found peculiar organs which he named the hypopleural organs (Fig. 190) between the first and second abdominal segments of some larvae thought to be female. In these organs there are cuticular pits containing either hyphal filaments or oidia in dense masses of *Stereum sanguinolentum*. Parkin, who found no fungus in the pupae or the newly ecloded female adult, could not see the reason why the larvae should store the fungus when they were virtually surrounded by it in the larval gallery.

Host trees

Attack by SIRICIDAE (Fig. 191) is confined to the PINACEAE. Notes for the recognition of trees likely to be attacked by *Sirex* are given by Chrystal. He says that generally speaking, trees while still green, having less than one-third of their crown remaining, are most likely to be selected by the *Sirex* for oviposition and that *S. cyaneus* cannot be considered a primary enemy of healthy green trees. In larch, the cerambycid *Tetropium gabrieli* often occurs simultaneously with *Sirex cyaneus* and larvae of the latter species have been found in company with those of *Urocerus gigas* in the same tree of silver fir.

Fig. 191. Damage by larvae of *Sirex*.

It now seems fairly certain that, in spite of many statements to the contrary, the SIRICIDAE found in Britain will attack all species of PINACEAE, no wasp having any particular preference. On the other hand *Urocerus* may usually be found in larger trees, this being accounted for by the longer ovipositor being able to penetrate the thicker bark, whilst *Sirex cyaneus* is more likely to attack trees in the pole stage. *Sirex* larvae have been found in the timber of *Cedrus deodara* in the Forest of Dean.

Siricids in Australia

The first record of the occurrence of siricids in Australia is that of Levick, who in 1926 reported *Urocerus gigas* in 'white deal' building timber imported from Danzig. No evidence was given that it had become established. Previous to this in 1909 the Quarantine Act of the Commonwealth of Australia listed all insects belonging to the family SIRICIDAE as prohibited noxious insects, but the 1926 introduction stimulated the authorities to list them in the Vegetation and Diseases Act 1915 which gave power to Quarantine Officers to seize and burn all infested timber.

In 1951 consignments of Central European whitewood heavily infested with *Sirex* arrived both in Sydney and Melbourne. This caused great concern to the authorities and resulted in more stringent regulations. During 1962 a number of records of *Sirex* damage have been made in Victoria.

In Tasmania *Sirex noctilio* has been established for many years first in one locality in the south but now it occurs in about half the island, mainly in plantations of *Pinus radiata*.

Siricids in New Zealand

Somewhere around the turn of the century *Sirex noctilio* became established in New Zealand where it is known as the Steel-blue Horntail Borer. In the year 1900 this insect was reported from North Island in Wairarapa. For many years it was rare, but with widespread planting of exotic conifers such as *Pinus radiata* and their subsequent non-thinning and general low standard of management, *noctilio* became a common insect as reported by Miller and Clark. In New Zealand *noctilio* attacks not only *Pinus radiata* but has been recorded from *P. ponderosa*, *P. laricio*, *P. muricata*, *P. austriaca*, *P. pinaster*, *Larix europea*. *Urocerus gigas* has also reached New Zealand on a few occasions. A notable and historic entomological event was the successful shipment of the parasites *Ibalia leucospoides* and *Rhyssa persuasoria* from Farnham House Laboratory in England to New Zealand, in order to effect some control of *Sirex*.

Notes on the species

Tremex

The genus *Tremex* contains about twenty species of which two are European. *Tremex columba* (L.) has been introduced into Britain from North America. The larva bores into various hardwood trees; *Acer*, *Quercus* and *Ulmus* having been recorded. Adults vary between 15 and 40 mm in length. The female has the thorax mostly grey-brown with the abdomen pitchy-black but with tergite 1 yellow and the following six each with a yellow band. The wings are more or less smoky. The males are mainly pitchy-black in colour.

Sirex

Sirex areolatus (Cresson). This species is occasionally introduced into Britain from the Rocky Mountains and Pacific coast of North America.

Fig. 192. *Sirex noctilio* (F.), female. Length: 29 mm.

Sirex noctilio (Fabricius) (Fig. 192). Although a holarctic species, this insect probably only has the status of an established alien in Britain. In Scotland it is found in Linlithgow.

Sirex cyaneus (Fabricius). This is another holarctic wood wasp but is almost certainly American in origin. It is generally established in southern Britain and is found also as an established insect in a number of localities in Scotland. Host trees recorded in Britain are Scots pine, silver fir, larch and spruce.

Sirex juvencus (L.). Another holarctic insect which exists in two distinct forms or races. The typical form has the base of the antennae red and in Britain is confined to England where it has become established after frequent introductions in imported timber. Benson considered that the form with all black antennae is confined to the northern coniferous forests. Many of the early records of this species in Britain most probably refer to *S. cyaneus*.

Urocerus

Urocerus gigas (Linnaeus) (Fig. 193). Three races occur in Britain: *gigas gigas* (L.) is endemic to Europe but is found from North to South Africa. It is now established in Britain. The subspecies *gigas taiganus*

Benson is considered to be native to the northern coniferous belt in Eurasia and also the Caledonian forest in North Britain. In addition it is introduced into Britain in timber from time to time. The third subspecies, *gigas flavicornis* (F.), is native to the northern coniferous belt in North America, and is introduced in Britain from time to time but is not established.

Fig. 193. *Urocerus gigas* (L.). Centre, normal female, length (including ovipositor) 37 mm; right, diminutive female, length 20 mm; left, diminutive male, length 18 mm.

Host trees recorded in Britain are Scots pine, silver fir, larch and spruce. *Urocerus gigas* is generally distributed throughout Britain; in Scotland it is found everywhere except in the far north and certain islands. There is great variation in the size of the adult. Evans records the largest as 38 mm and the smallest 18 mm, and Chrystal records both males and females 10 mm in length only and discusses the probable reasons—variability in fungal decay and low moisture content. In the British Museum collection is a female of length 51 mm.

Urocerus californicus (Norton). This species has been recorded only once in Britain. It is native to the Pacific coast of North America.

Urocerus albicornis Fabricius. Occasionally introduced into Britain from North America, where it attacks spruce, hemlock, balsam fir and silver fir.

Urocerus fantoma fantoma Fabricius. Previously known as *Urocerus augur augur* Klug, this species is native to Central and South-east Europe.

Urocerus tardigradus Cederhjelm. Native to Central Europe and western Siberia but has been found in Britain on one occasion.

Xeris

The genus *Xeris* contains four species, of which one is holarctic, two are confined to North America and one to the Himalayas. The holarctic species, *X. spectrum* (L.), is occasionally introduced into Britain in timber but has not succeeded in establishing itself here. It varies in length from 15 to 30 mm (excluding ovipositor) and is mostly pitchy-black in colour.

XIPHYDRIIDAE

The two British species of *Xiphydria* may be separated as follows:

1. Abdomen black with a red band in the middle (usually covering tergites 3–5 in ♀ and at least 4 in ♂) and a row of small white lateral spots; legs of ♀ red with the bases of all tibiae and the hind basitarsus white, and apical hind tarsal segments infuscate; legs of ♂ piceous. Antenna with second segment about as long as fourth. Hind basitarsus much longer than the three following tarsal segments together. Abdomen of ♂ with tufts of brown hairs on the sternites, especially on the fifth and sixth. 6–18 mm . *prolongata* (Geoffroy)

– Abdomen with lateral white spots but without any red middle segments. Legs of ♀ entirely red except for the infuscate apical tarsal segments; ♂ legs piceous. Antenna with second segment much shorter than fourth. Hind basitarsus only about as long as the three following tarsal segments together. Abdomen of ♂ without tufts of hair on the sternites. 10–21 mm. White stripes bordering post-ocellar area not reaching to back of head . . . . *camelus* (L.)

Notes on the species

Xiphydria prolongata (Geoffroy).

This is a rare and local insect found in south-eastern England but has been recorded as far north as Nottinghamshire. Although in continental Europe it is recorded in *Populus* and *Ulmus*, in Britain it appears to be confined to *Salix*.

An account of the biology of this insect has been given by Chrystal and Skinner. These authors state that the adults emerge from about the middle of June to the middle of August and the males, emerging first, tap the bark with their antennae in an audible manner and females subsequently emerge from situations which have been 'tapped'.

The authors observed oviposition in the bark of fresh willow logs where the females select the longitudinal grooves where the bark is thinnest. The ovipositor of this species was found to vary from 3·8 to 5·6 mm in length and the boring action was similar to that of *Sirex*, slight vertical movements of the abdomen accompanying the operation until the ovipositor is completely sunk into the bark. A liquid secretion flows down the ovipositor during the process, the function of which is probably lubrication and transport of symbiotic fungal oidia.

The egg of *Xiphydria prolongata* differs considerably from that of the SIRICIDAE. It consists of a globular egg body and a pedicel, the latter being longer than the former. The egg body is extremely elastic and although something like 0·5 mm in width can be compressed to approximately 0·16 mm, which is the internal diameter of the egg channel in the ovipositor. This has the effect of causing considerable elongation of the egg body, a length of 2·5 mm being observed when in this condition.

The ovipositor is made up of three parts, a dorsal fixed portion and two setae and when at rest it is protected by two sheaths. The cutting teeth on the movable setae extend from the tips for about one-third of their length. The female lays from two to nineteen eggs at each oviposition and the clutch of eggs can be detected just under the bark by the discoloration of the surrounding tissue. The authors were uncertain as to the mechanism by which the fungal mycelium or spores are introduced from the female to the larvae but it is certain that a fungus lives in symbiotic relationship with *Xiphydria prolongata*. Cartwright working with this insect considered that the fungus resembled *Daldinia concentrica* (Bolt.) Ces. and de Not. *Xiphydria prolongata* is thought to be a vector of *Bacterium salicis*, the cause of watermark disease in cricket bat willows.

About 16 days after laying, the young larvae hatch and they commence to bore in the cambial layer at first, then in the superficial sapwood and later at an average depth of 3 cm in the sapwood. The tunnels are packed with frass, irregular and winding, but the gallery is always clear for a distance of two to three times the length of the larva.

Xiphydria camelus (L.).(Fig. 182)

Another rare insect found locally in England, Wales and Scotland. The larvae attack *Alnus* and *Betula*. Adults emerge from May to August but in a heated room they emerge from branches in March, the males appearing 10 to 14 days before the females.

Knowledge of the biology of this insect and its parasite complex is due to Skinner and Thompson, and it is probably unique that the information is not available in a conventional scientific paper but as a colour film.

After mating, the female cuts a hole with her ovipositor through the bark of a recently dead branch of alder or birch—a site on the bank of a stream seems to be preferred. When the cambial layer is reached by the ovipositor, a thick, orange-coloured gelatinous fluid is exuded before the clutch of eggs is deposited. The egg has a long stalk, and during the passage of the egg down the ovipositor the egg substance is extruded into the stalk in order to decrease the egg's diameter. On emergence from the tip of the ovipositor the original shape is regained; the larva (Fig. 189c) tunnels in the wood for about 10 months, by which time it has returned to the cambium. It then pupates with the head adjacent to the bark and at right angles to it, and two weeks later the adult emerges by biting its way through the bark making circular flight holes.

Xiphydria is parasitized by four hymenopterous insects. The first, *Aulacus striatus* Jurine (family EVANIIDAE), reaches its host's eggs by passing the ovipositor down the already cut channel. It then places an egg in those of the *camelus* eggs which it can reach with its ovipositor. The *striatus* larva lives in the body of its *camelus* host but does not kill it until shortly before it is due to pupate. The adult *striatus* bores its own way out of the wood.

The second, *Rhyssella curvipes* (Gravenhorst), is an ichneumonid closely related to *Rhyssa persuasoria* (Linnaeus) the parasite of *Sirex* and *Urocerus*. The female *curvipes* is able to detect the position of a nearly fully grown *camelus* larva in the wood and it bores down into it, often going as deep as 20 mm and then lays an egg on it. The larva of *curvipes* is full grown in 16 days but remains in a state of suspended animation for 10 months before pupating. The pupal stage lasts 14 days, the adult then biting its way to the outside.

The third parasite, *Pseudorhyssa alpestris*, using the egg shaft of *curvipes*, lays an egg on the *camelus* larva alongside that of *curvipes*. When it hatches however the larva of *alpestris* is able to kill the *curvipes* larva by means of its large jaws. When it has fed on the *camelus* larva it also remains motionless for 10 months before it completes its life cycle.

The last parasite, *Xiphydriophaga meyerinckii*, is a chalcid. It is only about 3 mm in length and again a very different method of reaching the *camelus* larva is employed. The female enters the wood through an old *camelus* flight hole and she then bores through the packed bore-dust until in some unknown way the presence of a *camelus* larva is sensed. The chalcid then bores through the solid wood until the larva is reached; she paralyses this with her sting, before laying up to thirty eggs on it. The following year the larvae pupate, and on final emergence they reach the outside by the same route as that used by their mother to reach the host.

The Wood-boring Ants

Amongst the ants (FORMICIDAE) is the species *Camponotus herculaneus* L. which damages outdoor woodwork including external building timbers and sometimes living trees. It does not occur in Britain but imported timber is sometimes seen bearing the unmistakable signs of *Camponotus* attack (Fig. 194). It is often of economic importance in North America. Large irregular cavities are gnawed out, the softer spring wood being

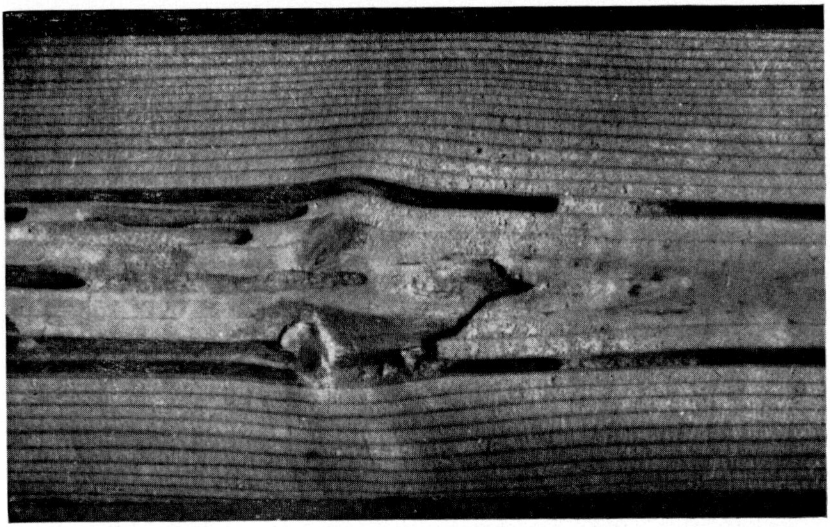

Fig. 194. Softwood damaged by the ant *Camponotus herculaneus* L.

extracted and leaving the stronger summer wood. Bore-dust, faecal pellets and fungal growth are all absent. The wood substance is not consumed but is removed to the outside where piles of fibrous borings are often an indication of the presence of *Camponotus*.

The Wood-boring Bees

The large Carpenter Bees, XYLOCOPIDAE, contain some species in the genus *Xylocopa* which bore into dry seasoned wood in some parts of the world. *X. orpifex* Smith is found throughout the eastern States of North America whilst *X. virginica* Drury is found in the West. The borings produced by these insects are about 13 mm in diameter and from 13 to 45 cm in length. In Britain a number of so-called carpenter bees classified in the APIDAE occur, but none bore into sound wood, although the larval cells are often found in short rows in tunnels excavated in fungally decayed wood. The soft wood merely acts as a receptacle for the cells which in some genera, such as *Megachile*, are made of pieces of leaf cut from rose bushes and other plants. One of the most common species in this country, boring into rotting woodwork (Fig. 195), is *Osmia rufa* but it will place its larval cells in anything convenient—key-holes, loose sand, etc.

The Wood-boring Moths

The order LEPIDOPTERA comprises the well-known butterflies and moths, which are characterized by the possession of four well-developed wings covered with small scales. The latter are often brightly or even

Fig. 195. Window sill partially decayed and containing larval cells of a carpenter bee.

brilliantly coloured, often with distinctive and attractive patterns and markings. Two families in the order contain species with wood-boring larvae— the COSSIDAE and the SESIIDAE (AEGERIIDAE).

COSSIDAE. This family has a world-wide distribution being especially common in Australia, and the adult moths are often of exceptionally large size. *Cossus cossus* (Linnaeus), the Goat Moth, is found throughout the British Isles and is given this name on account of the strong odour of the larva. The latter (Fig. 10) is large and fleshy with a pinkish tinge, the head being almost black with a partially sclerotized area on the pronotum. It has been recorded as boring into a number of different tree species, but is more usually found on elm, poplar, ash and willow. Older, over-mature trees are most often attacked and such trees become well known as 'Goat Moth Trees', the infestation continuing for many years. The indidivual larvae, however, bore for three or four years before vacating the burrow and searching for a site for their cocoon. The abdomen of the pupa is furnished with backwardly directed hooks which force the pupa out of the cocoon on emergence of the adult. The adult moths emerge in June and July and measure 7–10 cm across the outstretched wings. The latter are pale grey, mottled and clouded with brown, with a meshwork of fine, dark-brown, transverse lines and with two thicker ones near the hind margins. The hindwings are brownish and are less distinctly reticulated. The Goat Moth is of no significance in timber indoors.

Zeuzera pyrina (Linnaeus) is known as the Wood Leopard or Leopard Moth on account of its blue-black spotted wings. The long, narrow wings are whitish and only thinly scaled. The thorax is patterned with six blue-black spots and the abdomen is long. Whilst the male measures about 5 cm across the outstretched wings, the female similarly measures from 6–8 cm. The sexes are easily identified by the antennae which are entirely filiform in the female, but in the male they are pectinate basally. The larvae are whitish, sometimes tinged with yellow or pink and spotted with black. Two or three years are spent in the larval stage boring into the branches of trees or shrubs. It has been recorded in elm, ash, chestnut, beech, birch, hazel, lime, apple and pear and in addition the present author has found it in willow.

It is more generally found in southern England especially the London district. It is a swift flier at night. The larvae, and more rarely the moths, are sometimes brought into the laboratory for identification where they have usually been found in logs for the fire, but it is reported as sometimes causing economic losses in orchards.

SESIIDAE. The species of this family are known as the 'Clearwing' moths on account of the scaling of the wings being confined to the outer margins and the cross bars of the forewings. Fifteen species in seven genera are found in Britain, where the larvae bore into the stems, branches and sometimes roots of trees and shrubs. The life cycle is thought to be two years in the majority of cases. Two species are remarkable for the resemblance to hornets. None of the moths are likely to be found in timber indoors. Coloured illustrations of all the species are given by South, where also will be found illustrations of *Cossus cossus* and *Zeuzera pyrina*.

The Wood-boring Mayflies

In the order EPHEMEROPTERA—the mayflies—there is a species, *Povilla adusta* Navas, in East and Central Africa whose aquatic larvae burrow in the stems of reeds and similar plants and also cause considerable damage by tunnelling into the bottoms of boats and wooden harbour works. Such damage is easily identified by a leathery brownish lining to the tunnel which is secreted through the anus as a viscid mass and spread into position by the mouthparts and the forelegs. The life cycle is of four months' duration and the species is remarkable for displaying a high degree of lunar periodicity.

One of the most specialized of insect orders, the DIPTERA, possesses no species with wood-boring habits, although flies in the families TIPULIDAE and XYLOPHAGIDAE are found in fungally decayed wood.

REFERENCES

THE WOOD WASPS

BENSON, R. B. 1950. An Introduction to the Natural History of British Sawflies. *Trans. Soc. Brit. Ent.* **10**: 45–142.

BENSON, R. B. 1951. Handbooks for the Identification of British Insects. HYMENOPTERA—SYMPHYTA **6**, pt. 2 (a): 1–49.

CARTWRIGHT, K. ST. G. 1929. Notes on a Fungus associated with *Sirex cyaneus*. *Ann. appl. Biol.* **16**: 184–7.

CHRYSTAL, R. N. 1928. The Sirex Woodwasps and their Importance in Forestry. *Bull. ent. Res.* **19**: 219–47.

CHRYSTAL, R. N. 1930. Studies of the *Sirex* parasites. The Biology and Postembryonic Development of *Ibalia leucospoides* Hochenw. (HYMENOPTERA-CYNIPOIDEA).

CHRYSTAL, R. N. and MYERS, J. G. 1928. Natural Enemies of *Sirex cyaneus* Fabr. in England and their Life History. *Bull. ent. Res.* **19**: 67–77.

CHRYSTAL, R. N. and MYERS, J. G. 1928. The *Sirex* Woodwasps and their Parasites. *Empire For. Journ.* **7**: 145–54.

CHRYSTAL, R. N. and SKINNER, E. 1932. Studies in the Biology of the Woodwasp *Xiphydria prolongata* Geoffr. (*dromedarius* F.) and its Parasite *Thalessa curvipes* Grav. *Scottish Forestry Journal* **46**: 36–51.

FISHER, R. C. 1955. Australian Quarantine and Woodwasp problem. *Timber Technology* **63**: 480–3.

MÜLLER, W. 1934. Untersuchungen über die symbiose von Tieren mit Pilzen und Bakterien. III. Über die Pilzsymbiose holzfressender Insektenlarven. *Arch. Mikrobiol.* **5**: 84–147.

PARKIN, E. A. 1942. Symbiosis and Siricid Woodwasps. *Ann. appl. Biol.* **29**: 268–74.

RAWLINGS, G. B. 1952. Progress in Establishment of *Ibalia leucospoides*. *New Zealand Science Review* **10**: 107–8.

RICHARDS, O. W. 1956. Handbooks for the Identification of British Insects. HYMENOPTERA. *Introduction and Keys to Families*. **6**. **Pt. 1**: 1–94.

SKINNER, E. R. and THOMPSON, G. H. 1960. The Alder Woodwasp and its Insect Enemies. 16 mm film in colour and sound. Oxford Scientific Films Ltd., Long Hanborough, Oxford.

THE WOOD-BORING MOTHS

KIRBY, W. E. N.D. *Butterflies and Moths of the United Kingdom*. Routledge. London.

SOUTH, R. 1948. *Moths of the British Isles*. Warne. London.

CHAPTER TEN

Termites in Britain

A brief account of termites—Notes on the European species—Termites in Britain: accidental importation: rearing colonies under laboratory conditions

---

A brief account of termites

Termites are primitive insects allied to the cockroaches and are classified as a distinct order of insects—the ISOPTERA. They are of profound importance as destroyers of wood. In many parts of the world unless termite-resistant timber species are utilized or unless the timber is protected chemically or mechanically it is only a question of time—sometimes only a short time—before the wood is consumed and destroyed.

Termites are pests not only of wood in buildings but many species are important forestry pests attacking living trees usually when young when the roots are often destroyed. Mature forest or ornamental trees are often attacked, although in the latter cases, the termites gain an entrance from some primary wound such as a fire scar. Termites are important pests of crops in many tropical countries.

Termites are remarkable for the complexity of their social organization. The form of the individual is modified according to its function. The female or queen of which there is usually only one or a very few, is an extreme of modification, the swelling of the abdomen to contain the highly reproductive ovaries virtually imprisons the individual in a cell in the ground. There she lays a very large number of eggs—in one species one every two seconds has been recorded. Some individuals are modified as nurses tending the eggs and young individuals, others carry out nest construction, foraging and nest maintenance whilst highly modified forms serve as guards or soldiers. These are furnished with large, sharp mandibles and are pugnacious, concentrating where the nest becomes broken. An extreme form—the nasute—secretes a strong-smelling sticky chemical, although to humans its aromatic smell is not unpleasant.

In the soldier form the shape of the head and size and complexity of mandibles is often characteristic of species so that in identifying termites the soldier form must be collected as it will often be that the species can most easily be identified from this caste or form.

Geographical distribution

Harris (1971) states that the number of different kinds of termites present in any locality is greatest in tropical rain forests but the general level of activity appears to be the highest in deciduous woodlands and where man has farmed. It is generally thought that termites are confined to the tropical and subtropical regions of the world, but that is not the case. Their distribution extends far into the temperate zone both in the northern and the southern hemispheres, several species being found in New Zealand and Tasmania in the south and as far north as Vancouver, La Rochelle in France, and Hamburg (an introduced species) in Germany.

Classification

About 1,800 species of termite have been described and they are divided into six families. They are characterized by possessing typical mandibulate mouthparts and two pairs of somewhat similar membraneous wings which are held flat over the abdomen when at rest. The wings are longer than the abdomen. The anterior veins are strongly sclerotized and there are no true cross veins. There is a suture near the base facilitating shedding. A complete metamorphosis is absent but polymorphism in both sexes is pronounced.

Family 1. MASTOTERMITIDAE

Although widely known in fossils, today it exists only as the single species *Mastototermes darwiniensis* Froggatt, found in northern Australia. In this species, anal lobes are present on the hindwings showing a relationship with the cockroaches and, moreover, eggs are laid in pods of about twenty. Over a million individuals are reported as being present in fully developed communities. The nests are below ground.

Family 2. KALOTERMITIDAE

These are the well-known drywood termites. The worker caste is absent, all the work being carried out by the nymphs. The community does not usually reach large proportions and is always housed inside the infested wood. Flagellate protozoa are present in the intestine. The more important genera are *Kalotermes*, distributed throughout the tropics and subtropics including *K. flavicollis* (F.) in southern Europe, and *Cryptotermes* which includes species spread across the world through commerce. *C. brevis*, the West Indian drywood termite, and *C. domesticus* attack wood in buildings in the Pacific and Oriental regions and *C. dudleyi* in Central America, South-east Africa and East Africa.

Family 3. TERMOPSIDAE

These are known as dampwood termites. Like the KALOTERMITIDAE they are found only in small communities and they possess intestinal flagellate protoza. There are three subfamilies. TERMOPSINAE includes the species *Zootermopsis angusticollis* which occurs along almost the whole Pacific coast of North America living in decaying logs and stumps of Douglas fir and Californian redwood. The other subfamilies are the STOLOTERMITINAE and the POROTERMITINAE.

Family 4. HODOTERMITIDAE

These are the Harvester termites; there are only three genera and they tend to colonize arid regions, found mostly in North Africa, Arabia, Iraq, Persia and North India.

Family 5. RHINOTERMITIDAE

This family contains six subfamilies, all species being rather small and subterranean in habit. A character of the family is the presence of the 'fontanelle' or opening of the frontal gland which in the soldiers is used as a defensive organ as, when alarmed, they secrete a sticky fluid from the gland.

Subfamily PSAMMOTERMITINAE

This subfamily which contains a few African and Madagascan species is remarkable in that there are two sorts of soldiers.

Subfamily HETEROTERMITINAE

This subfamily contains two genera, *Reticulitermes*, several species of which are found in the northern temperate zone, and *Heterotermes* which are found in tropical areas of America and Asia but only one from Africa. Certain species are pests of sugar cane in the West Indies where they appear to gain entrance to the cane through the exit holes of boring beetles.

Subfamily STYLOTERMITINAE

One species only is classified in this subfamily, found in southern India.

Subfamily COPTOTERMITINAE

The genus *Coptotermes* with many species is distributed throughout the tropics. The soldiers have pear-shaped heads with a large fontanelle and sharp, slender, toothless mandibles. Many are pests of considerable importance attacking timber used in constructional work.

Subfamily RHINOTERMITINAE

A large number of species in the genus *Schedorhinotermes* damage structural timber in Australia, Malaya, the Solomons and tropical Africa.

Family 6. TERMITIDAE

This is by far the largest family, the great majority of termite species being included in it. In general they are subterranean or are mound builders, but some, for example *Nasutitermes*, construct nests in trees or on posts.

Subfamily AMITERMITINAE

This subfamily contains twenty genera, the most important of which are *Microcerotermes* and *Amitermes*. In the former genus some species build their nests in trees. There are about sixty species in the latter genus found throughout tropical and subtropical regions. Several are found in deserts and cause damage to railway sleepers and installations in such areas.

Subfamily TERMITINAE

Thirty-two genera are classified in this subfamily, they are all humus feeders—consuming soil and extracting humus.

Subfamily MACROTERMITINAE

All species in this subfamily construct fungus gardens and are either subterranean or are mound builders. *Acanthotermes* and *Pseudacanthotermes* are African genera, whilst *Macrotermes*, including several large species, is African and Asiatic. They build large mounds and are important destroyers of wood where they occur. *Odontotermes* is also of pronounced importance in Africa and Asia. *Microtermes* are soil dwelling and often of economic importance on account of their habit of damaging crops. As their name implies they are small.

Subfamily NASUTITERMITINAE

The fontanelle, in the soldier, opens at the end of the pear-shaped head. In almost all species the mandibles of this caste are much reduced. *Nasutitermes* is found throughout the world in the tropical regions. In *Trinervitermes* there are two soldier sizes.

Notes on the European species

Reticulitermes lucifugus (Rossi) (Figs. 196 to 198)

There has been much confusion and misunderstanding concerning the specific determination of the termite found in la Rochelle and the surrounding area (Charente Maritime). As this area roughly corresponded with the old kingdom of Saintogne, the termite is often referred to as *le termite de Saintogne*. Feytaud in 1923 confused the latter with *R. flavipes* Kollar which, although it had been named from material found in the conservatories in the gardens of the Palace of Schönbrunn in Vienna, was a widespread North American species. It is now considered that *le termite de Saintogne* is a subspecies of *R. lucifugus*, and is referred to as var. *santonensis* Feytaud. The people of the Charente, however, firmly believe that *their* termite—*le termite de Saintogne*—came to them from America. Lash has confirmed that it is not referable to any North American species.

Although *lucifugus* in its typical form and its subspecies *santonensis* differ little from each other in morphological characters, they differ widely in behaviour and in other respects. For example *santonensis* is considerably more active and its colonies are larger than *lucifugus*, and

Fig. 196. Workers of *Reticulitermes lucifugus* (Rossi). Wood Preservation Ltd.

whereas the latter is found most commonly in dead stumps of maritime pine (*Pinus pinaster*), *santonensis* attacks many tree species dead or alive. Buchli states that *santonensis* is a scourge in orchards, attacking many species of fruit trees. Walnut trees are often attacked to complete destruction, and the roots of many garden plants are destroyed, especially vulnerable being dahlias, tulips, cyclamens and geraniums. With regard to destructive powers, however, *santonensis* is by far the greater danger to woodwork in buildings. Attack by *lucifugus* indoors hardly ever reaches the same proportions.

Another interesting point of difference between the two forms, pointed out by Buchli, is that the form *santonesis* is very much more robust—withstanding much more physical damage to its body than the more delicate typical form.

Damage by R. lucifugus *var.* santonensis *Feytaud*

It is a comparatively easy task to find damaged woodwork without entering buildings. In the dock area of La Rochelle are a number of fishermen's huts and termite attack can be seen where rainwater downpipes are obviously faulty, and excess water, leaking into the woodwork, gives the conditions which are necessary for attack by this species. In the small port of Rochefort a number of properties damaged by *santonensis* can be seen. The Ministere de la Construction maintains a small department at La Rochelle in order to deal with this problem.

During a termite-collecting trip, the author was taken to the vicinity of Rochefort where there are some small factories, and at the back of one of these was a large dump of sawdust covering about 12 square metres. The sawdust was old and damp, and was warm to the touch, and constituted a most remarkable entomological situation. One had only to remove a handful of the sawdust to find thousands of *santonensis*. Several million termites must have been present, and although it was very late in the season, a few alates were also found. Returning to La Rochelle the author was able to collect a good number of *santonensis* in a cut trunk of maritime pine lying in a builder's yard. As saw and bench were at hand it was possible to cut the trunk up into pieces suitable to fit into two open-top four-gallon metal drums.

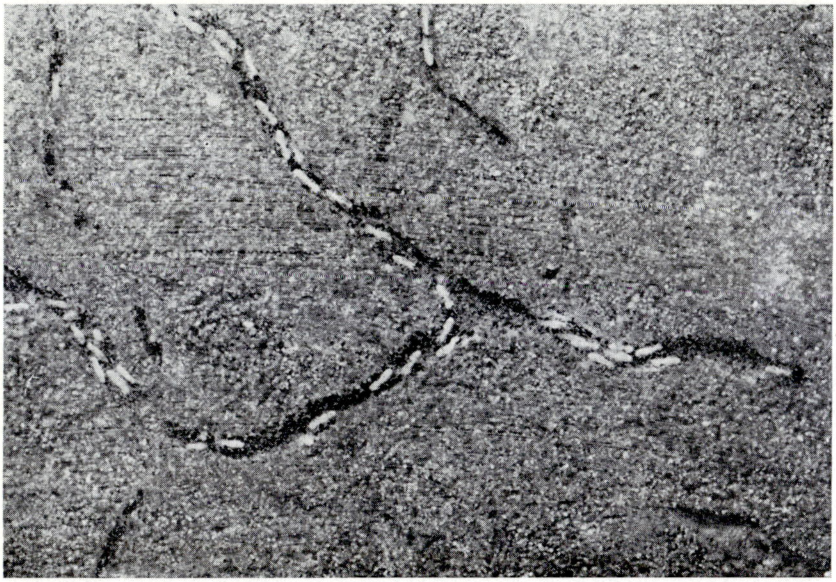

Fig. 197. Workers of *R. lucifugus* (Rossi) moving in their earth tunnels. Wood Preservation Ltd.

One factor was common to all *santonensis* colonies which were found, and that was the dampness of the wood infested. In this region the water table is extremely high, being only about a metre from the surface. This makes the conditions very favourable for *santonensis*. From such situations as wood lying on the surface of the ground, they are able to invade damp wood in buildings, especially by their ability to produce small auxiliary colonies in all directions. This is, of course, quite apart from their ability to produce flying nuptial swarms of winged adults.

Fig. 198. A swarm of winged adults of *Reticulitermes lucifugus* (Rossi). For a few minutes only in the whole life cycle, termites actively seek the light—this is when the winged sexuals are freshly emerged. Wood Preservation Ltd.

The typical form of *Reticulitermes lucifugus* occurs in certain parts o the Dordogne Valley, more particularly in the edges of woodlands containing maritime pine, and there is a regular collecting station for this termite two or three miles from Les Ezies. In this village there is a field station of the University of Paris, and the author is indebted to Professor Grassé who made available the excellent facilities of this laboratory. Here colonies of *lucifugus* were certainly much rarer than the author subsequently found in the south of France, a colony being found in about one in ten of the pine stumps. The author did, however, find a maritime pine which had been broken about 1½ m from the ground, in which was a very large colony of *lucifugus*. It was probable that the tree had been attacked by some other pest before *lucifugus* invaded.

Kalotermes flavicollis (Fabricius) (Figs. 199–201)

Banyuls-sur-mer on the Mediterranean coast of France is within a few kilometres of the Spanish border. There is situated the Laboratoire Arago at the water's edge, at the very foot of the Pyrenees, and although marine biology is its main preoccupation, the study of termites is undertaken by Dr. Harro Buchli. The Drywood termite, *Kalotermes flavicollis*, is only found in this region of France and Dr. Buchli took the author on two collecting trips. The nearby vineyards on the steep, stony slopes of the foothills were searched and termite colonies found in the stocks of the older vines, which were gnarled and twisted with longitudinal slits. On a hillside, the property of the laboratory, some of the older stumps were hacked up and opened with an axe—they are very tough—and in some of the stocks were small galleries containing *flavicollis* colonies. In addition,

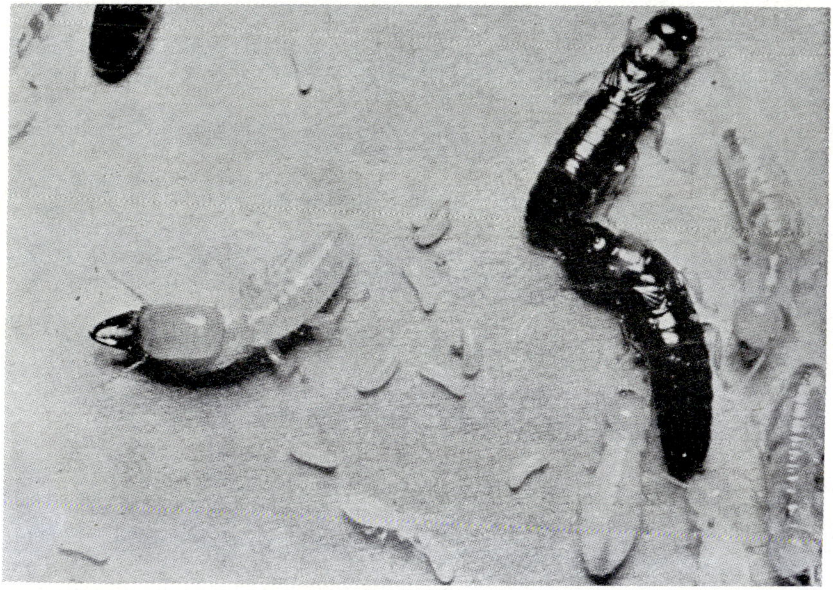

Fig. 199. The queen, king, soldier, workers and eggs of *Kalotermes flavicollis* (F.). Wood Preservation Ltd.

in some of the stocks were found *lucifugus* colonies. It was quite easy to differentiate between the two species in the field, as *flavicollis* is larger and not nearly so agile as *lucifugus*. Even when a probable gallery of *flavicollis* had been exposed it was not always easy to locate the termites. Half a dozen jarring raps on the stock with the back of the axe were often required to shake the first termite out. Altogether ten colonies of *flavicollis* were found. It is thought that an infestation by *flavicollis* cuts the length of life of a vine stock from 80 to 40 years. The vine stocks containing the colonies were trimmed and placed in polythene bags for the journey back to England. When a vine stump had been in the constant temperature room in the laboratory at Felcourt only two weeks, a specimen of the Cerambycid beetle *Chlorophorus varius* Müller emerged. This is a well-known pest of grape-vines and it would certainly appear that vine stocks infested by this beetle would give very much easier access to the inside of the stock to pairs of *flavicollis* searching for a brood site.

Fig. 200. Grape vine stock cut away to show rough bark—probably loosened and split by the cerambycid beetle *Chlorophorus varius* Müller and showing the galleries occupied by *Kalotermes flavicollis* (F.).

In the collecting trips around Banyuls every opportunity was taken of examining the cut stumps of maritime pine and in about a third of them *lucifugus* colonies were found in similar positions to those found in the Dordogne area—about 10 cm above the soil surface, in the partially rotted sapwood under the bark.

Termites in Britain

Accidental importation

There appears to be only five instances of accidental importation of living termites being observed, identified and recorded up to 1960. The identifications and summary of all the circumstances of all these occasions are given in table form below. These are only the *recorded* cases. It is highly probable that with the scale and variety of timber importation taking place today that many termites are accidentally brought into this country alive every year and even more arrive dead. W. V. Harris (1954) has discussed the chances of accidental importation of termites becoming of economic significance. He thought that the higher termites had too complex a social organization to allow them to become established in England—in addition, they are unable to withstand low temperatures. On the other hand, the primitive termites (KALOTERMITIDAE and RHINOTERMITIDAE) without a complex social organization are more likely to become established if introduced.

Fig. 201. Queens, kings and eggs of *Kalotermes flavicollis* (F.). Wood Preservation Ltd.

Fig. 202. Workers of *Zootermopsis angusticollis* Hagen.

SUMMARY OF RECORDS OF ACCIDENTAL IMPORTATION OF ISOPTERA INTO ENGLAND
(*See list of references for publication details*)

Date	Species	Family	Author	Place	Circumstances
1874	*Kalotermes* sp.	KALOTERMITIDAE	R. McLachlan	Kew, London	In a Copalwood trunk (*Trachylobium hornimanniana*) from East Africa.
1943	*Nasutitermes costalis* (Holmgren)	NASUTITERMITINAE	N. D. Riley	Spalding, Lincs.	In nest in packing material from Martinique.
1948	*Trinervitermes* sp.	NASUTITERMITINAE	R. C. H. Sweeney	Croydon, Surrey	Soldiers, from a hedge-bottom.
1955	*Zootermopsis angusticollis* (Hagen) (Fig. 202)	TERMOPSIDAE	W. V. Harris	London Docks or Bristol	In a parcel of softwood believed to be Douglas fir (*Pseudotsuga taxifolia*) from W. Canada.
1960	*Cryptotermes brevis* (Walker) (Figs. 203–205)	KALOTERMITIDAE	N. E. Hickin	Watford, Herts.	In small pieces of softwood and plywood in portable gramophone case from Trinidad.

Cryptotermes brevis (Walker), for example, has a wide distribution (in the United States it is known as the Rough-headed Powder-post termite). It occurs throughout the West Indies and Cuba; on the mainland it is found in Central and South America, but in North America appears to be restricted to Florida and Louisiana. It does a great deal of damage in Bermuda where a natural reservoir for widespread infestation exists in the extensive dead Bermuda cedars (*Juniperus bermudiana*). This species is found also in some of the very isolated Atlantic islands.

The KALOTERMITIDAE, including *Cryptotermes*, are known as Drywood termites. This is because the colonies are found entirely within the wood they are infesting and no communication with the ground or damp wood is necessary. Soldiers (Fig. 204) are relatively uncommon. There are no 'workers', all the work of the colony is carried on by the nymphs who will all attain sexual maturity unless death intervenes. Digestion of the wood particles cut off by the mandibles is carried out by microscopic protozoa. These can be easily seen under a microscope when the gut is dissected. This is a classical example of symbiosis and is one reason for the maintenance of living Drywood termite colonies for teaching purposes in universities.

Shelter tubes are rarely produced by Drywood termites, for only

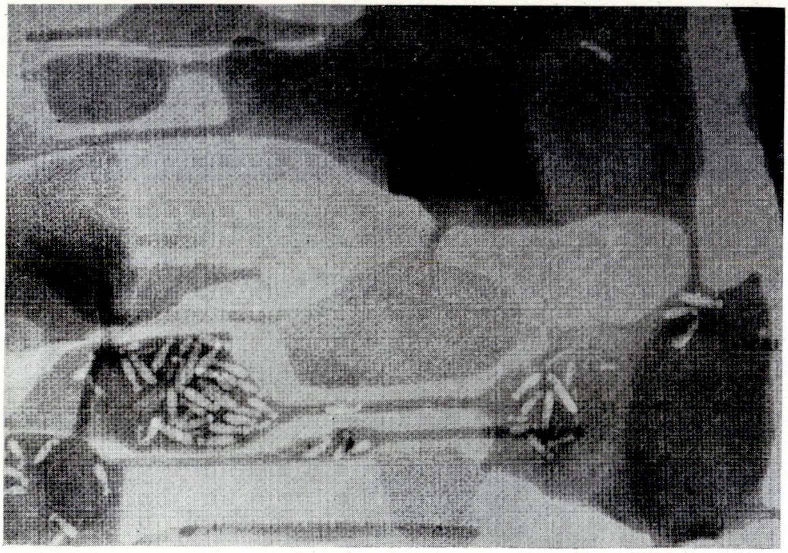

Fig. 203. X-radiograph of plywood containing nymphs of *Cryptotermes brevis* (Walker). Note the large chambers inter-connected by narrow tunnels. Some chambers (lighter grey) are in a different layer of the plywood.

occasionally do they find it necessary to bridge over from one piece of wood to another, when a cement made of saliva and faecal pellets is employed. Identification of Drywood termite attack is relatively easy because the excrement of the colony is ejected from the tunnels within the

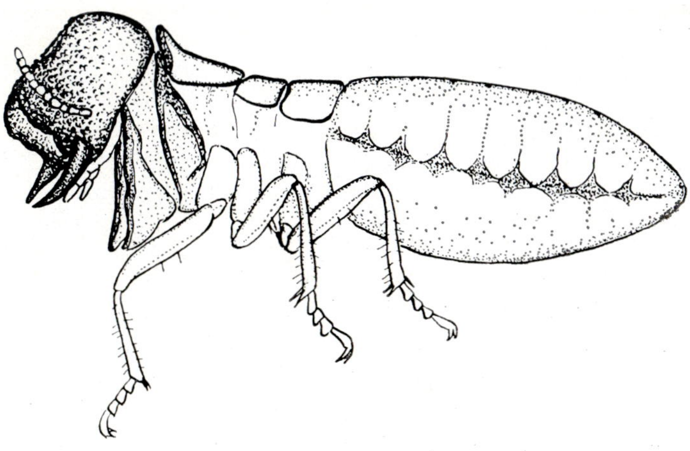

Fig. 204. Side view of a soldier of the Drywood Termite *Cryptotermes brevis* (Walker). Note the hard disc-like head used for plugging holes in the outer skin of wood in an emergency. Length: 4·5 mm.

wood through small holes made for the purpose. The excrement (the faecal pellets) is seed-like, always of constant form, hard and there are concave faces (Fig. 205). A small pile of 'poppy-seed' as it is often called, denotes Drywood termite attack. If the faecal pellets would be hard to observe on a rough bark background—say, in a roof space—then a sheet of newspaper could be put down for a few days and then taken up and examined.

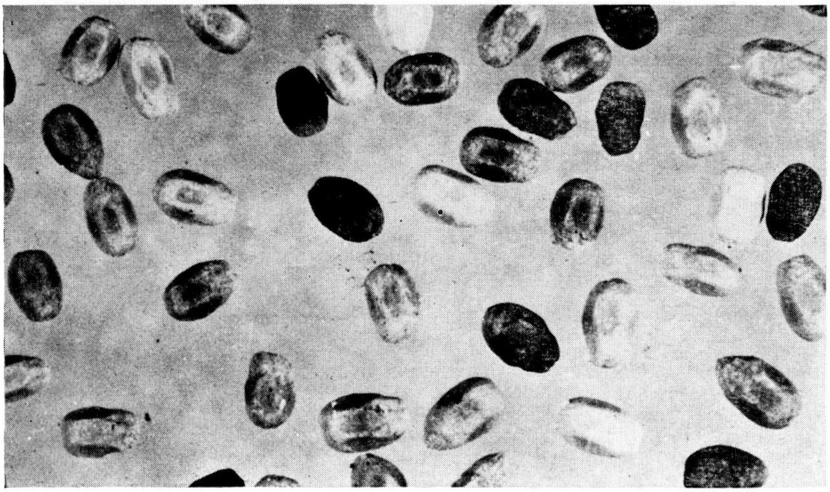

Fig. 205. The first sign of the activities of Drywood Termites is usually the pile of seed-like faecal pellets. The sides are concave and the general appearance is that of poppy seeds, by which name they are often called.

Legislation

Obviously there is little that can be done to prevent the accidental introduction of termites, say in imported timber, but legislation exists, the object of which is to prevent the importation of termites by unauthorized collectors. These are the various Destructive Insects and Pests Acts from 1877 to 1927, and more particularly the Destructive Insects and Pests Order of 1933. A collector, however, may be authorized to import living termites by a licence issued by the Ministry of Agriculture, Fisheries and Food. The licence is normally subject to the following conditions, although there have been some relaxations in recent years:

I. All experimental work with the termites shall be confined to the laboratories mentioned in the application, and all necessary precautions shall be taken to prevent the escape of the termites out of doors.

II. No living species of the named termites shall be removed from the said laboratories unless such removal has been authorized in writing by the Ministry or by an authorized officer of the Ministry.

III. An authorized officer of the Ministry shall be allowed to inspect the conditions under which the termites are retained.

IV. All specimens of the termites shall be destroyed as soon as the experimental work for which they are required is completed.

V. This licence, unless previously revoked, shall remain in force for one year only.

Rearing colonies under laboratory conditions in England

Recorded instances of termite colonies being maintained in England for purposes of demonstration and research are very few. Haviland and Sharp (1896) appear to record the first instance of specially importing and maintaining colonies of termites. A colony of *Cryptotermes domesticus* (Haviland) imported from Singapore was kept in a test-tube 15 cm by 15 mm wide. It was plugged with occasionally moistened cotton wool, and the termites were fed on a splinter of wood. The colony lived for nearly two years. It was exposed to a wide variation in temperature although it was kept near some hot-water pipes. Alates were frequently produced but were never removed from the tube.

Sweeney (1948) kept, for only seven weeks, a colony of fifty-two nymphs of *Cryptotermes brevis* (Walk.) which he had obtained from Chile. He described an artificial termitarium consisting essentially of a large glass jar in which was fitted an electric light bulb within a metal cylinder in order to obtain the necessary warmth. An apparent serious disadvantage possessed by this apparatus was that the termites were exposed to light. Sweeney states that 'the termites have little opportunity for concealment' as though this was an advantage. Termites in nature do not expose themselves to light except for a brief period when they possess wings, and it would seem prudent to simulate this condition as closely as possible when keeping termite colonies in captivity.

Colonies of six species were reared successfully for several years in the laboratories of Rentokil Limited, at East Grinstead, Sussex. The termites were kept in a constant temperature room at 25° C and 75 per cent relative humidity. Three types of containers were employed:

1. *Large general receptacles.* These consisted of conventional type aquaria of metal angle strip and glass, and all glass battery jars.

A most successful colony of *Reticulitermes lucifugus* var. *santonensis* was contained in a battery jar, 30 by 30 by 45 cm in height. Glass plates were used to cover the tops and a layer of pure washed silver sand covered the bottom. The cultures of subterranean species were kept moist by damping the sand and spraying the wood as necessary. The tropical drywood termites such as *Cryptotermes* were damped only very occasionally and the blocks of wood containing the colonies were not resting on sand. The temperate drywood termites (*Kalotermes flavicollis*) were damped from time to time, but were not given anything like the amount of water given to the subterranean species.

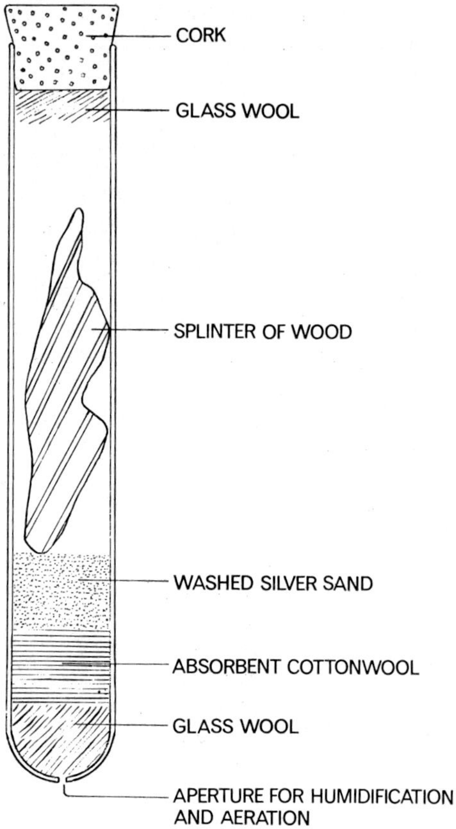

Fig. 206. Jucci-Grassé tube for experimental culture of termite colonies.

2. *Jucci-Grassé tubes.* These were of simple test-tube or boiling-tube pattern with a hole at the bottom (Fig. 206).

Buchli has given a table showing the dimensions of three tubes which he used and the maximum number of individuals in the colonies of *Reticulitermes lucifugus* var. *santonensis* which the tubes would contain, and which would continue to thrive but not increase in numbers.

The tubes were damped by placing the bottom in a beaker of water for a

Dimensions of Jucci-grassé Tubes

	Small	Medium	Large
Length	18 cm	25 cm	28 cm
Diameter	17 mm	25 mm	40 mm
Volume	35 cm	100 cm	300 cm
Maximum population	80 individuals	250 individuals	1000 individuals

few seconds. Water enters the aperture at the base of the tube and is absorbed by the cotton wool, but care must be taken to see that excess water is not allowed to enter. The object must be to make the cotton wool just damp.

3. *Lüscher plate-type termitarium.* This design of termitarium is better adapted to observational and display purposes than perhaps for experimental purposes, at least where wood preservation work is in progress. One small colony of *Kalotermes jouteli* was maintained in a termitarium of this type, but did not increase in size over a period of seventeen months. The termitarium is 15 by 8 by 1 cm in dimensions (Fig. 207).

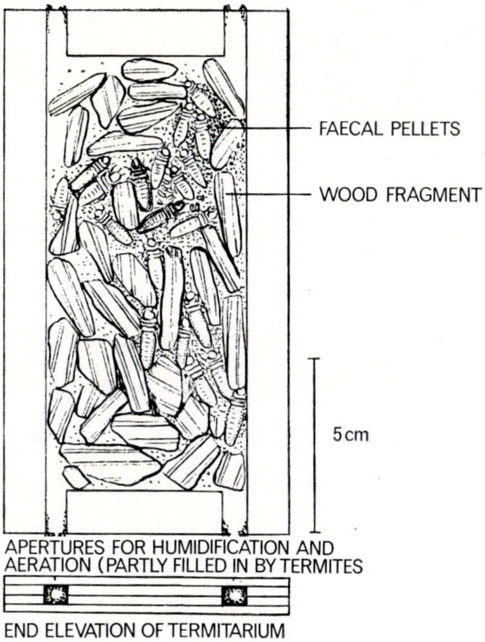

Fig. 207. Lüscher plate-type termitarium for experimental culture of colonies for observational purposes.

REFERENCES

Buchli, H. H. R. 1958. L'origine des castes et las potentialités ontogéeniques des Termites Européens du genre *Reticulitermes* Holmgren. Thèses présentees à la Faculté des Sciences de L'université de Paris.

Chaboussou, F. 1954. Les Termites de France. *Nature, Paris.* **3233**: 347–52.

Harris, W. V. 1954. Termites in Europe. *Ent. mon. Mag.* **90**: 194–7.

Harris, W. V. 1955. An American Termite in Imported Timber. *Wood* **20**: 366–7.

Harris, W. V. 1956. Termites Destructive to Timber. *B.W.P.A. Convention Record*, pp. 145–77.

HARRIS, W. V. 1958. More about Dry-wood Termites. *East African Agricultura Journal* **23**: 161–6.

HARRIS, W. V. 1971. *Termites, their recognition and control.* 2nd Edn. Longmans. London.

HICKIN, N. E. 1960. Termites in France. Collecting Species found in Europe. *Timber Technology* **68**: 459–61.

HICKIN, N. E. 1961. Isoptera in England. A Record of Accidental Importation. *Timber Technology* **69**: 26–7.

HICKIN, N. E. 1961. Maintenance of Termite Colonies in England for Research Purposes. *Pest Technology* **3**: 84–6.

HICKIN, N. E. 1971. *Termites: a World Problem.* Hutchinson. London.

KRISHNA, K. and WEESNER, F. M. 1969. *Biology of termites, Vol. I.* Academic Press, N.Y.

MCLACHLAN, R. 1874. A brood of White-ants (Termites) at Kew. *Ent. mon. Mag.* **11**: 15–16.

RILEY, N. D. 1943. An Occurrence of *Nasutitermes costalis* Holmgren in England (ISOPTERA). *Proc. R. ent. Soc. Lond.* (*A*) **18**: 95.

SNYDER, T. E. 1948. *Our Enemy the Termite.* New York.

SWEENEY, R. C. H. 1948. Soldiers of *Nasutitermes* (*Trinervitermes*) (ISOPTERA) in England with a Note on Artificial Termitaria. *Ent. mon. Mag.* **84**: 164–6.

CHAPTER ELEVEN

Incidence and Control

The relative incidence of the various wood-boring insects found in buildings in the United Kingdom—Methods for woodworm control: use of pre-treated timber: use of smoke generators and dichlorvos strips: fumigation of houses

---

Wood-boring insect damage is the major factor in the deterioration of wood in buildings. The need to assess the importance of this biological hazard to unprotected building timber has been appreciated for some time. At first, only a rough idea of the comparative importance of the various wood-boring insects was known. However, a statistical examination has been made of all surveys carried out by Rentokil Timber Infestation Surveyors in the United Kingdom, during the period 1 July 1960 to 31 December 1965. In this 5½ year period, 142,270 buildings were surveyed; constructional and joinery timbers being closely examined and reported upon.

Analysis of the data gives a measure of the actual number of infestations, as well as the relative incidence of the major wood-boring insects. The results obtained during the 5½ year period are reported in full elsewhere (Hickin, 1967); a brief summary is given below.

Types of building surveyed

Each survey report gave details of the type of building. The majority of these, as can be seen from Table I, were private dwellings. The other buildings surveyed gave a representative cross-section of all types of buildings in Britain.

TABLE I

Summary of composition of total survey by type of building in the United Kingdom over 5½ years.

Category	Number	Percentage of total number
Private dwelling-house	120,627	84·8
Commercial, factory, shop	9,236	6·5
Church, historic building	6,340	4·5
Licensed premises	1,452	1·0
Farm, country estate	4,615	3·2
Total number of all surveys:	142,270	

Age of buildings

When surveyed, a note was made of the approximate age of the premises. They were grouped into three categories, as shown in Table II.

TABLE II

Summary of age of buildings surveyed in the United Kingdom over 5½ years

Category	Number	Percentage of total number
Pre-1919	76,013	53·4
1920–1939	57,643	40·5
1940–present	7,464	5·2

Relative incidence of the various wood-boring insects

Insects were found in most of the buildings surveyed. Their incidence, relative to one another, is shown in Table III.

TABLE III

Summary of comparative incidence of various wood-boring insects as a percentage of total surveys, and actual numbers of infestations involved in the United Kingdom over 5½ years.

Insect	Actual number of infestations	Percentage of total number of surveys
Anobium punctatum	106,620	74·9
Xestobium rufovillosum	6,626	4·7
Weevils	7,433	5·2
Lyctus	1,295	0·9
Hylotrupes bajulus	533	0·4
Other insects	3,382	2·4
Nil infestation	5,276	3·7

Geographical distribution of the insects

Although Table III indicates that about three-quarters of the buildings surveyed contained an infestation of *Anobium*, there is considerable variation in this figure, depending on the geographical area. The lowest incidence was in North Wales and Lancashire, where less than 60 per cent were infested. At the other extreme, south coast counties from Dorset to Kent showed a percentage infestation of over 85.

Wood-boring weevils, at the time of the survey, were particularly common in London and East Anglia, with infestations reported in 13 to over 15 per cent of the properties examined. It is thought that weevils may now (1974) be more widespread, at least in the southern half of England.

Limitations of the $5\frac{1}{2}$ years survey

When considering the results given above, one must take into account the fact that the buildings examined were not a 'random sample'. In most cases, surveys were made only where the Surveyor was called upon to do so, usually because the owner believed wood-boring insects were present. Thus, the figure of 75 per cent for buildings infested with *Anobium* must be higher than would be expected for *all* buildings.

A second statistical analysis was made in 1961, when surveys were made of 694 properties whose owners were applying for woodworm insurance. It could be argued that this sample, to a great degree, consisted of house-owners who were under the impression that their properties were *not* infested by wood-boring insects.

In this survey, it was found that 60 per cent of buildings contained an active infestation of *Anobium punctatum*—perhaps a more realistic figure than that given by the $5\frac{1}{2}$ year survey. The number of nil infestations, however, was nearly 39 per cent, compared with only 3·7 per cent in the main survey.

Methods of Woodworm Control

The situation which has just been described is obviously a grave one which leads, in the individual householder, and, indeed, in some local authorities, to a feeling of disquiet and anxiety which has to be allayed.

It has been remarked on occasion that woodworm attack in buildings occurs at such a slow rate that no material damage is effected, but such is very far from the truth. It is true that an attack proceeds at a very slow rate at the beginning, and several reasons have been given to account for this. The average number of eggs laid is low, about twenty-eight; parasitization by the chalcid *Theocolax formiciformis*, the spathiine *Spathius exarator* and the mite *Pyemotes ventricosus*, and predation by the clerid *Korynetes caeruleus* often take place to a high degree. In addition, the length of the life cycle is long, averaging about three years. But in a very large number of cases it happens that at some stage in the attack a great increase takes place over a short period, as shown by the crops of new flight holes. The attack can then proceed until almost complete destruction of the sapwood has taken place (Fig. 208) and sometimes damage extends deep into the heartwood.

This is now a common occurrence in older buildings (Fig. 209), but as hitherto such patches of bad attack could be easily repaired by stripping out and replacing, it has escaped serious attention. Now, the price of replacement timber and the necessary labour has become an economic burden.

It happens occasionally that the flight holes in a piece of infested timber all appear old—the wood is greyish or dirty just inside and often the hole is rounded. This is in contrast to new flight holes where the wood is cream

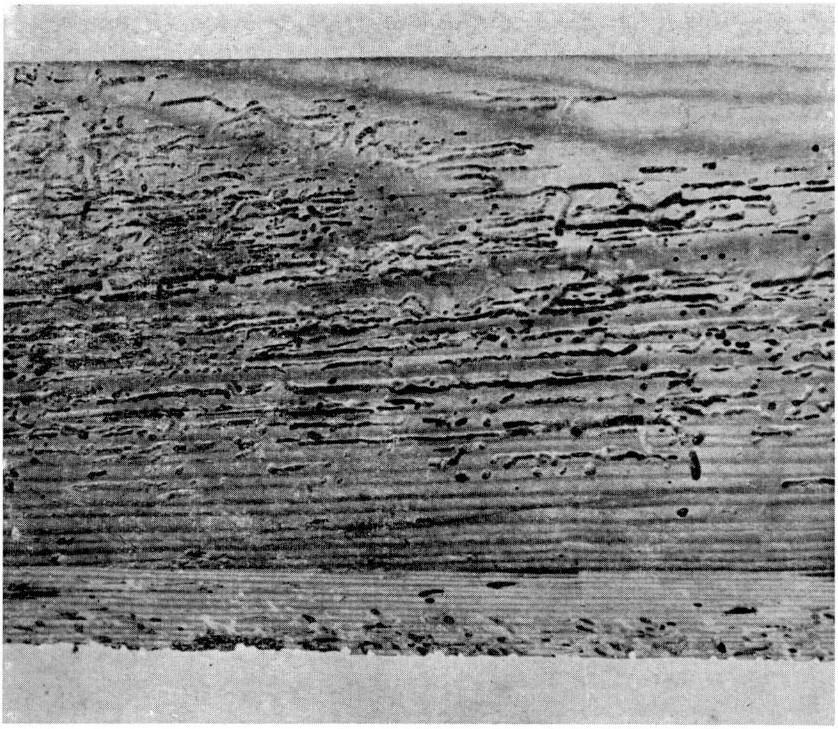

Fig. 208. Deal floorboard with the surface planed to show the tunnelling inside by *Anobium punctatum*. Damage equal to this can seriously impair the load-carrying properties.

in colour and clean, and the edge of the flight hole is often quite sharp, or sometimes with little plates of fibres still in position. In the former case it is usual to describe it as 'apparently inactive' although at present there are no means of verifying this. It will very often be found, however, that if such a piece of timber is carefully cut up, larvae will be found within. Old flight holes are frequently used by the female beetles for egg-laying sites. It is known from experience that reports of 'apparent inactivity' should be treated with reserve.

It is disquieting to find that from the insurance survey statistics (1961) already referred to, of the 166 properties in the 1940–61 group no fewer than 36 or 21·7 per cent had an active attack of woodworm. On account of timber shortages, licence arrangements and high price, the amount of timber in many of these buildings was much reduced compared with the other two age groups. This almost always meant smaller timber sections. Furthermore, such timber, most probably grown on a faster rotation, would have a greater sap/heartwood ratio, again compared with timber used in buildings in the other two groups. This inevitably means that when

Fig. 209. Destruction of softwood flooring by *Anobium punctatum* is a common incident in older properties.

woodworm attack occurs it will bring about the collapse of the wooden members much more speedily.

It is fortunate that there are several very important advantageous factors in the mechanics and chemistry of wood-boring insect control. In the first place the eggs of *Anobium punctatum* are not scattered or laid haphazardly but are carefully attached to the particular piece of wood that will be attacked. The whole of the larval and pupal periods are spent in the same piece of wood and even after the adult has metamorphosed, some days elapse whilst the cuticle of the newly emerged beetle hardens and it subsequently bites its way out. Thus, during the whole of the life cycle, with the exception of the flight, mating and egg-laying periods (some two or three weeks at most), the insect is virtually imprisoned in the wood for on average about three years.

Thus it is possible to aim certain forms of control accurately to within a very short distance of the insect although it cannot be seen.

Another great advantage which woodworm control shows over other forms of insect control concerns the nature of the wood itself. Wood, especially sapwood, absorbs chemical substances in oil solution very readily (to a lesser degree it absorbs solutions in water although if the wood is wet the absorptive capacity is higher). Many chemical substances of value in woodworm-killing formulations appear to possess a longer lasting effect if carried fairly deeply into the wood in an oil solution than if they were contained in a purely superficial layer on the wood surface. Perhaps the lessening of the ageing effect of sunlight and probably a slightly reduced oxygen tension bring about this very favourable effect. Certain oily solvents of low volatility can also be shown to be highly toxic to woodworm larvae and their non-volatility not only allows considerable 'creep' even after the lapse of many months but can take the insecticidal chemicals in with them even if, as can be shown by chemical analysis, the materials are strung out in an ever-decreasing concentration as they penetrate deeper into the wood (see also page 364).

It is not proposed to go into the complexities of formulation detail except to state two positive points, firstly that consideration of a woodworm-killing fluid to be widely used in all possible circumstances must entail a study of the human toxicity factor. The effect of accidental drinking by children and possible criminal use must be established accurately and guarded against by the possible use of an emetic 'built-in', or by other means. Cheap formulae have sometimes been published in professional journals which would not have stood the test of this elementary humane principle.

The second point concerns the 'effective dose' i.e. the woodworm-killing fluid must be at such a concentration that when a certain amount is supplied evenly over a given area it will be effective in 100 per cent of cases. Preferably from the economic standpoint such a concentration should be capable of application at one coat without 'run off'. This has been a very difficult exercise in view of all other essentials of formulation, but was achieved some years ago.

The final result of this is that provided certain rules of formulation are followed in the make-up of the woodworn-killing fluid and certain rules of fluid application followed rigidly, the elimination of *Anobium punctatum* from wood which it is infesting is virtually certain. What is more, the wood can be made impervious to future attack for many years to come.

Control specifications and methods

At this stage we must return to a consideration of the biology of *Anobium punctatum* and for that matter all other species of wood-boring beetle with which we have to contend, and reiterate that the flight holes observed on attacked woodwork occur only after the larva has spent some three years boring in the wood, pupating, then finally emerging as a beetle into the outside world by biting its way out through the wood. In the case of Death-Watch Beetle, this period is four or five years or longer; usually in the case of *Lyctus* one year, but could be a few months, or two years; and in the case of House Longhorn Beetle the period may be up to 15 years. Thus from a piece of wood showing no flight holes, one cannot infer that there is therefore no woodworm present. The legal case colloquially known as 'Boland's Piano' in Northern Ireland some years ago, put this question beyond doubt from the legal point of view. For this very reason a company carrying out woodworm control work can obviously only give a warranty concerning the effectiveness of their work on timber which has been included in a contract. The futility, therefore, of applying woodworm-killing fluids only to wooden surfaces showing flight holes is apparent, although of course such areas must be treated in addition to adjacent areas likely also to harbour larvae but where the infestation is of such duration that the first flight holes have not yet been made. These latter timbers, or areas, at present cannot be identified. It follows, therefore, that the treatment should be carried out over the widest possible area in the building—preferably all the timber susceptible to attack, where the infestation is active.

All surfaces of individual timbers which it is practicable to reach with equipment should be treated. These must be the guiding principles in control work, which have led to the outline specification most widely used for some years in the United Kingdom. In this, a building is divided into what may be described as simple separate building units with special reference to timber, and the treatments specified accordingly after a thorough examination by a trained Timber Infestation Surveyor.

The unit brief specifications are given here.

1. Wood-boring insects

The basic treatments have a common application to the following species of wood-boring insects: Common Furniture Beetle *Anobium punctatum*, Death-Watch Beetle *Xestobium rufovillosum*, Powder-Post Beetle *Lyctus* spp., and House Longhorn Beetle *Hylotrupes bajulus*. The treatments are detailed below and will, where appropriate, be specified in the report as complete treatments. Any variation from these treatments will be specified in the reports. House Longhorn treatment is specially dealt with in paragraph 2 below.

(*a*) *Roof Voids*. Lift floorboards as necessary in the roof void, clean down all exposed roof timbers and apply the insecticidal fluid to all such exposed surfaces of the rafters, purlins, roof boarding, struts, tie beams, ceiling joists, etc., and all surfaces of any roof floorboards before relaying.

This treatment is referred to as . . *Full roof void treatment*

(*b*) *Ground Floors and Upper Floors*. Lift the floorboards as necessary and clean down surfaces. Clean all accessible surfaces of the joists and floor framing timbers and apply the insecticidal fluid to all exposed surfaces of the floorboards, joists and floor framing timbers. Relay existing floorboards, renewing as specified with pretreated flooring.

This treatment is referred to as . . *Full floor treatment*

(*c*) *Joinery Timbers*. Timbers which are painted or varnished are less susceptible to insect infestation than sawn structural timbers or flooring.

Where evidence of infestation is observed in joinery timbers, the visibly infested timbers require to be treated by one of the three following alternative methods which will be specified in the report.

This treatment is referred to as . . *Full joinery treatment* Numbers 1, 2 and 3 respectively

Full Joinery Treatment No. 1. Carefully remove the visibly infested skirting, picture rail, architrave, door lining, panelling or other similar joinery timber as particularly described in the report and apply the insecticidal fluid to the unpainted reverse side and end grains, as also to all timber fixing plugs, grounds, etc. Replace and fix in position as before, leaving any necessary touching up and making good of decorations to match existing for client's decorator.

Full Joinery Treatment No. 2. (Where F.J.T. No. 1 is not economically justifiable.) Using a bradawl form holes each of a depth two-thirds the thickness of the infested timbers to supplement existing flight holes as means of pressure injection of the insecticidal fluid, the holes to be placed as far as possible in the least conspicuous positions, and (together with flight holes) at 7–10 cm centres. After injection extend holes to full depth and give low pressure trickle treatment through each hole to rear surface of the infested timber. Any necessary stopping of holes and making good of decorations to be left to client's decorator.

Full Joinery Treatment No. 3. (where redecoration of infested timbers is already proposed). Strip off existing paint from visibly infested timbers using NITROMORS or similar stripper to maker's directions or blow lamp as necessary and apply the insecticidal fluid to all stripped surfaces, also injecting into flight holes at 7–10 cm intervals.

(*d*) *Staircases*. Apply the insecticidal fluid to all exposed timber surfaces, ensuring particularly so far as possible the treatment of all accessible unpainted rear or undersurfaces of timbers painted or varnished on one side. Painted or varnished timbers not accessible for treatment on their reverse sides, e.g., wall strings, to receive F.J.T. No. 2, or No. 3 as may be specified in the report which will also indicate any need for the removal and, where necessary, the subsequent renewal of plaster or other soffites.

This treatment is referred to as . *Full staircase treatment* which term will be used in conjunction, where necessary, with F.J.T. Nos. 2 or 3.

(*e*) *Half-timbered Houses.* Carefully wire brush all accessible surfaces of such external hardwood timbers as are not sealed by varnish or paint films, and which are recommended for treatment, and apply the insecticidal fluid, also injecting into flight holes at 7-10 cm intervals. Any painted or varnished and infested timbers requiring treatment are, to extents to be specified in our report, to be treated as for F.J.T. No. 3.

This treatment is referred to as *Half-timbered houses treatment* (*ext.*) which term will be used in conjunction, where necessary, with F.J.T. No. 3.

2. House Longhorn Beetle

Hylotrupes bajulus. In the special case of Longhorn infestations the basic treatment specified above will be supplemented as follows: After cleaning, carefully cut away all heavily tunnelled sapwood down to sound timber, and examine all apparently sound accessible timber surfaces with a strong pointed instrument to determine the extent of sub-surface destruction of timber. Carefully collect, remove, and burn all frass. Reinforce and/or renew affected timbers as necessary to restore structural strength. Apply one liberal coat of the insecticidal fluid to all such defrassed areas, before applying the fluids provided for in the basic treatments specified above.

The present writer welcomes this opportunity of suggesting that the definitions or titles of the separate treatments are used universally. This would be of immense benefit to all concerned and would be a substantial step in the direction of standardization and a full understanding of the treatment processes involved.

It is obvious that a balance has to be struck between the widest possible application of the treatment in the affected building, and economics of the situation, or what, in the circumstances, can be called common sense. As an example, in the 1961 surveys already referred to, infestations of *Anobium punctatum* in roof voids only, occurred to the extent of 40·9 per cent. In almost all the instances 'full roof void treatment only' was recommended and carried out. In all other cases various combinations of the unit specifications were recommended.

If a householder has reported to him that there exists an active infestation of woodworm, say in the roof void of his property, surely the most reasonable and most economic method of dealing with the situation is to contract for the full roof void treatment to be carried out by a servicing company, and to insure against attack in the timber in the remaining parts of the property with a woodworm-insurance company. The writer, from his experience as a scientist working full-time on the problems of wood-boring insect and wood-rotting fungus control in buildings, considers that this system, now fully confirmed, is wholly in the public interest, and should be adopted by property owners as expeditiously as possible and on the widest possible scale.

Use of pre-treated timber

The control methods described above apply to timbers already incorporated into a building. If left untreated, the timber used in *today's* buildings will require the same expensive treatment in 10–30 years' time. To prevent this loss to property owners, it is recommended that all new buildings be constructed from pre-treated timber. This can be done at a fraction of the cost of the *in-situ* treatment described.

Wood preservation is rapidly becoming a major part of the timber trade, with many different treatment methods being employed (Hickin, 1971). Timber may be pre-treated by pressure impregnation, dipping, sap displacement, etc. All these methods are designed to get insecticides (and fungicides) into the timber *before* it is exposed to attack. Young wood-boring larvae are then quickly killed before they can damage the wood, and the property owner is thus relieved of all worries concerning the possibility of woodworm attack in the future.

Use of smoke generators and dichlorvos strips

A large proportion of *Anobium punctatum* infestations occur in the relatively confined areas of roof spaces. In the past, lindane smoke generators have been used in an attempt to kill invading beetles and those which emerge from infested timber already present. This method is not particularly effective, since the small particles of insecticide settle out on only horizontal or sloping, upward-facing surfaces. This leaves most roofing timbers completely unprotected. Also, of course, the insecticide does not penetrate the timber to kill woodworm larvae.

More recently, Baker (1972a) has carried out experiments with dichlorvos impregnated resin strips. In initial tests in an unheated, ventilated room, dichlorvos at concentrations of 0·2 and 0·5 μg/l were found to prevent adult beetles from emerging from pieces of infested plywood. Introduced adults were knocked down within 40 hours, and introduced eggs failed to hatch. *Larvae already established within timber were not affected by the insecticidal vapour.*

Field trials were later carried out using one dichlorvos strip per 1,000 ft^3 (14 m^3) of roof space. In roofs with a bituminous felt lining (typical of most modern buildings), dichlorvos levels of 0·015 to 0·045 μg/l were obtained, and a high proportion of adult beetles were killed before they could complete their exit holes. Any eggs laid on the surface were killed before hatching, but eggs laid deep in old flight holes produced viable larvae. In unlined roof spaces (typical of pre-1939 houses), dichlorvos levels were usually below 0·005 μg/l due to excessive ventilation, and few beetles were killed.

Baker concluded that, 'Such strips cannot replace liquid insecticides formulated to kill the larvae in wood. However, in modern, lined roof spaces their annual use throughout June to August could effect progressive control of an existing infestation and entirely prevent attack starting in sound wood. They might most usefully be used in support of local applications of liquids'.

This conclusion may be correct from the scientific point of view, but in practice, the method would be quite expensive when extended over 20 or 30 years. Also, no written guarantee can be obtained. Thus, when the house owner comes to sell his property, treatment with dichlorvos strips gives the vendor no documentation to show to a prospective purchaser, solicitor or Building Society. There is no knowledge that the job has been done properly at the right time every year.

On the other hand, when insecticidal fluids are applied by skilled operators, all existing larvae are killed, and any larvae hatching from eggs laid subsequently will also be killed. When completed, the owner obtains a written guarantee indicating the extent of the treatment work. The purchaser then knows exactly, by specification, which timbers have been treated, and the documentation retains its value against the rising price of property.

Fumigation of houses

Controlling drywood termites by fumigation under gas-proof sheets is a well-established procedure in some countries, but the use of fumigants for wood-boring beetle control is rarely practiced. The principal reason for this is the high cost and the fact that most companies do not have suitable equipment available. Nevertheless, the method is of value, particularly where the infestation is widespread or where valuable decorations may be damaged by fluids. Any type of building can be fumigated, including churches (see photographs in Hickin, 1971), and all types of wood-boring beetles are killed.

Rasmussen has observed the effect of methyl bromide when used in houses for the control of *Hylotrupes bajulus* in Denmark. He found that about 1 kg of methyl bromide was required for every 10 m^3 of space, and that all the gas should be delivered to the *top* of the building (i.e. the roof space or attic). The violent movement of the methyl bromide downwards through the house created sufficient turbulence to mix gas with air. This resulted in the stabilization of the gas/air mixture throughout the house in one or two hours. Although there was a tendency for the concentration of gas to be higher in cellars and basements, there was still ample gas in upper rooms and in cavities to effect control.

It should be borne in mind that fumigation kills only those insects which are in the house timbers at the time. The premises are open to re-infestation as soon as the gas has dispersed.

REFERENCES

BAKER, J. M. 1972a. The Insecticidal Action of Some Components of Wood Preservatives. *Pestic. Sci.* **3**: 229–239.

BAKER, J. M., 1972b. The Pest Status of British Wood-boring Insects. *Rec. Ann. Conv. B.W.P.A.*, pp. 3–16.

GOODBODY, R. G. 1962. Control of Woodworm. *Municipal Journal*. 20th July, 1962. No. 3622, pp. 2203–07.

HICKIN, N. E. 1951. Which Species does Most Damage? *Cabinet Maker*. March 1951.

HICKIN, N. E. 1958. Woodworm and its Control. *New Scientist*. 19th June, 1958.

HICKIN, N. E. 1960. The Problem of Wood Damage by *Anobium punctatum* in the

United Kingdom With Some Notes on the Current Pratice of Control. *XI. International Entomological Congress. Vienna.*

HICKIN, N. E. 1961. Die wirtschaftliche Bedeutung holzzerstörender Insekten in Grosbritannien. *Vorträge der 7. Internationalen Holzschutz-Tagung*. pp. 41–2. Berlin. In German with English summary.

HICKIN, N. E. 1967. *The Conservation of Building Timbers*. Hutchinson. London.

HICKIN, N. E. 1971. *Wood Preservation: A Guide to the Meaning of Terms*. Hutchinson, London.

HICKIN, N. E. 1972. *The Woodworm Problem*. 2nd Edn. Hutchinson. London.

RASMUSSEN, S. 1967. Fumigation of Houses with Methyl Bromide against the House Longhorn Beetle, *Hylotrupes bajulus. Material & Organismen* **2**: 65–78.

CHAPTER TWELVE

Research Methods

A description of the laboratory techniques used in wood preservation research: *eilarven* tests: larval transfer: emergence tests: use of X-rays: egg-laying and boring test: tests for permanence, accelerated ageing: termite tests

---

A number of methods have been described for assessing the efficiency of wood preservatives and wood presevative chemicals in respect of their ability to protect wood from wood-boring insect attack or to exterminate or control already existing infestations.

Two types of wood preservative, therefore, the one for protection and the other for extermination, are postulated. More simply perhaps the former is to prevent the young hatched larvae from boring in. The latter are to prevent the adults from boring out, as in practice it is only the presence or absence of new flight holes which can give an indication of success or failure of the treating fluid. Bletchly and Taylor have stated that it is desirable that preparations used for eradication should also afford protection against subsequent reattack, but, on the other hand, it is not necessary for a preservative to have curative properties. Whilst the former is certainly true, in the latter case it would appear to be unwise to rely on the absence of flight holes to determine whether a 'preservative' or 'curative' fluid should be used. The timber to be 'protected' could already sustain two or even three overlapping life cycles of *Anobium punctatum* without any visible external evidence of such an attack. Most tests have been devised in connection with wood-boring beetles.

Since the first edition of this book two methods of test have been issued by British Standards Institute for testing the toxicity of wood preservatives to *Anobium punctatum* and *Hylotrupes bajulus* by larval transfer (B.S. 3651 : 1963) and to *Anobium punctatum* by egg-laying larval survival (B.S. 3652: 1963). Standardized methods of testing using wood-boring insects as the test organisms have been official in several European countries for some years, and have largely come about through the work of Becker in Berlin. The following account is given as a brief guide only, the appropriate official standard should be consulted before any tests are commenced.

Tests for evaluating the efficiency of insecticidal fluids in protecting wood from attack by wood-boring beetles.

The test usually employed for this purpose is colloquially known as the *eilarven* test. This is the German word for freshly hatched larvae which are employed and is particularly described in the German Standard (Deutsche Norm.) DIN 52621/52163. It is more especially useful when using House Longhorn as the test insect.

Test blocks. These consist of standard pieces of planed air-dry Scots pine sapwood (*Pinus sylvestris*), 5 cm in length, 2·5 cm wide and 1·5 cm deep. They are so cut that the growth rings are parallel to the narrow side, not deviating from this by more than 30° (Fig. 210). The transverse surfaces of the blocks are treated with a paraffin wax to seal them and are then stored at 20° C and 60–70 per cent relative humidity for at least a week. This size and type of test block are standard in other tests. The blocks are treated with the preservative being employed and to the manufacturer's instructions. After treatment the test blocks are again stored, this time for four or five weeks at 20° C and 60–70 per cent relative humidity.

Number of test blocks. Six test blocks are used for each treatment variation, two of these blocks being untreated to act as controls, but if permanence of the treatment is to be assessed then further blocks are used for each treatment variation and stored for periods of one, three, six and ten years. For the one- and three-year periods the storage conditions are $20 \pm 2°$ C and 60–70 per cent relative humidity but for the longer storage times they are kept in a 'normal' attic, the conditions appertaining to which are described in the DIN.

Preparation of the boring slit. Plates of glass the same size as the large face of the blocks (5 cm × 2·5 cm) are placed on all the blocks and a gap or boring slit 1 mm wide allowed to remain along one edge by inserting a glass strip 1 mm thick and for convenience 20 mm long (Fig. 210). The two end slits—along the transverse faces of the block—are sealed by dipping

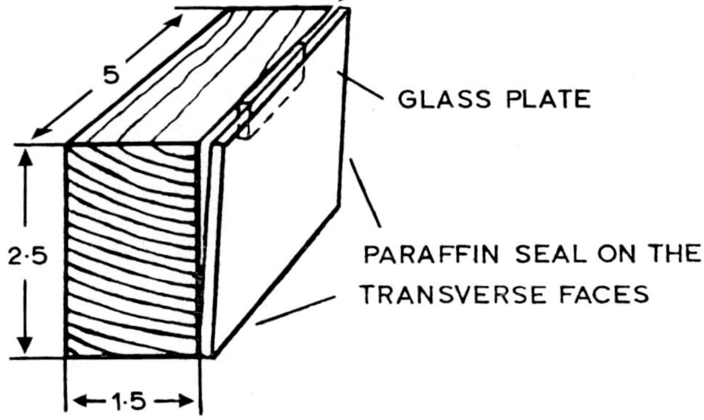

Fig. 210. Sample test block of sapwood of *Pinus sylvestris* giving dimensions and showing position of glass plate.

in molten paraffin wax and when the latter has solidified the small glass strip is removed. The test sample blocks are placed with the slit uppermost (as in the illustration) and they are then ready to receive the newly hatched larvae.

Method of rearing Hylotrupes bajulus to obtain newly hatched larvae. Wood removed from infested buildings known to contain living larvae is maintained in an outdoor insectary completely gauzed to prevent escape of emerging beetles. The latter are collected in single pairs isolated in containers. Sections of Scots pine sapwood 10 cm in diameter and about 7 cm thick are roughly split and the two segments are then held together rather loosely with an elastic band. This, with a piece of paper underneath, is introduced into the container with the mated pair. The eggs are laid in the crevices in the wood, often quite deeply by extension of the abdominal segments and the ovipositor and some will be laid on the paper. The eggs are kept at 28° C and 97–98 per cent relative humidity and when the young larvae hatch they are removed and immediately used in the test. Young larvae from different parents are mixed in order to attain uniformity.

Insertion of larvae. The newly hatched larvae (within three days of the actual hatching) are inserted into the boring slit on each block and the blocks are then stored at $20 \pm 2°$ C and at 70–75 per cent relative humidity. If there is any chemical in the treatment solution which acts as a fumigant then the controls must obviously be kept separately from the treated blocks.

Evaluation of results. X-radiographs of the blocks are taken after four weeks. The numbers of living and dead larvae are noted and the extent of tunnelling activity if any. A satisfactory test must show all larvae to be dead in the treated blocks after 12 weeks.

Tests for evaluating the efficiency of insecticidal fluids in exterminating wood-boring insects already present in wood

The most widely used test of this type is undoubtedly that known as the larval transfer method. One of the main advantages is that it can be carried out at all times of year. It is the method described in the German Standard DIN 52623/52165 and in B.S. 3651 : 1963 but is used in a number of European countries with variations. Very generally it is carried out as follows:

Test blocks. Similar standard blocks are employed as already mentioned, but when the test insect is *Anobium punctatum,* blocks of spruce and beech in addition to *Pinus sylvestris* are used in some laboratories.

Application of treatment fluid. This can be carried out in two ways. In DIN 52623/52165 and in B.S. 3651 : 1963 the treatment fluid is prepared in a series of dilutions with a suitable solvent and the blocks of *Pinus sylvestris* sapwood completely impregnated. Transfer of the larvae conducted after a suitable period has elapsed to enable the solvent to evaporate. In the second method the larvae are transferred and after a period suitable for establishment the drilled face and the two end faces are covered with paraffin wax and the treatment fluid is applied at a fixed dosage rate. This latter variation has some value in giving a more practical assessment of treating fluids, but has grave disadvantages.

Preparation of test blocks. Before the larvae are 'transferred' it is necessary to drill holes in the test blocks to receive them. For larvae of *Anobium punctatum,* ten holes are drilled equally spaced as shown in the illustration (Fig. 211). This is best carried out with a nail of 12 S.W.G. with the head removed to fit in the drill chuck and with the pointed end filed for about 1 cm to give four flat faces. The hole should be drilled of

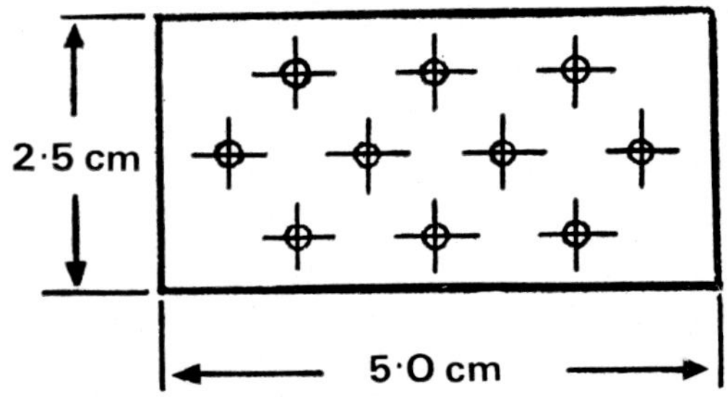

Fig. 211. Sample test block showing positioning of ten holes for larval transfer.

such a depth so that when the larva is in position there is a free space above it equal to about half its length.

Test insects. Medium sized larvae of *Anobium punctatum* are removed from naturally infested wood, plywood being convenient for this purpose. Larvae weighing from 1·0 to 3·0 mg are suitable, above this weight the larvae may pupate when inserted into the test blocks, although Bletchly and Taylor consider that a weight range of 1·5–5·5 mg is satisfactory when using intermediate storage blocks. When the larvae are taken from infested woodwork, Bletchly and Taylor recommend keeping the larvae in small dishes which have been treated with a 2 per cent solution of β napthol in 97 per cent alcohol—allowing the latter to dry off before use—as a precaution against infestation by the mite *Pyemotes ventricosus.* The larvae are examined next day for injuries or mites and if suitable in every respect they are inserted head first into holes drilled in intermediate storage blocks. The latter are in every respect similar to the untreated test blocks used for the controls and their purpose is to store the larvae until required, allowing a period when they will be under constant temperature and humidity control and standard conditions as far as nutrition is concerned. In addition, they can be kept away from the attentions of parasites but perhaps more important from an economic point of view, larvae smaller than the minimum size allowable may be 'stored' and used when they develop into the specified weight range.

The transfer. The larvae are stored in the intermediate storage blocks for at least two months under the same conditions of temperature and relative humidity as in the the test itself, namely 22° C and 75 per cent R.H. and are then removed from the storage blocks and transferred to the test and control blocks. This is done very carefully with a camel-hair brush or, preferably, with a vacuum tweezer (Figs. 212 and 213).

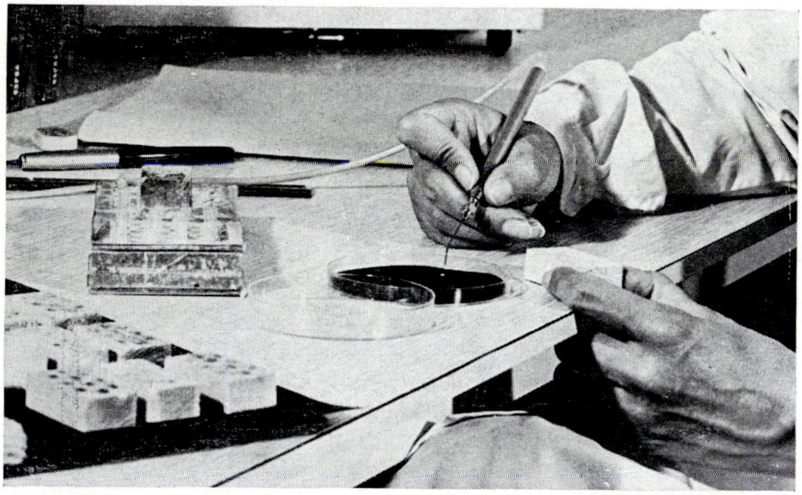

Fig. 212. Transferring *Anobium* larvae to holes drilled in Scots Pine blocks. Each larva is picked up with a vacuum tweezer.

Interpretation of results. The test blocks are allowed to remain under the standard conditions for one, three, six and twelve months. X-radiographs (see below) are taken each time, and the number of dead larvae is recorded. With this test, therefore, it is possible to determine the dilution which will kill all the larvae and the next greater dilution allowing some larvae to live. The toxic threshold will lie between these values.

Emergence tests

All the above tests show the effect of an insecticidal fluid on developing larvae. In practice, however, the applied insecticide may not penetrate the timber completely, leaving untreated wood in which some larvae may survive and mature. Therefore, the timber fluid should also prevent emergence of the adults from these survivors.

Taylor (1971) devised a test method in which a core of untreated timber was sandwiched between two treated veneers. *Anobium* larvae were introduced into the core and the blocks stored in the laboratory and outdoors over the winter, until adult beetles emerged.

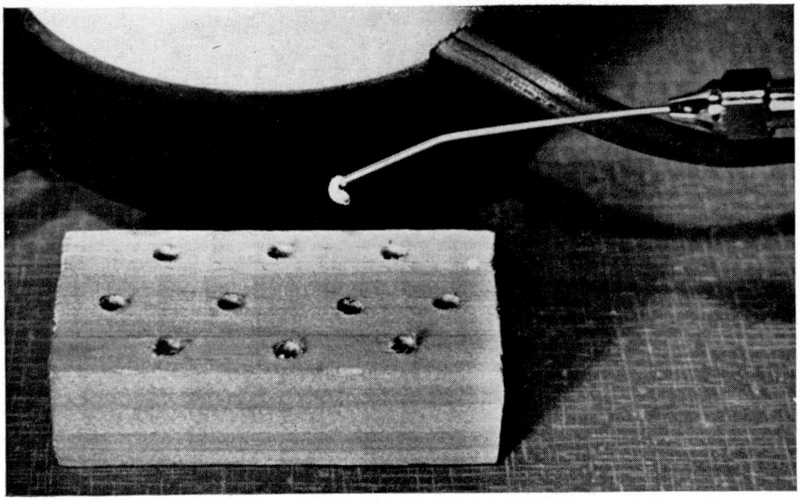

Fig. 213. *Anobium* larva held at the tip of a hypodermic needle by vacuum, prior to placement in a test block.

These tests showed that lindane and dieldrin at 0·25 kg/m^3 in veneers 3 or 5 mm thick were successful in preventing emergence of adult *Anobium*. Nevertheless, veneers treated with lindane permitted more larvae to survive in the core than was the case with dieldrin-treated veneers. At the concentrations used, five other stomach poisons failed to prevent emergence through 5 mm veneers: pentachlorophenol, zinc napthenate, tri-butyltin oxide, polychlornapthalene and copper-chrome-arsenate. The last two treatments, however, markedly reduced the number of beetles able to penetrate the treated veneers.

The use of X-ray techniques

One of the most important techniques developed during the last few years for facilitating the testing procedure is the use of X-rays. Apparatus was first suggested by the work of Fisher and Tasker and it has been improved and reported on satisfactorily by Jacquiot. Commercial equipment is now available for routine use. This is fitted with a beryllium window and supplies 'soft' X-rays up to 30 kV. If the larvae are kept at a minimum distance of 60 cm from the source they are not harmed during the exposure for the production of a radiograph.

The immediate benefits accruing from the equipment are first of all in examining infested woodwork to see whether a suitable number and size of larvae are present to make worthwhile extracting them. Secondly, the test blocks can be examined from the radiographs to ascertain the fate of the contained larvae (Fig. 214). This not only saves the time of splitting up the blocks but means that an experiment can be ended when all the larvae

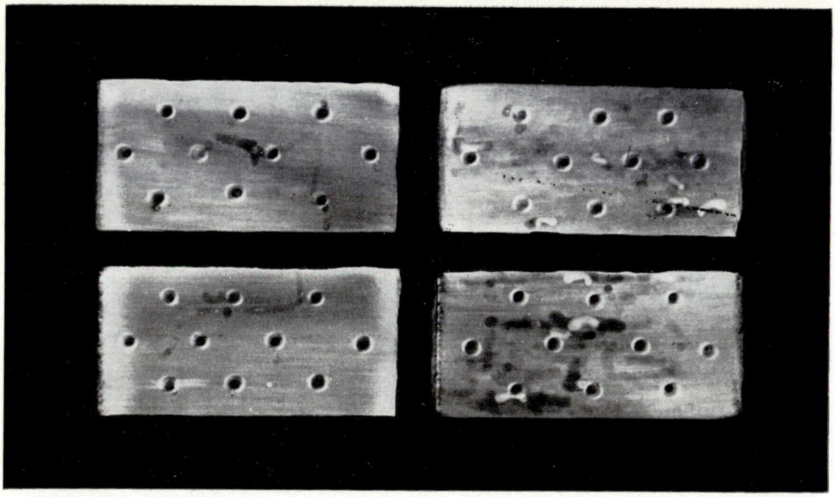

Fig. 214. X-radiography has proved a great help in seeing the fate of larvae when transferred to blocks treated with woodworm killing fluids. Here, treated blocks on the left show only a few short tunnels and no live larvae, compared with untreated controls on the right, where extensive tunnelling and live larvae can be clearly seen.

in the test blocks are dead and there is no need to wait until the specified period has been reached.

The Anobium egg-laying and boring test

Bletchly and Taylor have described a method of evaluating the efficiency of wood preservatives in respect of attack by *Anobium punctatum*, in use at the Building Research Establishment, Princes Risborough. This has now been issued as a British Standard (B.S. 3652: 1963). In this test a range of treated standard test blocks and controls is exposed to pairs of the adult beetles. The end grain of the blocks must be rough enough to be attractive to the female as an egg-laying site and the ends must be specially roughened if this is not the case. Three months after the commencement of the test the numbers of eggs laid and hatched are recorded. After a further nine months the blocks are X-rayed. The toxic threshold of the insecticidal fluid under test is found by determining the highest concentration in which living larvae are found and the lowest concentration in which no larvae survive.

As there are no satisfactory methods of culturing *Anobium punctatum* so that adults can be made available at any predetermined date, this test can only be started in the months of June and July—during the period of natural emergence. Against this must be set the undoubted advantage that this test stimulates quite closely natural conditions of infestation.

Contact tests with adult beetles

Taylor has described a test which she has employed in order to assess the effect of contact insecticides on the adult beetles. Various concentrations of contact insecticides in solution are applied to pieces of filter paper and groups of beetles were obliged to come into contact with it for various periods. The effect of this on the subsequent length of life and on the numbers of eggs subsequently laid were recorded. Taylor was able to show that after the beetles had been in contact for certain periods with specified concentrations of certain contact insecticides co-ordinated movement was lost and subsequently egg-laying was either greatly reduced or ceased completely. The species *Anobium punctatum* and *Xestobium rufovillosum* were used in the tests which were carried out out-of-doors.

This type of test is of value in determining the effect of the new contact insecticides. Most insecticdes used for *in-situ* protection or extermination of the two species named above contain such materials, but of course, it is usual to effect as deep a penetration into the wood as possible, one good reason being that such insecticides are much longer lasting when in wood than when exposed to light and air currents on the wood surface. In *Lyctus* beetle control in the timber yard, however, a superficial insecticidal film of high adult killing properties is of value during the flight season.

Tests for permanence

An essential requirement of a wood preservative is that when applied to the wood it is desired to preserve, the effect is long-term or permanent. Definitions of 'permanent' in this regard vary from 20 years to the lifetime of the building in which the timber is situated. This requirement, of course, concerns the situation in the United Kingdom with regard to the use of building timber. In West African countries short-term insecticidal wood preservatives are used to protect logs from Ambrosia beetle attack.

The tests which have already been described in this chapter are likely to be employed in permanence tests but the test blocks will have undergone an accelerated ageing treatment carefully evaluated to give the effects of oxidation and volatilization likely to be encountered during many years under natural conditions of use. Apparatus designed to subject the test blocks to accelerated air flow and increased and fluctuating temperature have been used in Germany and in Britain.

Figure 215 shows the wind tunnel in the Rentokil laboratories: equipment on the left controls fans, heaters and humidifiers so that a known air speed, temperature and humidity can be selected. The stream of air passes through the pair of tunnels round three sides of a small room and leaves through a duct on the right. (Cornwell & Hawkes, 1975.)

In one test, *Pinus sylvestris* blocks were sprayed with 1 per cent lindane solution at the rate of 1 gallon /200ft^2 (1 litre/4 m^2), and allowed to dry for one week. Half were transferred to a roof void for 'normal ageing', the other half were subjected to 'accelerated ageing' in the wind tunnels. Chemical and biological assays were made after one week, three months, one year and three years.

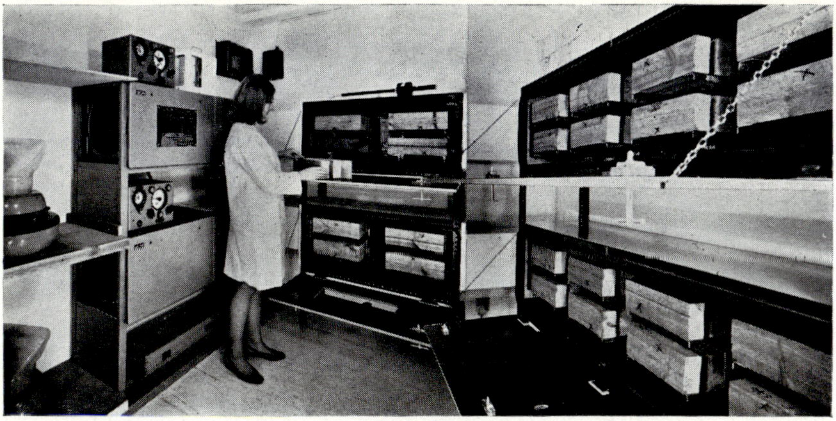

Fig. 215. Wind tunnel consisting of upper and lower chambers used for accelerated ageing of treated blocks.

Results of chemical assays (Fig. 216) showed that 50–70 per cent of the lindane was quickly lost from the outer 2 mm. Losses were much less at greater depths, and gains (indicating movement of lindane into the timber) were registered at 6–7 mm from the surface after three months normal ageing and at 10–14 mm after three years in both situations.

The acceleration factor varied with the depth: losses in the outer 2 mm with three years normal ageing were very similar to those after three months of accelerated ageing—a factor of x12. In the 2–7 mm layers, however, three years in the roof void was equal to about nine months in the wind tunnel—an acceleration factor of x 4.

Since reinfestation by newly hatched larvae occurs by penetration from the outer surface of the timber, an acceleration factor of at least x10 for the wind tunnel is appropriate in judging the ability of *Pinus sylvestris* to resist attack.

Bioassay was carried out by keeping *Anobium punctatum* larvae in compacted sawdust taken from successive 1 mm layers of the treated blocks (Doughty, 1974). Samples taken from the 1–2 mm layer gave 100 per cent kill irrespective of type or period of ageing. Lindane content in this layer was found by analysis to be about three times the minimum required to kill all *Anobium* larvae.

These results suggest that when sprayed on timber surfaces, lindane should be capable of preventing reinfestation for a period of up to 30 years. This allows the commercial treatment companies to offer a 20 year warranty with confidence.

Termites in tests

A number of investigators in Europe and the tropics have tested wood preservatives by exposing treated wood to termites. A large number of

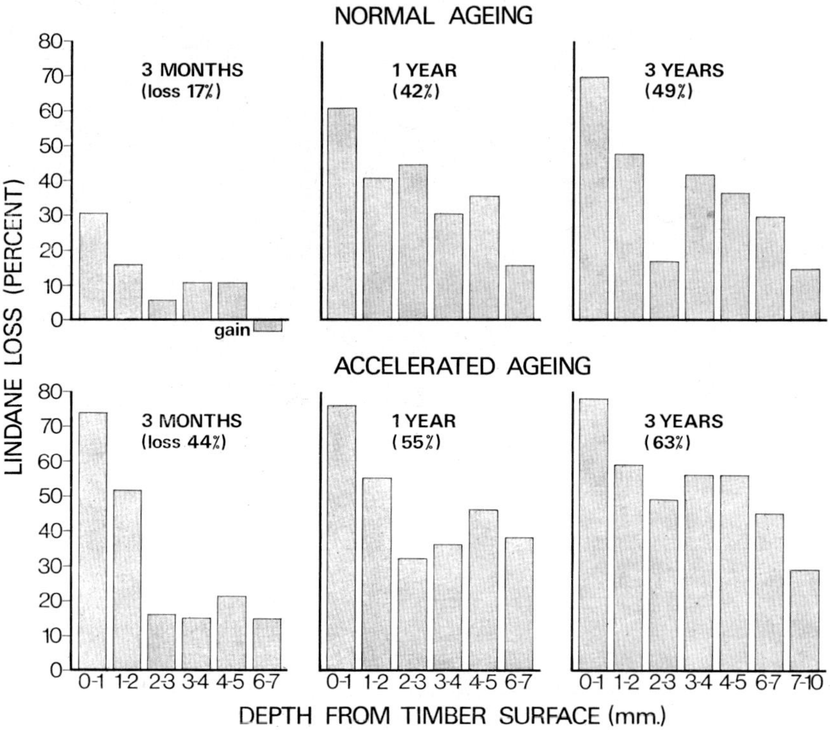

Fig. 216. Percentage loss of lindane from various depths of rough sawn *P. sylvestris* after normal and accelerated ageing.

field experiments have been carried out in tropical countries in which slabs of treated wood and controls have been partially buried in the ground where termites are known to be present (Hickin, 1971). Obviously such methods have their limitations and can really only supply negative evidence.

It is perhaps to be expected that the European species *Kalotermes flavicollis* and *Reticulitermes lucifugus* and its form *santonensis* are the species usually employed in laboratory testing in Europe. Although a number of standard tests have been proposed in Germany and France in fact none have yet found universal favour. It does seem evident, as Becker has suggested, that several species of termites should be used in tests and there seems no reason why, with the increasing availability of termite-rearing facilities, widespread and highly damaging species such as *Cryptotermes brevis* should not be used.

REFERENCES

BECKER, G. 1942. Prüfung der vorbeugenden wirkung von Holzschutzmitteln gegen termiten. *Wiss. Abh. Deut. Materialprüfungsanstalten* **2**: 40–55.

BECKER, G. 1961. On the examination and estimation of the natural resistance of wood to termites. *Holz.* **19**: 278–90.

BLETCHLY, J. D. and TAYLOR, J. M. 1961. Methods of Testing Preservatives Against the Common Furniture Beetle (*Anobium punctatum* Deg.) and the House Longhorn Beetle (*Hylotrupes bajulus* L.) at the Forest Products Research Laboratory. *J. Inst. Wood. Sci.* **8**: 4–11.

CORNWELL, P. B. and HAWKES, C. 1975. Penetration and Persistence of Organic-solvent Preservatives, Spray Applied to *Pinus sylvestris*. *Holzforschung* **29**.

COUDREAU, FOUGEROUSSE, BRESSY, and LUCAS 1960. Recherches en vue de déterminer une nouvelle méthode destinée à apprécier las résistance d'un bois aux destructions causées par les termites (*Reticulitermes lucifugus* Rossi). *Holzforschung* **14**: 40–51.

DOUGHTY, K. 1974. A novel circular saw technique for producing sawdust samples for use in wood preservative determinations. *Int. Biodetn Bull.* **10** (4).

FISHER, R. C. and TASKER, H. S. 1940. The Detection of Wood-boring Insects by Means of X-rays. *Ann. appl. Biol.* **27**: 92–100.

GROSSWALD, K. 1958. Zum Wirkungsmechanismus on Thiodan. *Z. angew. Zool.* **45** (2): 129–51.

HICKIN, N. E. 1971. *Termites—A World Problem*. Hutchinson. London.

HRDY, I. 1961. A Quick Laboratory Method of Determining the Termite Resistance of Materials. *Beitr. Ent.* **11**: 546–556.

JACQUIOT, C. 1961. Some Aspects of Wood Preservation in France. *Record of Annual Convention of the British Wood Preserving Association*, pp. 151–6. London.

SCHULZE, B. and GASDA, G. 1954. Die Verdunstbarkeit on Holzschutzmitteln und ihre Prüfung. *Holzforschung* **8**: 78–87.

TAYLOR, J. M. 1961. The Use of Contact Insecticides for the Control of Woodboring Insects. *Record of the Annual Convention of the British Wood Preserving Association*, pp. 5–17. London.

TAYLOR, J. M. 1971. Testing the effects of liquid eradicants on emergence of *Anobium punctatun* (Deg.). *Ann. appl. Biol.* **67**: 201–10.

Index

All timbers are indexed under their scientific names; common names are cross-referenced. Most insects are indexed under both names, the exceptions being the common species which are broken down into sections under the scientific name.
Figures in bold type indicate the page on which the insect occurs in an identification key.
Figures in italics indicate illustrations.